Estuarine and Coastal Hydrography and Sediment Transport

A practical guide to the latest remote and in situ techniques used to measure sediments, quantify seabed characteristics, and understand physical properties of water and sediments and transport mechanisms in estuaries and coastal waters. Covering a broad range of topics from global reference frames and bathymetric surveying methods to the use of remote sensing for determining surface-water variables, enough background is included to explain how each technology functions. The advantages and disadvantages of each technology are explained, and a review of recent fieldwork experiments demonstrates how modern methods apply in real-life estuarine and coastal campaigns. Clear explanations of physical processes show the links between different disciplines, making the book ideal for students and researchers in the environmental sciences, marine biology, chemistry and geology, whose work relies on an understanding of the physical environment and the way it is changing as a result of climate change, engineering and other influences.

R. J. Uncles has worked on estuarine and coastal projects at the Plymouth Marine Laboratory and overseas since gaining a PhD in physics from Imperial College, London (1972) and has published numerous research articles. He is an Associate of the Royal College of Science, a Fellow of the Institute of Physics and was President of the Estuarine and Coastal Sciences Association (ECSA) between 2009 and 2012.

S. B. Mitchell lectures at the School of Civil Engineering and Surveying at the University of Portsmouth, having gained a PhD in civil engineering at the University of Birmingham in 1998. He serves as editor-in-chief for the *Estuarine, Coastal and Shelf Science* journal and is a Fellow of the Institution of Civil Engineers. Results of his extensive estuarine and coastal field work have been published widely.

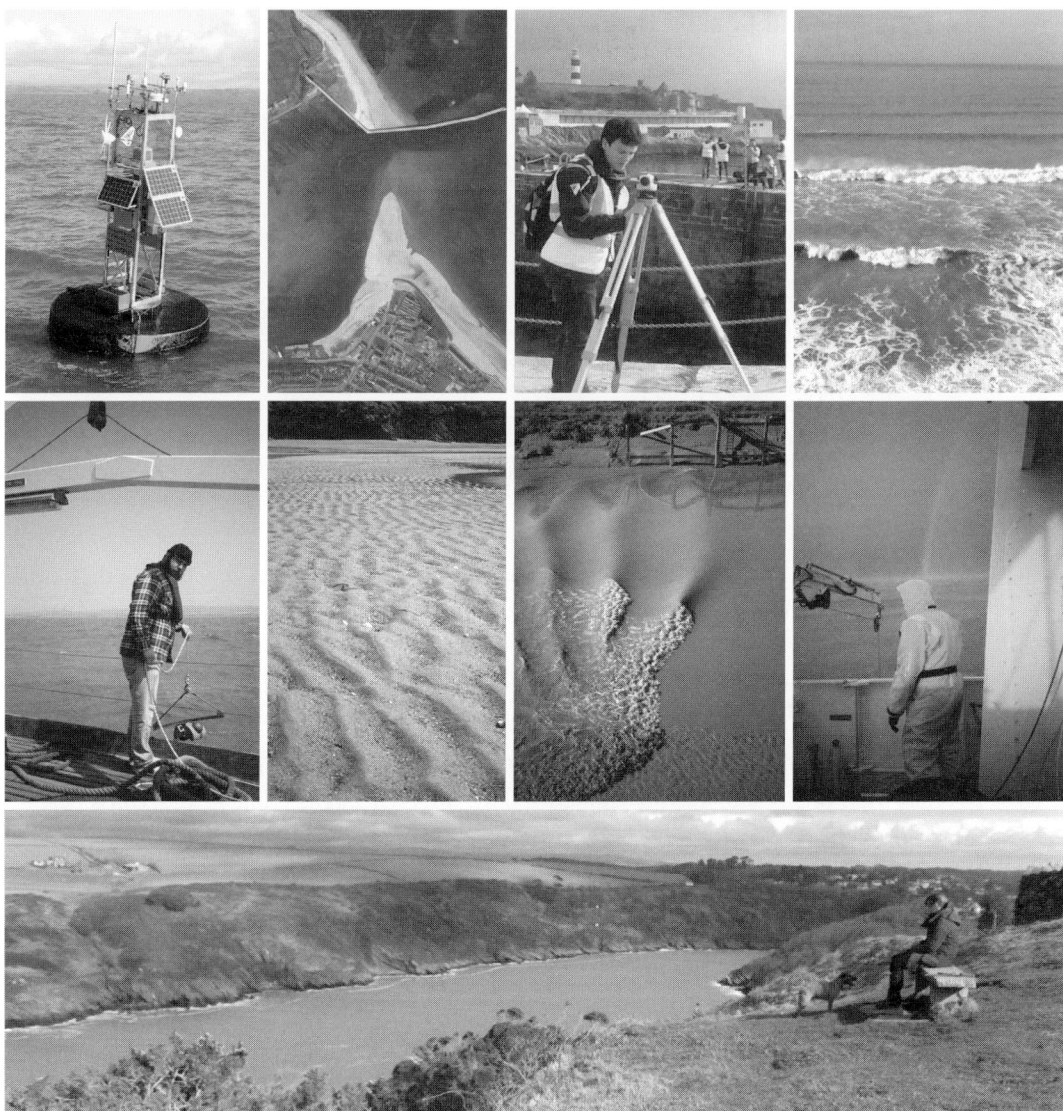

R. J. Uncles and S. B. Mitchell have edited this book on behalf of the Estuarine and Coastal Sciences Association. Above photographs: R. J. Uncles, except upper left three by, respectively, J. R. Fishwick; NERC Airborne Research Facility, UK; and V. J. Abbott; with permission.

Estuarine and Coastal Hydrography and Sediment Transport

Edited by

R. J. UNCLES
Plymouth Marine Laboratory

S. B. MITCHELL
University of Portsmouth

CAMBRIDGE
UNIVERSITY PRESS

CAMBRIDGE
UNIVERSITY PRESS

University Printing House, Cambridge CB2 8BS, United Kingdom

One Liberty Plaza, 20th Floor, New York, NY 10006, USA

477 Williamstown Road, Port Melbourne, VIC 3207, Australia

4843/24, 2nd Floor, Ansari Road, Daryaganj, Delhi – 110002, India

79 Anson Road, #06–04/06, Singapore 079906

Cambridge University Press is part of the University of Cambridge.

It furthers the University's mission by disseminating knowledge in the pursuit of education, learning, and research at the highest international levels of excellence.

www.cambridge.org
Information on this title: www.cambridge.org/9781107040984
DOI: 10.1017/9781139644426

First published 2017

Printed in the United Kingdom by TJ International Ltd. Padstow Cornwall

A catalogue record for this publication is available from the British Library.

Library of Congress Cataloging-in-Publication Data
Names: Uncles, R. J., editor.
Title: Estuarine and coastal hydrography and sediment transport / edited by R.J. Uncles,
Plymouth Marine Laboratory, S.B. Mitchell, University of Portsmouth.
Description: Cambridge, United Kingdom ; New York, NY : Cambridge University Press, 2017.
Identifiers: LCCN 2017005281 | ISBN 9781107040984 (Hardback)
Subjects: LCSH: Hydrography. | Sediment transport. | BISAC: NATURE / Ecosystems &
Habitats / Oceans & Seas.
Classification: LCC GB661.2 .E78 2017 | DDC 551.46/18–dc23
LC record available at https://lccn.loc.gov/2017005281

ISBN 978-1-107-04098-4 Hardback

Contents

J. A. Stephens
Plymouth Marine Laboratory, Plymouth, UK

J. D. Turton
Met Office, Exeter, Devon, UK

R. J. Uncles
Plymouth Marine Laboratory, Plymouth, Devon, UK

R. J. S. Whitehouse
HR Wallingford, Wallingford, Oxfordshire, UK

J. Wolf
National Oceanography Centre, Liverpool, UK

M. R. Wright
Partrac Ltd, Newcastle Upon Tyne, UK

Preface

Some aspects of estuarine and coastal hydrography and sediment transport have experienced revolutionary change since the publication of the original Estuarine and Coastal Sciences Association (ECSA) Handbook in 1979, much of which is still inevitably in progress. The dedicated and deep-thinking scientists researching estuaries and coastal systems in the 1970s would have been well aware of the possibilities of advanced measurement and logging – humans had, after all, reached the surface of the moon – but they might not have foreseen the challenges in terms of data requirements for high resolution, three-dimensional modelling, data management and interpretation. They understood, as Professor Keith Dyer did in the original preface written in 1978, that the most important progress lies in adopting a multidisciplinary approach, and that handbooks of this type are a necessary means of softening the divide between the traditions that separately accompany the biology, the water quality and the physics.

As an example of the scope and ambition of the changes, the advent during the early 1970s – and subsequent refinement of – acoustic methods to collect data on water motion, and related efforts similarly using advances in marine acoustic technologies to improve subsea positioning accuracy and underwater communications, produced a sea change in marine measurement capabilities. This ECSA handbook considers estuarine and coastal physical processes research, focusing on many of the same areas considered in the 1979 volume, as well as some entirely new ones. Clearly, there have been considerable advances in the ability to record and log data at high temporal and spatial resolutions, leading in turn to challenges in assimilating and synthesising large amounts of data prior to its subsequent storage and archiving.

Of course, all technological developments have a tendency to be driven by research needs, which in turn are driven by the needs of society to understand the world, develop and grow economically. To take one example, an ability to predict sediment transport paths, rates and processes in estuaries is fundamental to their management, whether this be from an ecological or an engineering viewpoint. The major expansion in port activity since 1979 has meant that dredging technologies and their related impacts have come to the fore as a key area of interest in scientific policy making. European and other directives related to ecosystem management also clearly drive all port development and determine the need to provide new and sustainable habitats. The threat of sea-level rise implies a critical importance associated with understanding the response of all coastal systems to a variety of rates and types of environmental change. There is thus no doubt of the continuing need of dredging programs and other estuarine and coastal management activities to be properly informed by good scientific understanding and the availability of the best and most modern measurement techniques and technology, and it is hoped that this handbook will furnish a guide to these activities.

1 Estuarine and Coastal Hydrography and Sediment Transport

R. J. Uncles, S. B. Mitchell

1 Introduction

This chapter provides an introduction to the handbook and summarises some of the more recent fieldwork on estuarine and coastal research topics and the instrumentation deployed. The handbook deals with numerous methodologies that are currently in use to study the physical behaviour of estuaries, their coastal plume areas and the coastal zone in general. It is clearly impossible to provide detailed accounts of all the methods used and their various laboratory and field procedures; rather, what we hope to provide in this book is a source of material that can be used both as an introduction to the methodologies and a starting point from which to follow up, in much greater detail, the various techniques that are briefly described here.

The chapters deal with physical properties such as bathymetry, water circulation and waves, and sediment properties and behaviour. Other ECSA handbooks are planned that will deal with the biology and chemistry of estuarine and coastal systems. Holistically, many, if not all, of the 'life' systems and processes and their interactions with suspended and deposited sediments and water-column and particulate chemistry within estuaries and coastal waters are strongly dependent on physical processes, whether they are associated with horizontal transport, vertical and horizontal mixing or the erosion and deposition of sediments. Although the handbook is largely focused on physical processes, biological effects are not ignored. For example, Chapters 6 to 9 consider several properties of the suspended and deposited sediment, including its particulate organic content and other biologically relevant variables, and Chapter 10 outlines methods that are used to deal with and mitigate the biofouling of moored instrumentation.

The estuarine and coastal-waters system is of great importance to us; most of the world's major cities and most of our most populous areas are near the coast (e.g., Huntley et al., 2001). Water-borne effluents and wastes derived from industries, cities and river catchments are transported by rivers and estuaries to the coastal zone. Sediment movements are particularly difficult to predict because of the complications of resuspension, settling and deposition, and particle flocculation. Observations of the pertinent processes are an important part of our effort to model these processes and provide essential data for model validation, interpretation and understanding. The instrumentation and methodologies described and presented in this handbook form a key part of the armoury of estuarine and coastal oceanographers in their attempts to obtain these data.

Salinity, Stratification and the ETM

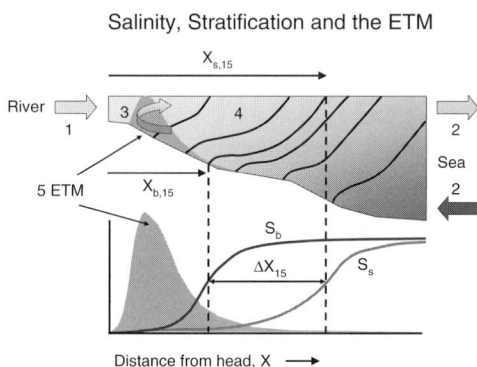

Figure 1-1 A qualitative, indicative schematic illustrating a partly mixed estuary and showing: (1), river inflow at the head; (2), the residual exchange-flow of waters from, and to, the sea at the mouth; (3), the residual estuarine circulation null-point near the bed and close to the head; (4), some salinity isohalines that exhibit stratification due to freshwater buoyancy, which leads to the surface isohalines, S_s, located at distance X_s, being displaced down-estuary relative to the bed isohalines, S_b, located at distance X_b ($X_{s,15}$ and $X_{b,15}$ and their separation ΔX_{15} are shown for the 15 isohaline); (5), SPM concentrations that maximise near the head within the ETM and then fall dramatically progressing closer to the head and the nontidal river. Tidal oscillations are superimposed on this depiction of residual properties, leading to intratidal variations in salinity intrusion, mixing, stratification and ETM location and magnitude.

In his introduction to the earlier handbook, Dyer (1979) summarised the tidally averaged (residual) estuarine physical processes as they were known at that time. These focused on the classic concept of estuarine circulation (or buoyancy or gravitational current) and the classic classification scheme of tidally averaged circulation and salinity stratification. We do not attempt to update this knowledge here, although it is worthwhile to mention these topics very briefly and illustrate some of the most fundamental concepts (a partially mixed estuary is shown schematically in Figure 1-1). The inflow of freshwater and buoyancy to an estuary generates an estuarine circulation (reviewed by MacCready and Banas, 2011), which, in a tidally averaged sense, is down-estuary near the surface and up-estuary near the bed. This in turn leads to an exchange flow at the mouth; a null-point in the residual circulation in the upper estuary; greater or less salinity stratification (surface salinity less than bed salinity, dependent on freshwater inputs and mixing); and, depending on the nature and extent of sediment supply, the potential for an estuarine turbidity maximum (ETM) to form in the upper reaches (Figure 1-1). This simple picture is greatly complicated by tides, winds and sediment behaviour, and one of the consequences of the development of new technologies and associated instrumentation in the intervening years since the first handbook was published has been the ability to study, in great detail, the influences of tidal variability and turbulence on estuarine and coastal dynamics.

We start with the handbook's raison d'être and then provide a preview of the chapters and their contents. In an attempt to highlight the current state of knowledge and general direction of travel of research efforts in the topics covered by the handbook, a summarised description of some of the latest (generally 2010–2016) research work that has been undertaken using the methods and instrumentation described in these chapters is then given. We conclude with some final comments on the relevance of the topics covered and a note of caution about ensuring the timely use of the knowledge base.

2   The Handbook's Raison d'Être

To give just a few examples, the technology behind the acoustic Doppler current profiler (ADCP), the HF (high frequency) ocean surface current radar (OSCR) and, more recently, X-Band radar, the optical backscatter (OBS) sensor, and airborne and space-borne remote-sensing technology, were either not available or in their infancy when the previous handbook was published (Dyer, 1979). Since that time, these and other advances in instrumentation, such as accurate, handheld global positioning system (GPS) readout units, have produced a huge advance in our ability to observe spatial and temporal variations in estuaries and coastal waters.

Satellite and airborne remote sensing are continually providing new insights that are invaluable to research programmes that investigate estuarine and coastal phenomena, e.g., understanding key processes, as well as aiding studies of estuarine and coastal water quality. Airborne remote sensing, utilising light aircraft, can provide very high spatial resolution images, typically of the order of metres or less, and OSCR has provided highly resolved maps of surface currents in the coastal zone.

Given these huge advances since 1979, we felt that it would be valuable to provide an updated version of the handbook, generally following the earlier version in overall coverage of the research topics considered therein, but concentrating on the instruments and methods that have evolved since. This does not render the earlier version obsolete because many of the methods described there are still useful, although there are inevitably some overlaps with the updated handbook when older methods have been considered too important to omit; e.g., the fundamental technique of gravimetric determination of suspended solids concentration is described in Chapter 7.

3   Preview of Chapters

The handbook begins with a description of the modern techniques used in hydrography, followed by those utilised in hydrodynamics for the measurement of waves, conductivity, temperature, salinity and density, depth and water velocity. The instrumentation and methodology required to measure properties of the deposited sediment and the suspended primary and flocculated sediment and transported sediments and particulate matter within estuaries and the coastal zone are then covered. Finally, there are discussions on the methodologies currently used to capture remotely acquired data from moored observation platforms and from satellite and aircraft images.

3.1   Hydrography

Abbott describes the modern methodologies that are used to measure estuarine and coastal bathymetry and tides in Chapter 2. The chapter includes topics such as the recording of vertical tidal movements and the processing of data; global reference frames and positioning on the Earth's surface, including satellite positioning; the technologies, methodologies and types of survey platform utilised for the determination of bathymetry and the associated uncertainties; and finally the approach taken to plan survey field work, including safety aspects. In Chapter 3, Jones, Abbott, Manning and Jakt continue this theme and describe the applications of sidescan sonar to seabed imaging, geological mapping and habitat mapping, which developed from early 12-kHz systems through to the modern 900-kHz and 1600-kHz systems. Chapter 3 covers swathe bathymetric systems, using multiple echo sounder transducers, interferometric bathymetric swathe sounding and Multiple Beamforming Echo-Sounder (MBES), Lidar, subbottom profilers, and survey planning and seabed classification.

3.2   Hydrodynamics

3.2.1   Temperature, Salinity, Density and Velocity Measurement Techniques

Technologies for velocity measurements in estuaries and coastal waters, especially acoustic Doppler instruments (velocimeters – ADVs, and profilers – ADCPs), are described by Souza in Chapter 4. He describes the underlying physical principles of operation of these instruments and highlights their great operational flexibility; e.g., their use when mounted on moving platforms, for which the ship's velocity is removed using either the bottom-tracking of the same ADCP in shallow waters or else calculated from the ship's GPS. He then goes on to describe the use of ADCPs and ADVs for the estimation of turbulence production and dissipation, pointing out the simplicity and versatility of the ADCP instrument, which allows the user to obtain full water-column estimates of turbulence production and dissipation and even directional wave spectra and suspended particulate matter (SPM) concentrations with appropriate calibrations. Nevertheless, turbulence measurements from bottom-mounted moorings are usually obtained using the ADV, which rapidly samples the three components of velocity in a small sampling volume. Remote sensing techniques are also mentioned because of their sea-surface descriptions of ocean winds, waves, temperature, ice conditions, suspended sediments, chlorophyll, eddy, and frontal locations.

Also described in Chapter 4 are modern measurement techniques for the determination of temperature, conductivity and hence both salinity and, with SPM concentrations, water-column density, along with an historical review of these measurement methodologies. Although modern measurements of temperature and salinity are usually carried out using conductivity-temperature-depth (CTD) sensors, temperature alone is also measured using thermistors. The main operational difference between CTD sensors, aside from accuracy and stability considerations, resides in whether they use inductive or conductive sensors and the configuration of sensors on the instrument. Other advances include reliable self-recording systems that can be either moored or mounted on ships and AUVs (automated underwater vehicles) and gliders and the availability of communication technologies that provide real-time estimates of density. Measuring sea-surface temperatures from space is well established although the measurement of salinity from space is still not routine, despite the fact that techniques using passive microwave radiometry have been under development for more than 20 years.

3.2.2 Wave Measurement Techniques

Tides and buoyancy due to freshwater inflows and surface heating generally dominate the hydrodynamic behaviour of estuaries and coastal systems, although wind is often an important influence, both as a driver of currents and surface waves. A description of linear wave theory and its various definitions in coastal and estuarine waters is given by Wolf (Chapter 5) to provide an introduction to modern measurement techniques for the observation of waves in shallow waters and the determination of surface wave properties. Several types of instrumentation are documented, including directional wave buoys, arrays of current meters and pressure sensors, and satellite, aircraft and land-based, remotely deployed instrumentation, which sense wave characteristics such as surface roughness using microwave radar or optical systems.

Wolf points out that wave-following buoys can be difficult to deploy in shallow estuarine and coastal waters, so that combined current meter and pressure-sensing instrumentation are frequently used (the PUV – pressure and two orthogonal components of horizontal velocity – method). ADCPs, adapted to capture wave data, have been used for this PUV method. The ADCP's vertical profile configuration is used for wave measurements and high-frequency data are recorded at several Hertz to resolve the periodicities of surface gravity waves. Remotely sensed instrumentation includes ground-based radar systems using radar ranging from structure-mounted instruments, X-Band radars that measure waves via analysis of the backscattered radar energy from the sea's surface, and HF radar that can simultaneously observe waves and currents over a spatial grid of many kilometres. Other methods covered include ultrasound (>20 kHz) tide and wave gauge instruments and the laser altimeter (LiDAR), which can be used from a satellite, plane or helicopter.

Chapter 5 also includes an overview of wave modelling techniques for the nearshore zone, which includes the spectral (phase-averaged) approach and the SWAN model, which has been a landmark development in spectral wave modelling for the coastal zone.

3.2.3 Sediment Measurement Techniques

Solving important practical problems that involve sediments, e.g., estuarine and coastal erosion and accretion, accretion within harbours and ports and the maintenance of navigational channels, requires a quantitative understanding of the transport and behaviour of sediment. Fine sediments, which are an attribute of many large, strongly tidal estuaries, cause an additional water quality issue because of their high adsorptive capacity for dissolved pollutants. Currently, the numerical prediction of sediment transport phenomena largely involves the computation of physically based relationships, although the relevance of biological effects is increasingly acknowledged. These physical relationships are predominantly empirical in nature and rely on the accurate measurement of sediment properties and processes in the field and in the laboratory.

The sampling and analysis of estuarine deposited sediments are discussed by Spencer in Chapter 6. The nature and composition of the surface layer of deposited sediments (the surface or surficial sediments) within an estuary are characteristic of present conditions, whereas analysis of sediment cores can provide information on the deposited sediment at depth within the sediment column, which may indicate past estuarine conditions.

Various methods of collecting surface sediment samples are described in Chapter 6. These range from the manual deployment of scoops or lightweight grab samplers in very shallow waters, to the use of ship-borne mechanical winches for the deployment of heavy grab samplers. Similarly, core samplers range from handheld polycarbonate or PVC tubes that are pushed into the sediment, suitable for use on, e.g., exposed salt-marshes, to gravity and piston corers, vibro- and percussion corers, and box corers. References are made to the instruments and methods that attempt to sample sediment at the thin sediment–water interface.

Other topics covered in Chapter 6 include survey design and frequency of sediment sampling, the storage, preparation and pretreatment of samples, which may include washing, removal of biogenic calcite and organic matter, dispersion and drying. Frequently, particle sizes of the coarse-grained sediment fraction are determined by sieving. The fine-grained sizes are usually determined by sieving, gravimetric techniques or automated methods such as laser diffraction (granulometry). Other methods of size analysis are described, including electro-resistance using the commonly employed Coulter Counter. A discussion is given of other key sediment properties, such as particle shape, sediment structure, porosity and density.

Methodologies for the sampling and analysis of SPM, which encompasses suspended minerogenic sediment, are described by Mitchell, Uncles and Stephens in Chapter 7. They discuss various kinds of sampling platforms, from ships to bridges, and various sampling methods, from pumping to bucket collection, as well as land-based, moored platform-based, and boat-based surveys to determine SPM concentrations; the benefits of working tidal-cycle stations and utilising moored instrumentation are mentioned. Chapter 7 provides a description of the various methods that are currently used for measuring SPM concentrations, including gravimetric analysis and optical and acoustic measurements. The optical sensors that are discussed include the older Secchi Disc and Transmissometer instruments, as well as newer sensors such as the OBS sensor and those operating remotely from satellites and aircraft; acoustic sensors include the ADV and ADCP as well as the acoustic backscatter (ABS) sensor. Also discussed are the calibration requirements of these various instruments.

In addition to the sampling process, Chapter 7 also covers techniques to measure several important properties of the SPM; these include particle mineralogy, the particle size distribution and the effect of organic material and aggregation on particle size, in situ settling velocity of the SPM, and its specific surface area (SSA). Other properties covered are loss-on-ignition (LOI) and the particulate organic carbon (POC), chlorophyll-a, and extracellular polymeric substances (EPS) components of the SPM.

Fine, cohesive SPM plays an important role in muddy, strongly tidal estuaries and is greatly affected by floc formation, which is induced by turbulent mixing and stresses. Measurement methodologies that quantify these processes are described by Manning, Whitehouse and Uncles in Chapter 8. Flocs may settle as individual aggregated particles, or when suspended sediment concentrations are large, they may be subjected to 'hindered' settling. Chapter 8 begins with a discussion of clay mineralogy and the composition of biological material within muddy sediments, followed by the influences of SPM concentration and turbulence, salinity and organic matter on the flocculation of muddy SPM as well as the behaviour of mixed-sediment SPM.

Chapter 8 then provides a review of instrumentation and devices for floc measurements and floc sample collection. These include settling tubes and sensing instruments such as the Lasentec (which uses rapidly scanning laser light), the LISST (laser in situ scattering and transmissometry), and InSiPid (in situ particle imaging device). Simultaneous measurements of floc size and settling velocity are made using instruments such as INSSEV (a video-based floc camera system), LabSFLOC (a laboratory-based derivation of INSSEV), and PICS (a particle imaging camera system). The analysis and processing of floc data are described, together with theoretical results for turbulence and shear stresses and sediment deposition.

Sediment transport methodologies are discussed by Black, Poleykett, Uncles and Wright in Chapter 9. The topics include the direct measurement of sediment fluxes, using both classical vertical profiling techniques as well as ADCPs that estimate the SPM concentrations from acoustic backscatter intensity. Direct quantification of sediment bedload transport is difficult to achieve, and the use of mechanical traps and bedform migration rates for measuring this transport are described. An alternative methodology, particle tracking, is discussed as a method with which to determine the movement of suspended and bedload sediments through space and time. Recently, a dual signature sediment tracer has been introduced that combines a fluorescent colour signature with a paramagnetic signature. Sediment erosion methodologies are discussed, including the use of high-resolution recording altimeters for fine-scale deposition–erosion measurements and benthic flumes as a means of assessing seabed stability in terms of erosion thresholds and erosion rates.

3.3   Autonomous Sampling Platforms and Remote Sensing

The use of autonomous sampling platforms is discussed by Fishwick and Turton in Chapter 10. The emphasis of their chapter is on moored data buoys, although recent autonomous technologies are also covered. They include the platform design of moored data buoys and stress the legal obligations associated with the deployment of such buoys in the sea. Power generation and battery considerations are discussed, including solar charging, power generation from buoy motion and wind turbines. An example is given of such systems, using the Western Channel Observatory data buoy in the English Channel. Data handling and control considerations are discussed together with buoy communications via radio, GSM (Global System for Mobile communications), satellite and wireless Ethernet communications. Various types of data-buoy measurement capabilities are covered in Chapter 10, including those for temperature, conductivity, salinity, density, turbidity, optical (apparent and inherent optical properties) and ADCP-derived variables, as well as meteorological and sea-state measurements.

Satellite and aircraft remote sensing, primarily focusing on derived variables such as surface water level, temperature, salinity and suspended particulate matter is discussed by Lavender (Chapter 11). A discussion is given of the use of the electromagnetic spectrum for optical remote sensing and the topics of atmospheric correction, optical theory applications to the derivation of water constituents, and the use of algorithms to derive quantitative variables, as well as the remote sensing of bathymetry. Thermal imagery is discussed together with, as examples, the derivation of biogeochemical constituents and the mapping of water surface temperatures from satellites and aircraft. In addition, microwave remote sensing is discussed, and its use for water level measurement is illustrated.

4   Some Recent Observational Work, Its Instrumentation and Results

In support of the general rationale behind the handbook, we now seek to place it in the context of current ongoing research work in the field. It is clearly not possible to provide an exhaustive review here, but it is nevertheless important to stress the relevance of the handbook within the ongoing scientific debates on coastal and estuarine matters. Some of the more recent work on estuarine research topics and the instrumentation deployed are briefly summarised here, generally for the period 2010–2016. An attempt is given to provide a 'flavour' of the research conclusions for each topic; however, it is recognised that this brevity cannot do full justice to the work discussed. The types of instruments used and their product names are given for completeness, although this does not imply recommendation of these products by us or the publisher.

In the following text, the abbreviation SPM refers to suspended particulate matter, which includes suspended, inorganic mineral sediment; however, the abbreviation SSC is widely used to refer to suspended sediment concentration or suspended solids concentration, without reference to the organic content, which can vary widely within and between estuaries (Uncles et al., 2015). In this section the concentrations of SPM, which includes inorganic suspended sediment, e.g., suspended sand grains, are referred to by the abbreviation SSC (suspended solids concentration(s)).

4.1   Currents: Tides, Density, Winds and Turbulence

Generally, estuarine tides are forced by those in the adjacent coastal sea, and the associated currents are mainly horizontal because tidal wavelengths are very long compared with water depths. The shallow depths of most estuaries can also lead to the development of overtides and tidally induced residual currents (reviewed by Li, 2011) and, occasionally, tidal bores (Figure 1-2a and b). Cross-estuary spatial variations in the longitudinal tidal and residual currents are influenced by the cross-estuary channel shape (reviewed by Valle-Levinson, 2011); for example, the estuarine circulation

Figure 1-2 Some examples of fronts and tidal bores: (a), a spring-tide tidal bore in the upper Humber-Ouse Estuary, UK; (b), a spring-tide tidal bore in the upper reaches of the Dee Estuary, northern English–Welsh border region, UK; (c), a front separating waters with different colours between the Tamar Estuary's main channel (to the left) and the shallows closer to the mouths of the two subestuaries (to the right); (d), an ebb-tide spring-tide front separating shallower, near-shore (smooth surface) and deeper (ruffled surface) waters caused by flow separation at the end of a pontoon (right of centre) in the Plym Estuary, UK; (e), a tidal intrusion front during a late flood-tide spring tide in a small creek off the Dee Estuary, UK; (f), an ebb-tide river plume off Penang, Malaysia. Photographs: R. J. Uncles, except (c), which is cropped from an aerial photograph that is © Cornwall Council and is reproduced with permission. Photograph (f) is reprinted from Uncles (2011) with permission from Elsevier.

(Figure 1-1) has both vertically and horizontally sheared flows, with inflow at depth and outflow along the estuary sides (not depicted in Figure 1-1) and at the surface.

Wind stress on an estuary's surface generates waves (Chapter 5), surface currents and turbulence (reviewed by O'Callaghan and Stevens, 2011). Remote, nonlocal wind stresses generated by along-shelf winds cause an Ekman setup of water levels that can then propagate into an estuary. Currently, there are few reported measurements that highlight and quantify the interaction between waves and currents at estuary mouths and in tidal inlets. A review of the theoretical and modelling aspects of wave dynamics is given by Wolf et al. (2011).

Stacey et al. (2011) review sheared, stratified turbulence with applications to estuaries, plumes and coastal seas and explain that water-column stratification acts to decrease turbulent mixing, current shear acts to increase it, and turbulence acts to reduce both shear and stratification. They show that the total kinetic energy (KE) can be separated into the mean KE and the turbulent KE (TKE) using the Reynolds' decomposition (Chapter 4).

4.1.1 Tides, Overtides and Bores

O'Donncha et al. (2015) used HF radar (Coastal Ocean Dynamics Applications Radar – CODAR) measurements to describe surface flows in Galway Bay, Ireland. Two HF radars continuously measured radial water velocity components at a frequency of 25 MHz; the effective depth of measurement was 0.48 m, the sampling range 25 km, and the spatial resolution 0.3 km. Two radar systems were used to determine the water velocity because one system measures only the radial component of velocity along a line originating from its antenna (Chapter 4). Barotropic model simulations and HF radar data demonstrated good agreement for tidal currents, although the comparison with residual currents was much less satisfactory.

Sound transmission experiments, designed to measure tidally influenced river currents, were carried out by Zhu et al. (2012) in a freshwater, tidal-bore reach of the Qiantang River, China. Two coastal acoustic tomography (CAT) transceiver nodes were installed on opposite tidal-river banks to form a 3-km-long section diagonally across the Qiantang River. The broadband transducer (Neptune Sonar, Model T170) – operated for sound transmission and reception – was suspended at a depth of 3 m. The transducer's central frequency and bandwidth were 5 kHz and 2 kHz. The CAT was able to successfully monitor the abrupt fluctuations in river flow that resulted from the passage of a tidal bore. It also estimated steady river flows, in the absence of bores, which were well correlated with measured water-level data.

Bonneton et al. (2015) investigated the physics of tidal bores. In the Garonne Estuary, France, pressure sensors were deployed in shallow water, and two 1.2-MHz ADCPs (RDI), with vertical bin sizes of 0.2 m and 0.05 m, and one acoustic wave and current profiler (AWAC, Nortek) with a 0.5 m vertical bin size, were deployed along the estuary axis. The pressure sensors sampled at 10 Hz and other instruments (OBS-3A, Campbell Scientific; ALTUS Altimeters, IFREMER and MICREL, France) at 2 Hz. A seabed-mounted 1.2-MHz ADCP with a 0.1-m bin size, and an ADV (Nortek), which sampled at 2 and 32 Hz, respectively, were deployed in mid-channel, and three pressure sensors that sampled at 10 Hz were deployed over the river cross-section. Field data showed that the dimensionless tidal range was the parameter that governed bore intensity and that the waves associated with a bore in a rectangular channel were significantly different from those associated with undular bores in natural estuaries (e.g., Figure 1-2a and b).

4.1.2 Vertical Current Structure

De Vries et al. (2015) measured the vertical structure of main-channel currents in the Marsdiep basin estuarine system (Dutch Wadden Sea). A 1.25-m-high seabed frame was deployed that held an upward-looking, four-beam 1.2-MHz Workhorse Monitor ADCP with a bin size of 0.5 m, a ping rate of 0.43 Hz, and an ensemble-recording interval of 30 s (storing 10 pings), and a SBE 37-SM MicroCAT (Sea-Bird Electronics) CTD that recorded data every 30 s. Together with tidal-cycle, water-column profiling, the data showed that a flood-ebb asymmetry in bed friction led to greatest vertical shears in the lower part of the water column during the ebb, and that these shears led to reduced stratification (produced by along-stream tidal straining); in contrast, stratification was enhanced by cross-stream tidal straining during the late flood.

4.1.3 Transverse Current Structure

Measurements were made by Valle-Levinson and Schettini (2016) in Mossoró Estuary, northeastern Brazil. CTD and velocity profiling over estuarine transects was undertaken for full tidal cycles at spring and neap tides, and the data used in conjunction with those from moored instrumentation. A 1.2-MHz ADCP, with a vertical bin size of 0.25 m and recording every 0.4 s, was towed over a 200-m-long cross-estuary transect (approx. 2 km from the mouth) for 12.5 h during both wet and dry seasons; a YSI Castaway CTD (SonTek) operating at 5 Hz was used in the dry season and a Rinko CTD (JFE Advantech) in the wet season. Seabed-mounted instrumentation included a 1.0-MHz Aquadopp ADP (Nortek) with a bin size of 0.35 m and an Infinity-CTW (JFE Advantech) CTD sensor logging at intervals of 5 min. The data showed that the spatial structure of residual flows changed dramatically from neap to spring tides during the dry season. A residual surface inflow of coastal seawater occurred above a residual outflow of hypersaline estuarine waters at neap tides (the reverse of that shown in Figure 1-1), whereas spring-tide residual currents exhibited a laterally, rather than vertically, sheared structure, such that the residual flow was up-estuary in the middle of the section and down-estuary over the flanks. The spring-tide structure remained qualitatively similar during the wet season, whereas the neap-tide structure reverted to the classical estuarine exchange flow (Figure 1-1).

Waterhouse and Valle-Levinson (2010) measured the transverse variations of estuarine exchange flows and lateral flows in Ponce de Leon Inlet, US. In conjunction with bottom-mounted pressure sensors and ADCPs, vertical current profiles over cross-estuary transects were measured using a vessel-mounted 1.2-MHz ADCP with bottom-tracking capability. A SBE 19plus CTD profiler was used to measure vertical density profiles, and a SBE 37 CT sensor, recording at 0.2 Hz, measured surface density. Results identified a residual current inflow over the shoals and an outflow in the channel with small wind effects; residual inflows were increased and outflows decreased with onshore winds, whereas the reverse was true with offshore winds. Similar measurements were made by Waterhouse et al. (2013) in St. Augustine Inlet, US. Residual flow across the inlet mouth had outflow over the shoals and inflow in the channel. Reversal of this channel residual flow only occurred immediately following the strong winds associated with tropical storms.

Collignon and Stacey (2012) measured cross-channel variations in physical properties every 30 min throughout a tidal cycle over a shoal-channel interface in south San Francisco Bay, US. In addition, instrumented moorings were deployed on the slope and in the channel with ADCPs sampling at 1 Hz with a vertical resolution of 0.5 m in the channel and 0.25 m on the slope that were adjacent to CTD sensors and turbidity sensors sampling at 3- or 6-min intervals. Mooring lines were used to install near-surface CTD sensors. Other deployed instrumentation included ADVs (Sontek or Nortek) sampling the local current for 10 min every hour at 4, 8 or 10 Hz and a LISST sensor. Transect surveys across the shoal–channel interface measured velocity with a boat-mounted 1.2-MHz ADCP, recording 1-s ensembles with 0.25-m vertical bins, and vertical profiles of salinity, temperature, and turbidity. Cross-channel circulations over the slope had pronounced intratidal variability, and their directions greatly influenced the convergence fronts that occurred at the shoal edge (e.g., Figure 1-2c).

Other examples of this type of transverse survey include work by Garel and Ferreira (2013) in the lower Guadiana Estuary, Portugal; work by Valle-Levinson et al. (2015) at Destin Inlet, US; and work by Blain et al. (2015), who investigated a nonbuoyant shear-front (e.g., Uncles, 2011 and Figure 1-2d) at the confluence of Nanjemoy Creek and the main Potomac River channel, US; sets of two sequential remote-sensing images (Terra ASTER and Landsat ETM+) were used to estimate surface velocities, calculated according to the Global Optimal Solution (Chen and Mied, 2013).

Becherer et al. (2015) measured the circulation in a curved, well-mixed, tidal inlet in the German Wadden Sea; a moored, bottom-mounted 1.2-MHz ADCP operating in mode 12 (0.25-m bin size), and a vessel-mounted 1.2-MHz ADCP (0.5-m bin size) were used (all data averaged over 1 min). An SBE 37 MicroCat CTD sensor recorded data at 0.05 Hz at a depth of 1.5 m. An MSS90 (ISW, Germany) free-falling microstructure profiler with a precision CTD sensor (Sea & Sun Technology), a pressure sensor, a fast response temperature sensor (FP07) and two airfoil shear probes (PNS06 from ISW) were deployed (all sampling at 1024 Hz). Results showed that lateral straining of the salinity field led to significant stratification during the late flood tide. The lateral circulation comprised a single-cell structure that was strongest at late flood and largely absent during the ebb. This flood-ebb asymmetry was an important contributing mechanism in the generation of the estuarine circulation. Similar measurements were made by Russell and Vennell (2014) in a curved channel within Otago Harbour, New Zealand. They demonstrated the occurrence of secondary flows that were as much as 20 per cent of the primary flow and a vertical velocity, inferred from the secondary flow, as much as 1 per cent of the primary flow. The vertical velocity was downward on the outside of the bend and upward on the inside.

4.1.4   Wind Effects on Currents

Muscarella et al. (2011) analysed winds and currents in the mouth region of Delaware Bay, US. Measurements of surface current were made using two standard-range 25-MHz SeaSonde (CODAR) radars. Radial velocities, defined on a polar-coordinate grid with 5° azimuth cell spacing and 1.5-km range cell spacing for each system, were measured every hour. Results showed that surface currents, although dominated by tidal forcing, responded rapidly to changing winds. In a similar experiment, Piedracoba et al. (2016) analysed data from HF radar systems that operated for one year in the Ría de Vigo, Spain. HF radar data from two short-range SeaSonde (CODAR) radar antennas were used to derive surface currents in the lower third of the ría. The radar systems were separated by approx. 7 km. They operated at frequencies of 46.2 and 46.8 MHz and measured continuous radial current components every 30 min with a 10 km (maximum) horizontal range. Two oppositely directed eddy structures were identified at subtidal frequencies that were largely caused by wind forcing.

Currents in the inlets that connect the Venice Lagoon to the Adriatic Sea were analysed by Mancero-Mosquera et al. (2010) using 600-kHz Workhorse Sentinel ADCPs. They were used to profile currents in the inlets with a 1-m vertical bin resolution, a ping every 10 s, and an output, formed by averaging an ensemble of 60 pings, every 10 min. Analyses showed that, generally, the inlet inflows or outflows were strongly related to winds.

Whitney and Codiga (2011) analysed the effects of wind events on Long Island Sound, US. Current velocities were measured using a 600-kHz ADCP mounted on the ferry M/V *John H* as it steamed along its cross-estuary transect. Currents were averaged in 2-m vertical bins. Tidal residual currents were most strongly modulated by spring–neap variations and winds. Based on the wind-affected component of the velocity (velocity anomaly), down-estuary wind events drove downwind, surface-enhanced anomalies, flanked by deeper, central, upwind anomalies, whereas up-estuary wind events drove weaker, broader and deeper downwind anomalies.

4.1.5 Wave Effects on Currents

Bolaños et al. (2014) assessed the importance of wave–current interactions in the hypertidal Dee Estuary, UK. A tripod moored in a major channel held an upward-looking ADCP that measured velocity profiles with a vertical bin resolution of 0.5 m. A second, smaller rig was deployed in another major channel that measured near-bed currents (ADV) and pressure. Estimates were made of wave parameters (Chapter 5). Tidal variations in water depth largely controlled the significant wave height, and the tidal current-induced Doppler shift affected wave periodicities.

Wargula et al. (2014) quantified the residual, longitudinal momentum balance in a shallow, well-mixed coastal sea inlet (New River, US). Pressure sensors were deployed near, and in, the inlet mouth that measured at 2 Hz for 3072 s every hour. These were collocated with either ADVs that sampled at 2 Hz for 3072 s every hour or with ADCPs that were used to estimate wave directions and currents (Chapter 4 and Chapter 5). Moored ADCPs at shallow sites used 0.25-m vertical bins and 1-min averaging for the currents. One main-channel ADCP measured currents in 0.50-m vertical bins every minute for 12 min, repeating every 30 min. It also sampled near-surface flows and near-bed pressures at 2 Hz for 1024 s every 30min. Over the km-wide ebb shoal, data showed that the longitudinal pressure gradient, the bottom stress and the wave radiation-stress gradient were the dominant forcing terms in the local, along-inlet momentum balance; nonlinear advective terms were important in the channel. Breaking waves generated onshore radiation-stress gradients, especially during storms, which enhanced flows into the inlet during the flooding tide.

4.1.6   Estuarine Turbulence

Orton et al. (2010) used a small, self-orienting catamaran, anchored at a 5.1-m-deep site in the Hudson River Estuary, US, as a platform for measuring near-surface turbulence, wind velocity and air–estuarine fluxes of heat. A 1.2-MHz, bottom-mounted ADCP with CTD measured vertical profiles of velocity, acoustic backscatter and turbulence properties close to the catamaran site. The ADCP used rapid-sampling mode 12 and recorded one ensemble average every second, which was an average of 21 subpings (raw, unstored data) that were collected over 0.63 s; the vertical cell size was 0.25 m. The observations indicated that wind was the principal controlling factor for turbulent dissipation at 0.5 m beneath the surface. On a sunny, calm-weather day, solar heating and tidal straining caused similar buoyancy fluxes that promoted restratification during the ebbing tide.

MacVean and Lacy (2014) made measurements on a shallow estuarine mudflat in northern San Francisco Bay, deploying bed-mounted tripods over a cross-shore mudflat transect. ADVs were distributed among the tripods and recorded mean currents, waves and turbulence by sampling at 10 Hz for 412 s every 15 min. Multiple-height CTD sensors and OBS sensors sampled every 15 min. They observed energetic turbulence in the presence of strong stratification, which was reconciled by considering the TKE budget to show that, in general, dissipation and buoyancy flux were balanced by local shear production, and all of these were increased during wave events.

Talke et al. (2013) used an infrared (IR) remote camera with in situ instrumentation to measure turbulence in the Snohomish River Estuary, US. The equipment was mounted on the end of a retractable A-frame, which extended 6-m upstream of their deployment barge. Sequential IR images were used to estimate the characteristics of the surface flow using a particle-image-velocimetry technique that delineated the surface manifestations of bottom-generated coherent structures or boils in an unstratified reach of the estuary. Other instrumentation included ADVs, an upward-looking ADP (Acoustic Doppler Profiler; Nortek) at a depth of 1.05 m and recording at 8 or 4 Hz; a sideways-looking ADP at a depth of 0.3 m; a downward-looking 1.2-MHz ADCP deployed in mode 12 with a frequency of 1 Hz and 0.25 m bins; and an SBE 19plus CTD profiler recording at 4 Hz. It was concluded that coherent structures and their surface manifestations played an important role in the vertical transport of TKE and the water column distribution of TKE dissipation.

Zippel and Thomson (2015) used wave-following buoys designed for near-surface measurements (SWIFT drifters) to investigate wave breaking and turbulence in the presence of currents at New River Inlet, US. The drifters measured wind, position, drift speed, waves (using a Qstarz BT-Q1000eX GPS) and turbulent dissipation rate (using a Nortek 2-MHz Aquadopp HR); took images (using a GoPro Hero); and had local tracking capability (using Garmin Astro collars). Wave breaking was observed in deeper water with opposing currents during the ebb. Measured turbulent dissipation rates for the wave-breaking regions were used to evaluate two wave-turbulence models.

Pan and Gu (2016) used survey data and modelling to investigate turbulent mixing in the estuary and plume of the Pearl River, China. The instruments used in the cruise survey included a Sea & Sun Technology MSS 90 microstructure probe; a surface sampling SBE CTD; a Sea & Sun Technology CTD; and a ship-mounted 600-kHz ADCP. Underway and 25-h anchor station data were collected. They found that mixing was stronger during the ebb than the flood inside the estuary and in the near-shore waters, and that estuary mixing was controlled by tidal current speeds.

Huguenard et al. (2015) made spring- and neap-tide velocity and microstructure measurements in the James River, US. Vertically profiled data were obtained for one tide over a 2-km transect using a 1.2-MHz ADCP, sampling at 2 Hz with a 0.25-m vertical resolution mounted on a small catamaran (1.2 m). At four cross-estuary locations a Self-Contained Autonomous Microstructure Profiler (SCAMP, Precision Measurement Engineering, PME) was repeatedly deployed. Data showed that during the ebb tide, near-surface mixing was related to lateral circulations.

4.2 Salinity: Salt Intrusion, Fronts and Plumes

Estuarine circulation and exchange flows at the mouth of an estuary (Figure 1-1) are important to the estuarine salt transport (MacCready and Banas, 2011). The tidally averaged rate of salt transport through an estuarine cross-section can be divided into components (e.g., Uncles, 2002) that are driven by the estuarine circulation (and by other sources of spatial shear in the currents), by the freshwater discharge into the estuary, and by the so-called tidal pumping due to oscillatory water transport (sometimes referred to as tidally induced dispersion).

Salinity stratification is often a pronounced feature of estuaries and their coastal waters (Figure 1-1) that can exhibit strong intratidal variability, ranging from stratified to vertically mixed during a single tidal cycle (e.g., Uncles, 2002). The buoyancy that is associated with this stratification can lead to the formation of frontal systems between mixed and stratified waters (e.g., Figure 1-2e and f) or between horizontally distributed water masses of different density (Figure 1-2c). A review of surface fronts is given by Uncles (2011).

Plumes present a stratified and spreading outflow of buoyant waters and associated material to the coastal zone (Figure 1-2f) that is affected by tidally driven exchanges of water, momentum and materials (reviewed by Chant, 2011). Turbid plumes form at many river mouths and affect their coastal areas because of the low-salinity waters, sediments, pollutants and nutrients that they transport. These buoyancy-affected coastal areas are often referred to as ROFIs (regions of freshwater influence) and exhibit stratification that tends to vary with wind speed and the spring-neap cycle. A review of coastal physical oceanographic processes is given by Prandle et al. (2011).

4.2.1 Salt Intrusion

Field observations throughout a tide in the Rotterdam Waterway were undertaken by de Nijs et al. (2011). Vertical profiles of turbidity were measured with OBS-3 and OBS-3A (with integral CTD) sensors. The OBS-3 probes, together with pressure sensors and CT probes (Falmouth Scientific Inc.), were mounted at the same level on frames. The operating frequencies were 1 Hz and, for the OBS-3A, 2 Hz. Vessel-mounted 1.2-MHz ADCPs were used to profile velocities, using mode 1, a bin size of 0.5 m, and an ensemble averaging period of 5 s with 15 water pings. They concluded that the combined baroclinic and barotropic forcing, together with turbulence suppression at the water-column density interface, accounted for the observed mean and time-dependent flow structure, and that advection governed the salt-wedge structure and movement.

Aristizábal and Chant (2015) measured salt fluxes in the Delaware Bay, US. Instrumented moorings were deployed across a mid-reach section of the estuary (68 km from the mouth), together with a single, mid-channel mooring (at 54 km). Generally, bed-mounted 1.2-MHz ADCPs acquired velocity profiles at 1 Hz for 2 min every 10 min with a bin size of 0.25 m (0.5 m at 54 km). Some moorings had surface CT probes, and all had bottom CT probes that sampled once every 10 min. Cross-estuary boat surveys utilised a vessel-mounted ADCP and a tow-yo'd CT (RBR) sensor. Results showed that the section-averaged, tidally averaged advective salt fluxes exceeded those due to tidal oscillations (tidal pumping) and shear dispersion.

Behrens et al. (2016) measured salinity in the bar-built Russian River Estuary, US, following closure of its tidal inlet (the source of its ocean saltwater) by coastal sediments. Boat-based and moored instrumentation included an SBE 19plus CTD logger for profiling; HOBO pressure-temperature loggers (recording at 10-min intervals); bed-mounted, 1.2-MHz Workhorse Sentinel ADCPs with 0.5-m vertical bins, which operated in mode 12 and sampled at 12 subpings per second during 10-min bursts. Following closure, topographic sills in the central reaches prevented up-estuary movement of the trapped, dense, saline waters. Subsequently, wind-forcing events produced a one-way transport of high-salinity waters over the sills and into the deeper pools of the upper estuary.

4.2.2   Salinity Stratification and Fronts

Observations in the shallow, strongly tidal, salt-wedge Snohomish River Estuary, US, were made by Giddings et al. (2011). Moored equipment comprised bottom-mounted 1.2-MHz ADCPs with 0.25-m depth bins, operating in mode 12 and averaging 10 subpings per 1-Hz sample; Nortek ADVs (16 Hz in 10-min bursts every 50 min); SBE 16plus CTD sensors (one sample per minute); a Nortek Aquadopp Profiler (ADP) with 0.4-m depth bins, and a CTD and RBR pressure sensor, all sampling once per minute; and a 600-kHz ADCP, recording ensembles once per minute. Transect-survey measurements utilised a 1.2-MHz ADCP and an SBE 19 CTD. Results showed that vertical stratification was minimised during flood-tide springs due to greater bottom-generated TKE and the influences of tidal straining. During fast ebbing currents, advection and tidal straining increased stratification but could eventually lead to strong mixing at a sharpened density interface.

Scully and Geyer (2012) made spring-tide observations in the Hudson River Estuary, US, during a period of higher freshwater flows. Their moored instrumentation comprised a bottom-mounted, 1.2-MHz broadband workhorse ADCP that sampled every

2 s and recorded data in 0.25-m bins, and a profiling winch that was connected to an SBE 19 CTD. A vessel employed an instrument array that included a 25-Hz ADV (Sontek), an SBE 7 micro-conductivity probe (operating at 300-Hz) and a 6-Hz (RBR) CTD sensor. The vessel also employed a downward-looking 1.2-MHz ADCP that utilised a bin spacing of 0.25 m and sampled every 2 s. Results showed that along- and cross-estuary advection of horizontal gradients in the vertical salinity gradient, and tidal asymmetry in the magnitude of the vertical mixing, needed to be taken into account to explain the intratidal behaviour of salinity stratification.

Baroclinic surface fronts in Skagit Bay, US, were investigated by Mullarney and Henderson (2011) using vessel transects and deploying 2-MHz Nortek Aquadopp ADCPs and XR-420 (RBR) CT sensors and XR-620 CTD profilers. Moored ADCPs were used to record velocities over longer periods together with CT sensors that were located at 0.1 and 0.6 m above the bed and recorded every 10 s. The ADCPs used either 0.1-m bins and recorded every 20 s or operated in pulse-to-pulse coherent mode, with 8-Hz sampling and 0.025-m vertical resolution. Surface convergence fronts were observed to form along the edge of the channel in shallow water (<0.7 m) soon after the flood-tide inundated the adjacent tidal flat (e.g., Figure 1-2c) and moved away from the edge later in the flood tide.

4.2.3   Estuarine Plumes

Molleri et al. (2010) studied Amazon plume dynamics using satellite data from Sea-WiFS (Sea-viewing Wide Field-of-view Sensor, Chapter 11) ocean colour images, which were related to sea-surface salinity (SSS) via calibration with ARGO float salinity data (Chapter 4). Fournier et al. (2015) used SSS data derived from the SMOS mission (Chapter 11), and satellite data from SMOS, Aquarius, MODIS (Moderate Resolution Imaging Spectroradiometer) were used by Korosov et al. (2015), together with model data from the TOPAZ system, for studying the Amazon plume. The studies showed that freshwater flow was the dominant influence on the variability of the plume's surface area; winds controlled the plume's residence time close to the mouth of the estuary, and its coastal transport and surface currents had an important effect on the plume's dispersion patterns.

Long-term, medium-resolution, optical satellite imagery was used by Petus et al. (2014) to monitor the Adour River turbid plume (south-eastern Bay of Biscay, France). This study was based on the MODIS-Aqua surface reflectance data (MYD09). The data had a high frequency (1/day), had a spatial resolution of 250 m and facilitated the plume's monitoring over a multi-annual time period. The plume was shown to be mainly controlled by the river discharge rates and modulated by winds.

Chung et al. (2014) utilised images from the Formosat-2 satellite, which provided a 2-m resolution, together with data on environmental factors to study the physical behaviour of the Gaoping River plume, Taiwan. It was found that the plume's offshore and onshore movements were controlled, respectively, by the flood and ebb tides and that the freshwater inputs were directly correlated with the plume's core size. Remote sensing was also used by Fernández-Nóvoa et al. (2015) to produce images related to the turbid plume that formed at the mouth of the Ebro River, Spain; data from the

MODIS sensor onboard the Aqua and Terra satellites were utilised. The plume was mainly influenced by freshwater flows, followed by wind and regional Mediterranean Sea circulation.

Kilcher and Nash (2010) measured properties of the Columbia River plume, US. Velocity profiles were measured using a bottom-tracking, hull-mounted 300-kHz ADCP with a 1.5-s averaging time and a 1-m bin size between 4.5 and 100 m below the surface, and a bottom-tracking, 1.2-MHz ADCP mounted alongside the ship with a 1.5-s averaging time and 0.5-m bins to cover depths between 2.2 and 24.2 m. Acoustic backscatter from turbulence (and zooplankton) was profiled and recorded every 0.5 s in 0.017-m bins, using a 120-kHz Biosonics acoustic echo sounder. A Chameleon microstructure profiler was used to profile for velocity shear and other in-water physical variables. Surface roughness was derived, and surface frontal spatial properties determined, from images recorded every 2 min by the ship's X-Band radar. The plume lifted from the bottom as the front moved offshore and then behaved like a freely propagating gravity current; winds altered the front's shape, and local currents controlled intratidal variability in front speed.

Huguenard et al. (2016) made measurements in the Choctawhatchee River plume, US. Their acquisition systems were simultaneously sampled using a GPS-linked synchronization server. Velocity profiles were recorded at 0.8 Hz using a downward-facing, vessel-mounted 1.2-MHz ADCP with a 0.25-m bin size. Water levels and waves were monitored at 50 Hz using an ultrasonic distance meter. A CTD profiler (YSI Castaway) sampled at 5 Hz during profiling deployments. Two ADVs (Sontek), sampling at 20 Hz and located at 0.14 and 0.41 m beneath the surface, and a vertical microstructure profiler (Rockland Scientific), sampling at 512 Hz, were used to estimate TKE dissipation rates. It was estimated that frontal processes provided 59 per cent of the overall mixing.

4.3   Estuarine Suspended Sediments

The speed and flood-ebb asymmetry of tidal currents, wave activity (Chapter 5), the residual wind-driven and tide-induced currents, and the estuarine circulation can all have a strong impact on sediment dynamics, as can the sediment supply from an estuary's river systems and coastal sea (e.g., Winterwerp, 2011). Much of an estuary's fine, cohesive SPM exists in a predominately flocculated state, and this adds considerable complexity to the prediction of fine-cohesive and mixed-sediment sediment transport (Chapter 8). An ETM is often observed in the upper reaches of estuaries (Figure 1-1), although the processes that contribute to sediment trapping within the ETM are very variable and differ in dominance at different times within an estuary and for different estuaries (Figure 1-3). Satellite observations provide an increasingly useful tool for cataloguing this variability in SPM concentrations (SSC) within estuaries, their plumes and coastal seas (Chapter 11 and, e.g., Moreira et al., 2013).

4.3.1   Resuspension and Sedimentation

Instrumented tripods were deployed by Liu et al. (2011) in the North Passage of the Changjiang (Yangtze) Estuary. Each had an up-looking 1-MHz current profiler

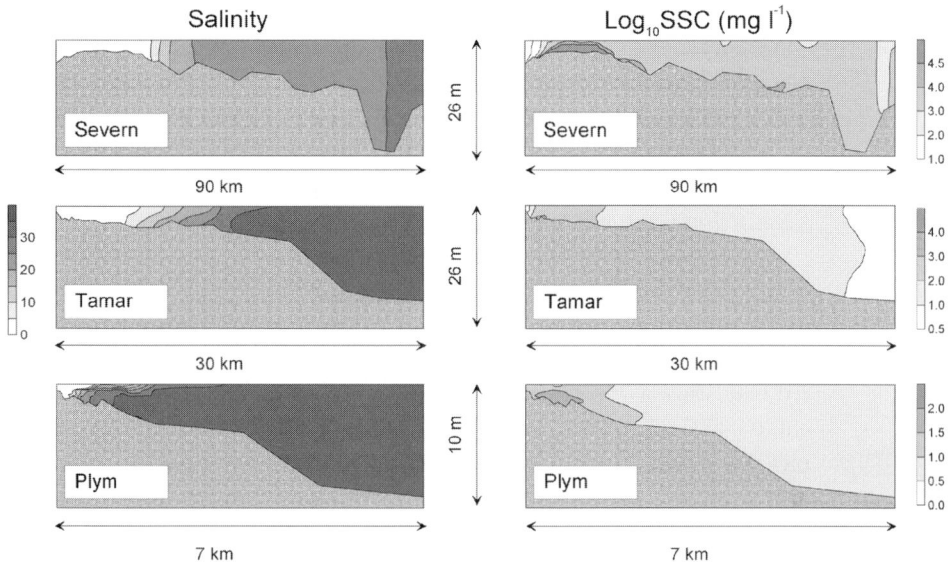

Figure 1-3 Contour plots that illustrate salinity intrusion and stratification (left panel) and SPM concentrations (SSC on a logarithmic scale, right panel) close to high water of spring tides for three strongly tidal UK estuaries having widely differing tidal lengths (the Severn data are plotted for the upper 90 km of its tidal length). All these estuaries exhibit an ETM in their upper reaches and have very different salt intrusion lengths, salinity stratification and ETM magnitudes.

(ACP, ADP or AWAC) and a down-looking 2-MHz current profiler (ACP), or an ADV Ocean (Sontek) for measuring velocity at a point, and OBS sensors for turbidity measurements. A strong salinity front was observed that induced sediment trapping and near-bottom convergence and deposition of sediment.

Shen et al. (2013) collected annually averaged SSC in, and monthly averaged sediment discharges to, the Changjiang (Yangtze) Estuary and utilised the MEdium Resolution Imaging Spectrometer (MERIS, Chapter 11) level 1b, top of the atmosphere radiances, to quantify the variability in SSC. The results showed that the upper estuary had substantial seasonal and annual variations in SSC, which were a response to annual and seasonal variability of the river flow.

The trapping of SPM by an estuarine plume's halocline throughout a spring-neap cycle was studied by Ren and Wu (2014) during the dry season in the Pearl River Estuary, China. Their bottom-mounted tripod held instruments that included a Pulse Coherent Acoustic Doppler Profiler (PC-ADP, Sontek) with a central frequency of 1.5 MHz, 100–200-Hz 3D point ADVs (Nortek), a 1-Hz XR-420 CTD and XR-620 OBS (RBR) sensors. A LISST-100B (Sequoia Scientific) was used for profiling, and a shipboard 2-MHz Aquadopp profiler ADP (Nortek) sampled every 5 min for an average interval of 90 s with a bin size of 0.1 m. Observations showed that during the neap tide, when the estuary was highly stratified, larger floc-sized particles (Chapter 8) were concentrated on the plume halocline.

Gensac et al. (2016) used MODIS data (MOD09Q1, MOD09A1, MYD09A1 and MYD09Q1) to study the dynamics of large mud banks that occur in the shallow coastal waters between the mouth of the Amazon River and the northern part of the Amapa region of Brazil and the eastern part of French Guiana. A single-band algorithm was used to compute SSC, based on the infrared (841–876 nm) MODIS 250-m reflectance band. The results were used to estimate mud-bank migration rates, which were as fast as 5 km y^{-1} in some regions, and their variability.

4.3.2   Suspended Sediment Fluxes and the ETM

Constantin et al. (2016) developed a water-turbidity retrieval methodology for the Danube Delta coastal area, Black Sea, using both MODIS Aqua and Terra satellite red-band data at 250-m resolution. The atmospheric correction was particularly important (Chapter 11). An HI-98713 turbidity meter (Hanna Instruments) was used to measure in situ turbidity near the Danube River mouth for calibration purposes. The data showed a strong relationship between water turbidity in the coastal area and the Danube's freshwater discharge.

Mariotti and Fagherazzi (2011) investigated sediment transport and currents in a small channel and its flanking mudflat in Willapa Bay, US using 2-MHz ADPs (Nortek). They measured profiles of velocity with a cell size of 0.10 m, a burst-averaging time of 60 s and a recording interval of 15 min. A pressure transducer (RBR) was located in the upper channel and burst-sampled 512 data at 2 Hz every 15 min. SSC was estimated using the acoustic backscatter (ABS) signal of the ADP. Sediment fluxes were calculated by combining current flows and the SSC estimates. Dominant ebb-tide transport resulted in a net export of sediment from the channel.

Song et al. (2013) measured sediment fluxes during the dry season in the Deepwater Navigation Channel of the Changjiang (Yangtze) River Estuary, China. A seabed quadrapod was deployed, instrumented with ADCPs, an ADV, and OBS 5+ sensors to measure turbidity. A XR-420 (RBR) pressure sensor with a CTD data logger was also used. An SBE 19 CTD with an OBS 3+ probe attached were used for vertical profiling. The down-estuary sediment fluxes were dominated by river flows, whereas the up-estuary fluxes were driven by tidal-pumping effects during spring tides and shear processes during neap tides.

Downing-Kunz and Schoellhamer (2013) measured fluxes of SPM in Corte Madera Creek, San Francisco Bay, US. The along-channel velocity was measured using a two-beam, side-looking 2-MHz EasyQ ADV (Nortek USA). The ADV recorded velocity in three bins in the horizontal plane, using a 4-min average every 15 min. The ADV also measured water level via a vertical beam that was reflected off the water surface, recorded as a 2-min average every 15 min. Instantaneous measurements of channel discharge were made using a boat-mounted, downward-looking 1.2-MHz Workhorse ADCP. Other instruments included an SSC-calibrated multiparameter water quality sonde (Model 6920 V2, YSI, Inc.) equipped with an optical turbidity sensor (Model 6136, YSI, Inc.). The residual flux of suspended sediment was seaward during wet periods and landward during dry periods. During summer, the landward flux was attributed to wind–wave resuspension of sediment in the estuary and subsequent up-estuary transport during flooding tides; during autumn, it was attributed to increased spring-tide flood current speeds, leading to local resuspension and increased up-estuary transport.

Nauw et al. (2014) used long-term, ferry-based measurements to determine the fluxes of SPM through the Marsdiep inlet (Dutch Wadden Sea). A 1.0-MHz ADCP (Nortek) was mounted beneath the ferry as it repeatedly crossed the 4.5-km wide inlet. After calibration, the ABS intensity was used to estimate SSC profiles, so that an SPM flux could be determined from these SSC and velocity profiles. Results showed that tidal asymmetry led to a residual import of SPM through the inlet.

Nowacki et al. (2015) measured velocity, salinity, and SSC for 25 h over three cross-sections in the tidal Song Hau distributary of the Mekong River, Vietnam. Ship-based surveys were made during high and low river-flow seasons, using a 600-kHz ADCP, an XRX-620 (RBR) CTD with integrated OBS-31 sensor, both sampling at 6 Hz. The vessel repeatedly traversed estuarine cross-sections. There was a seaward sediment export during high river flows and a landward import during low flows. Sediment export during high flows was mainly caused by fluvial advection of sediment. Tidal processes and local resuspension, together with exchange flows (Figure 1-1), were largely responsible for the residual sediment import during low river flows.

Carlin et al. (2015) described the interactions between SPM and the salt wedge in the lower reaches of the Brazos River, US, using vessel-based survey data. The instruments included an SBE 19plus SeaCAT Profiler CTD and an XR-420 (RBR) CTD with OBS sensor that were deployed for profiling. Bathymetry data were collected with a 0.05-m vertical resolution using a 200-kHz C3D-LPM (Teledyne Benthos) high-resolution sidescan sonar and swathe bathymetric system (Chapter 3). An SB-216S (EdgeTech) Full Spectrum Sub-bottom CHIRP seismic sonar system, operating on 2–16 kHz frequencies, was used to collect subbottom seismic data. Depending on the type of seabed, penetration depths for this system typically were between 6 and 80 m, with a 0.06- to 0.10-m vertical resolution. Results showed that the salt wedge modulated the export of sediment to the Gulf of Mexico by trapping sediment within the lowermost 9 km of the estuary to form a decimeter-thick mud layer. At higher river flows the salt wedge was not present, and the lower-estuary seabed changed from depositional to erosional.

Sommerfield and Wong (2011) measured SPM fluxes in the ETM zone (illustrated for three estuaries in Figure 1-3) of the Delaware River Estuary, US. The instruments used included Ocean Probe ADVs (Sontek), 1.2 MHz and 600-kHz ADCPs mounted on seabed frames with SBE 37 CT and OBS-3 sensors; SBE 37 and OBS-3 sensors were mounted on a mooring cable. The ADV and OBS-3 turbidity sensors sampled in 2-min bursts at sampling rates of 1–6 Hz every 15 min, whereas the SBE 37 CT sensors sampled in 3-s bursts every 5 min. To generate profiles of SSC, the ADCP echo intensity was calibrated with data recorded by the moored OBS-3 sensors. Results indicated that residual sediment fluxes were largely driven by the estuarine circulation (Figure 1-1). Tidal pumping was a contributing mechanism to the residual sediment flux in the channel near the ETM, although its direction and magnitude in the upper estuary varied with river flow and resident sediment-source budget. Over the subtidal flats, tidal pumping dominated, although residual sediment fluxes were weak.

Sediment trapping in the ETM of the Changjiang (Yangtze) Estuary was investigated by Wu et al. (2012) using longitudinal, profiling transects and 25-h anchor stations located up- and down-estuary of the ETM zone (e.g., Figure 1-3). Instruments included a ship-mounted, downward-looking 600-kHz ADCP with sampling intervals of between 1.6 and 30 s and a 0.5-m bin size; an OBS-3A in profiling mode to measure temperature, salinity and turbidity with sampling intervals of 1–10 s; and a LISST-100C (Sequoia Scientific) in profiling mode to measure the volume concentration of SPM with a sampling interval of 2 s. In the high river runoff (flood) season, the ETM appeared as a high-concentration hyperpycnal current that was thought to have been formed by flocculation and rapid settling. In the low river runoff (dry) season, the ETM appeared as a low-concentration sediment cloud throughout the water column, possibly caused by sediment resuspension.

An analysis of measurements made in the Hudson River Estuary, US, was presented by Ralston et al. (2012). They used seabed instrument frames that were deployed on the shoals and in the channel. Their instruments included CT and OBS sensors at multiple heights above the bed; ADVs; a PC-ADP that measured velocity and ABS in 0.01 m bins through the near-bottom 1 m – ABS sensors profiled at 1, 2.5, and 5 MHz over the same depth range; an upward-looking ADCP that profiled currents and ABS over the water column; and a surface buoy fitted with CT and OBS sensors. Their analysis showed that fluxes of SPM were down-estuary on the shoals and strongly up-estuary in the channel within the saline reaches of the estuary. Marked changes in width or depth were associated with the formation of seabed salinity fronts. Local maxima in SSC and sediment deposition were created near these fronts, due to sediment flux convergences, which led to the formation of secondary ETMs at marked changes in bathymetry.

4.3.3   Sediment Processes – Wave Influences

Brand et al. (2010) investigated sediment dynamics at shoals in South San Francisco Bay, US. Two stations were utilised on a line that was perpendicular to the axis of a deep channel. The instruments used included 10-MHz ADVs (Sontek Hydra) that recorded either 8-min bursts at 10 Hz every 12 min, or 10-min bursts at 8 Hz every 60 min, and a 6-MHz ADV (Nortek Vector) recording 10-min bursts at 8 Hz every 60 min; a LISST-100B with a particle size range of 1.2–250 µm; a downward-scanning Model 881A sonar (Imagenex) to monitor bed forms; and a Ponar grab (Chapter 6) that was used to sample bottom sediments at both stations. The majority of resuspension events were associated with flood-tide currents that followed low-water wave events.

Measurements of very small waves caused by light winds, with heights <0.1 m and periods between 1.0 and 1.8 s, and their accompanying resuspension of sediment, were analysed by Green (2011) for an intertidal flat in the microtidal Tamaki Estuary, New Zealand. Instruments used included a Druck ceramic pressure sensor (GE Sensing), with a sampling frequency of 5 Hz, 256 data per burst, and a burst interval of 4 min, which was mounted 0.25 m above the bed; a 10-MHz ADV (Sontek) with the measurement volume located 0.1 m above the bed and a sampling frequency of 16 Hz, with 2048 samples per burst and a burst interval of 4 min; an Infrared OBS (Seapoint) at

0.1 m above the bed with a sampling frequency of 5 Hz and 256 data per burst with a burst interval of 4 min. Results showed that sediment resuspension was controlled by wave period because of its effect on the wave friction factor. The top of the tidal flat experienced the greatest mass of wave-resuspended sediment; wave-induced resuspension was smaller farther shoreward because of wave dissipation and was smaller farther seaward because of the increased attenuation of orbital motions due to greater depths.

4.3.4 Suspended Sediment Flocculation

Observations were made by Braithwaite et al. (2012) in the Menai Strait, UK. An instrumented bottom-mounted frame included an upward-looking 1.2-MHz Sentinel Workhorse ADCP measuring in 0.3-m vertical bins at 1 Hz and a LISST-100C. The TKE dissipation rate was estimated using the structure function method (Chapter 4). Largest particles were observed at slack tide and the smallest at times of maximum flood and ebb, presumably due to floc aggregation and disaggregation during slow and fast flows, respectively.

Observations were made by Wang et al. (2013) in the macrotidal Jiulong River Estuary, China. A seabed tripod was deployed that held 6-MHz ADVs (Nortek Vector) using a burst-sample duration of 2.13 min at 20-min intervals with a 16-Hz sampling frequency; electromagnetic current meters (EMCM, JFE Advantech) that measured the horizontal components of flow with 2-min bursts at 20-min intervals and with a sampling rate of 1 Hz; and an OBS-3A and a SBE 26 pressure transducer. An anchored vessel was used to deploy a downward-looking 1.2-MHz ADCP recording every 2 s, with a bin size of 0.15 m, together with hourly profiling using a LISST-100X (Sequoia Scientific) and an SBE 25 CTD. The results exhibited intratidal variability in flocculation processes. The upper water column was associated with larger suspended particles and the bottom boundary layer with finer particles. Turbulence appeared to control particle aggregate size, which was inversely related to the turbulence dissipation (Chapter 8).

4.4 Bed Sediments

The erosion properties of bed sediments (especially muddy sediments – reviewed by Winterwerp, 2011) depend not only on their physical characteristics but also on the biological influences to which they are subjected. For example, water velocities and roughness lengths are affected by the presence of plants on the bottom, e.g., seagrass meadows, beds of kelp and other macrophytes (Nepf, 2011; Venier et al., 2012). Polysaccharide polymers (extra-cellular polymeric substances, EPS) are produced by benthic diatoms and other organisms that inhabit intertidal mudflats in estuaries. EPS influences the local stability of the muddy sediment by increasing its threshold for erosion (Andersen and Pejrup, 2011). EPS may also act as a source of organic material that coats the surfaces of suspended fine sediments and enhances their ability to aggregate and produce large flocs (Chapter 8). Biodestabilisation mainly results from the bioturbation caused by burrowing and deposit-feeding animals (Andersen and Pejrup, 2011).

4.4.1   Bed Sediment Erosion

The stability of cohesive sediments in the Venice lagoon, Italy, was measured in situ by Amos et al. (2010) using the benthic flume 'Sea Carousel'. The Sea Carousel provided a time series of sediment erosion, which usually lasted about 90 min and comprised 8–9 velocity steps, each of 5-min duration. Erosion thresholds and erosion rates were estimated, and it was found that the higher thresholds for intertidal stations were the result of air exposure, which influenced consolidation, density and organic adhesion.

Salehi and Strom (2012) made in situ measurements of the critical shear stress for the onset of erosion in the microtidal San Jacinto estuary, US. A 6-MHz ADV Vector (Nortek) was used to measure the velocity, SSC (*via* water sample calibration) and relative changes in bed height (sampling volume to the bed recorded at 1 Hz). Results showed that when estimating bed shear stress, the logarithmic profile method performed best in current-dominated regions, whereas the modified TKE method performed best in the wave-dominated zone.

Ge et al. (2015) estimated critical shear stress for erosion in the Changjiang Estuary, China. Time-series of sea surface SSC were derived from Geostationary Ocean Colour Imager (GOCI) satellite data (Chapter 11) at hourly intervals and combined with bed-property surveys and hydrodynamic modelling to estimate the critical shear stress for erosion in the clay-dominated bed region.

4.4.2   Bed Sediment Characterisation and Habitats

Changes in the thickness of the upper soft mud layer in the York River Estuary, US, were measured by Rodríguez-Calderón and Kuehl (2013) using a dual-frequency mobile hydrographic echo sounder (200/33-kHz Echotrac CVM system). A Smith-Macintyre grab sampler (Chapter 6) and a Gomex box corer were used to collect sediment samples. Water content trends showed no correlation with seabed stability, and observed changes in the soft mud thicknesses were interpreted to result from both biological and physical factors.

High-resolution imagery of seafloor terrain in the Bay of Fundy, Canada, was analysed by Todd et al. (2014). Their instruments included a Kongsberg EM3002 with 160–254 beams operating at 300 kHz (Chapter 3), an IKU (Institutt for Kontinental-sokkel Undersøkelser) grab sampler to obtain large-volume (approx. 1 m^3) sediment samples and a small van Veen grab sampler with a 0.2-m sampling depth capacity (Chapter 6) and high-resolution seafloor imagery using Campod, an instrumented tripod equipped with video and still cameras.

Carey et al. (2015) undertook sediment-profile imaging (SPI) and plan-view (PV) imaging surveys to map sedimentary habitats within the Milford Haven Waterway, UK. The instruments used included Ocean Imaging Systems (OIS) Model 3731 SPI and Model DSC16000 PV camera systems, both equipped with Nikon D7000, 16.2-megapixel SLR (single-lens reflex) cameras, and two OIS Model 400–37 Deep Sea Scaling lasers that were mounted on the DSC16000, attached to the sediment-profile camera frame and used to provide scaled PV photographs of the seabed. The PV rig included a housing and deployment system for the Nikon D-7000 camera. Stations were classified in terms of their location and their (inferred) sedimentary and biological properties, thereby providing information on landscape-scale habitats.

Acoustic data were collected on the Portuguese continental shelf in depths of 15–150 m by Mamede et al. (2015) to characterise sediment bed types for benthic habitat studies. The instruments used included a QTC VIEW, Series V, with automatic gain control, connected to a single-beam Hondex 7300II echo sounder operating at 50 kHz, with 600 W transmit power, a pulse duration of 265 μs, 7 pings s^{-1}, and a beam width of 28°. A Differential Global Position System (DGPS) was used for the determination of location.

4.5 Morphology

In addition to experimental and theoretical physical and morphological work at the saltmarsh, mudflat, sandflat and beach scales (e.g., Andersen and Pejrup, 2011; Nepf, 2011; Townend, 2012), there has also been much theoretical interest in estuarine morphology at the whole estuary scale and attempts to compare theoretically derived morphologies with observations (e.g., Uncles et al., 2015). Human influences are nonnegligible at these scales. For example, turbidity and sediment supply have decreased significantly in some Asian estuaries because of dam construction, whilst in others they have greatly increased because of forest clearance and other land clearance schemes in the estuarine catchment basins (Chen et al., 2010).

4.5.1 Waves

Callaghan et al. (2010) compared bed-sediment movement proxies for four tidal-mudflat/saltmarsh sites that were contrasting due to the extent of their wind exposure and their shrinking or expanding morphological development. At each site, eight instrumented locations provided data across the tidal flat to the mature marsh. Druck PTX1830 pressure transducers, positioned within 0.01 m of the bed, were deployed at all locations, and Valeport 2-axis electromagnetic current meters were used to measure horizontal velocities at 0.2 m above the bed at four locations. The burst-sampling interval was 15 min, during which 2048 data at 4 Hz were recorded, as well as logger-determined time-averages from 600 data (i.e., over 150 s). Measurements showed that the bed shear stress was largely determined by wind waves, with tidal and wind-driven currents being less important, and that these wave stresses greatly decreased landward, illustrating a hydrodynamic transition from the tidal flat and pioneer zone to the mature marsh.

Similar results were obtained by Yang et al. (2012) for the Yangtze Estuary, China. Measurements of waves and water depths were made from 50-m landward of a low marsh edge to 10-m seaward at nine locations. Site elevation was measured using a Real-Time Kinematic GPS (Ashtech, US). Waves and depths were measured using self-logging wave SBE sensors mounted 0.15 m above the bed, which recorded waves at 4 Hz over 128-s periods for each 10-min burst-sampled interval and water depths by averaging 4-Hz data over 60-s periods. Results showed that the wave attenuation was much greater over the marsh than over an adjacent mudflat.

Most of the wave attenuation was due to vegetation, and plant height was the dominant influence on this attenuation.

4.5.2   Tidal Flats, Saltmarshes, Beaches and Shorelines

Google Earth images have applications to morphology research. For example, Goudie (2013) used detailed images of saltmarshes across England and Wales, together with morphometric analyses, to investigate their pan (small shallow pools on marsh surfaces) and creek characteristics and relate these to tidal range, sediment types and other variables. The Google Earth ruler tool was utilised at the whole marsh scale to measure shoreline extent, marsh width and distance to the coast. Results showed that the tidal prism affected creek density and that lower creek densities occurred where there were coarser sediments; creek sinuosity tended to increase with larger tides. Pan densities were higher on back barrier and drowned valley marshes and where creek density was low. Maximum pan size was partly controlled by pan density.

Gade and Melchionna (2016) studied four intertidal areas on the German North Sea coast using SAR images acquired from sensors on the European ENVISAT and ERS-2, the German TerraSAR-X and the Canadian Radarsat-2 satellites. Pixel sizes ranged from 3 m to 1 m or less. Results showed that pairs of SAR images, if combined using straightforward algebraic manipulations, could yield indicators for the identification of oyster and mussel beds as well as morphological changes.

Beach morphological changes have been analysed by Guastella and Smith (2014) using webcam and Google Earth imagery and fieldwork. Low-water (low-tide) images were downloaded from a network of webcams that were deployed to monitor beaches in KwaZulu-Natal, South Africa. The images were used in conjunction with swell-wave height and wave direction data and Google Earth images, using the Google Earth ruler tool to measure distances, in order to illustrate seasonal variations in beach morphology and to investigate channel outlet migration.

Nagarajan et al. (2015) also used Google Earth images, together with downloadable high-resolution Landsat images, to investigate the development of a spit in the Baram River mouth, Malaysia, during 1998–2008. They found that its greatest expansion during 2005–2010 was related to sediment supply due to the erosion caused by deforestation and land-use changes in the estuary's drainage basin.

Almonacid-Caballer et al. (2016) used 270 Landsat-derived shorelines and 17 shoreline positions obtained from RTK-GPS and LiDAR (Chapter 3) surveys to study shoreline development at Saler Beach, Valencia, Spain. Other data were derived from two QuickBird satellite images and an orthophotograph (geometrically corrected aerial photograph). RTK (Real Time Kinematic) GPS from a roving beach vehicle and LiDAR surveys were used to measure shorelines with high precision.

4.6   The Use of Autonomous Vehicles

It is likely that future work in the coastal zone will increasingly utilise technologies such as gliders (Chapter 10), which in synergy with, e.g., satellite altimetry, are becoming increasingly important for work in deeper waters (Bouffard et al., 2010). Other developments include a portable tow-yo instrument for use in coastal seas – the Yoing Ocean Data

Acquisition Profiler (YODA Profiler), which can vertically profile, continuously, fine-scale in-water features (Masunaga and Yamazaki, 2014). The YODA has a small, memory-type CTD sensor, the RINKO-Profiler (JFE Advantech Co., Ltd.), which deploys (mounted at the bottom of the profiler) temperature, conductivity, pressure, fluorescence and turbidity sensors (0.2-s response time) and oxygen sensors (0.4-s response time).

Autonomous underwater vehicles (AUVs) have also been applied to plume studies. Rogowski et al. (2012) described the development of an AUV (REMUS) for ocean outfall discharge plumes, and Rogowski et al. (2014) conducted four surveys using a REMUS AUV to investigate the ebb-tide outflow from the New River Inlet, US. A SLOCUM glider (Teledyne Webb Research) survey was carried out by Many et al. (2016) in the vicinity of the Rhône River mouth, France. Ship-board cruise instrumentation was used to sample stations over 30-km long cross-shelf transects corresponding to the glider section. The glider moved in a sawtooth-shaped trajectory between 1 m below the surface and 2 m above the bed at an average horizontal speed of 0.2 m s^{-1}. For inner-shelf waters, data showed that the majority of SPM was flocculated (Chapter 8). Particles became progressively finer progressing seaward in the Rhône's plume and in the seabed nepheloid layer.

5 Final Remarks

As is evident from the preceding summary, the use of modern technology has been an essential element in the advancement of estuarine, coastal and shelf science and our attempts to answer fundamental questions related to, for example, the effects of urban development and climate change on these environments. The chapters in this handbook provide an introduction to the current technologies that are commonly available for the measurement of important physical variables in estuaries and their adjacent coastal seas. The continuing deployment of these technologies will enhance our understanding of estuarine and coastal processes and provide valuable data for the validation and application of numerical models, in order that these models may then be used to guide the management of our valuable estuarine and coastal ecosystems and infrastructure resources (e.g., Jennerjahn and Mitchell, 2013; Mitchell et al., 2015).

To end this introduction we paraphrase, and reword for our own purposes, Wright's Epilogue in his 1995 book on the morphodynamics of inner continental shelves (Wright, 1995): We acknowledge that the chapters offered in this handbook will soon need to be rewritten in order to accommodate the fast pace of technological progress in the field of estuarine and coastal science. However, it is stimulating to know that exciting new paradigms and understandings of our science are on the horizon.

Acknowledgements

We thank all of the contributing authors for providing us with excellent material for the handbook. RJU also thanks Professor Steve de Mora, Chief Executive of the Plymouth Marine Laboratory, for the award of a Senior Research Fellowship.

References

Almonacid-Caballer, J., Sánchez-García, E., Pardo-Pascual, J. E., Balaguer-Beser, A. A., Palomar-Vázquez, J., 2016. Evaluation of annual mean shoreline position deduced from Landsat imagery as a mid-term coastal evolution indicator. *Marine Geology* 372, 79–88. doi:10.1016/j.margeo.2015.12.015.

Amos, C. L., Umgiesser, G., Ferrarin, C., Thompson, C. E. L., Whitehouse, R. J. S., Sutherland, T. F., Bergamasco, A., 2010. The erosion rates of cohesive sediments in Venice lagoon, Italy. *Continental Shelf Research* 30, 859–870. doi:10.1016/j.csr.2009.12.001.

Andersen, T. J., Pejrup, M., 2011. Biological influences on sediment behavior and transport. *Reference Module in Earth Systems and Environmental Sciences: Treatise on Estuarine and Coastal Science* 2, 289–309. doi:10.1016/B978-0-12-374711-2.00217-5.

Aristizábal, M. F., Chant, R. J., 2015. An observational study of salt fluxes in Delaware Bay. *Journal of Geophysical Research: Oceans* 120, 2751–2768. doi:10.1002/2014JC010680.

Becherer, J., Stacey, M. T., Umlauf, L., Burchard, H., 2015. Lateral circulation generates flood tide stratification and estuarine exchange flow in a curved tidal inlet. *Journal of Physical Oceanography* 45, 638–656. doi:10.1175/JPO-D-14-0001.1.

Behrens, D. K., Bombardelli, F. A., Largier, J. L., 2016. Landward propagation of saline waters following closure of a bar-built estuary: Russian River (California, USA). *Estuaries and Coasts* 39, 621–638. doi:10.1007/s12237-015-0030-8.

Blain, C. A., Mied, R. P., McKay, P., Chen, W., Rhea, W. J., 2015. Bathymetrically controlled velocity-shear front at a tidal river confluence. *Journal of Geophysical Research: Oceans* 120, 5850–5869. doi:10.1002/2014JC010563.

Bolaños, R., Brown, J. M., Souza, A. J., 2014. Wave–current interactions in a tide dominated estuary. *Continental Shelf Research* 87, 109–123. doi:10.1016/j.csr.2014.05.009.

Bonneton, P., Bonneton, N., Parisot, J.-P., Castelle, B., 2015. Tidal bore dynamics in funnel-shaped estuaries. *Journal of Geophysical Research: Oceans* 120, 923–941. doi:10.1002/2014JC010267.

Bouffard, J., Pascual, A., Ruiz, S., Faugère, Y., Tintoré, J., 2010. Coastal and mesoscale dynamics characterization using altimetry and gliders: A case study in the Balearic Sea. *Journal of Geophysical Research: Oceans* 115, C10029. doi:10.1029/2009JC006087.

Braithwaite, K. M., Bowers, D. G., Nimmo Smith, W. A. M., Graham, G. W., 2012. Controls on floc growth in an energetic tidal channel. *Journal of Geophysical Research: Oceans* 117, C02024, doi:10.1029/2011JC007094.

Brand, A., Lacy, J. R., Hsu, K., Hoover, D., Gladding, S., Stacey, M. T., 2010. Wind-enhanced resuspension in the shallow waters of South San Francisco Bay: Mechanisms and potential implications for cohesive sediment transport. *Journal of Geophysical Research: Oceans* 115, C11024, doi:10.1029/2010JC006172.

Callaghan, D. P., Bouma, T. J., Klaassen, P., van der Wal, D., Stive, M. J. F., Herman, P. M. J., 2010. Hydrodynamic forcing on salt-marsh development: Distinguishing the relative importance of waves and tidal flows. *Estuarine, Coastal and Shelf Science* 89, 73–88. doi:10.1016/j.ecss.2010.05.013.

Carey, D. A., Hayn, M., Germano, J. D., Little, D. I., Bullimore, B., 2015. Marine habitat mapping of the Milford Haven Waterway, Wales, UK: Comparison of facies mapping and EUNIS classification for monitoring sediment habitats in an industrialized estuary. *Journal of Sea Research* 100, 99–119. doi:10.1016/j.seares.2014.09.012.

Carlin, J. A., Dellapenna, T. M., Strom, K., Noll, C. J., 2015. The influence of a salt wedge intrusion on fluvial suspended sediment and the implications for sediment transport to the adjacent coastal ocean: A study of the lower Brazos River TX, USA. *Marine Geology* 359, 134–147. doi:10.1016/j.margeo.2014.11.001.

Chant, R. J., 2011. Interactions between estuaries and coasts: River plumes – their formation, transport, and dispersal. *Reference Module in Earth Systems and Environmental Sciences: Treatise on Estuarine and Coastal Science* 2, 213–235. doi:10.1016/B978-0-12-374711-2.00209-6.

Chen, W., Mied, R. P., 2013. River velocities from sequential multispectral remote sensing images. *Water Resources Research* 49, 3093–3103. doi:10.1002/wrcr.20267.

Chen, Z., Yanagi, T., Wolanski, E., 2010. Editorial: EMECS8 – Harmonizing catchment and estuary. *Estuarine, Coastal and Shelf Science* 86, v. doi:10.1016/j.ecss.2009.10.018.

Chung, H. W., Liu, C. C., Chiu, Y. S., Liu, J. T., 2014. Spatiotemporal variation of Gaoping River plume observed by Formosat-2 high resolution imagery. *Journal of Marine Systems* 132, 28–37. doi:10.1016/j.jmarsys.2013.12.011.

Collignon, A. G., Stacey, M. T., 2012. Intratidal dynamics of fronts and lateral circulation at the shoal–channel interface in a partially stratified estuary. *Journal of Physical Oceanography* 42, 869–883. doi:10.1175/JPO-D-11-065.1.

Constantin, S., Doxaran, D., Constantinescu, S., 2016. Estimation of water turbidity and analysis of its spatio-temporal variability in the Danube River plume (Black Sea) using MODIS satellite data. *Continental Shelf Research* 112, 14–30. doi:10.1016/j.csr.2015.11.009.

de Nijs, M. A. J., Pietrzak, J. D., Winterwerp, J. C., 2011. Advection of the salt wedge and evolution of the internal flow structure in the Rotterdam Waterway. *Journal of Physical Oceanography* 41, 3–27. doi:10.1175/2010JPO4228.1.

de Vries, J. J., Ridderinkhof, H., Maas, L. R. M., van Aken, H. M., 2015. Intra- and inter-tidal variability of the vertical current structure in the Marsdiep basin. *Continental Shelf Research* 93, 39–57. doi:10.1016/j.csr.2014.12.002.

Downing-Kunz, M. A., Schoellhamer, D. H., 2013. Seasonal variations in suspended-sediment dynamics in the tidal reach of an estuarine tributary. *Marine Geology* 345, 314–326. doi:10.1016/j.margeo.2013.03.005.

Dyer, K. R., 1979. *Estuarine Hydrography and Sedimentation – A Handbook* (ed. K. R. Dyer). Cambridge, UK: Cambridge University Press.

Fernández-Nóvoa, D., Mendes, R., deCastro, M., Dias, J. M., Sánchez-Arcilla, A., Gómez-Gesteira, M., 2015. Analysis of the influence of river discharge and wind on the Ebro turbid plume using MODIS-Aqua and MODIS-Terra data. *Journal of Marine Systems* 142, 40–46. doi:10.1016/j.jmarsys.2014.09.009.

Fournier, S., Chapron, B., Salisbury, J., Vandemark, D., Reul, N., 2015. Comparison of spaceborne measurements of sea surface salinity and colored detrital matter in the Amazon plume. *Journal of Geophysical Research: Oceans* 120, 3177–3192. doi:10.1002/2014JC010109.

Gade, M., Melchionna, S., 2016. Joint use of multiple Synthetic Aperture Radar imagery for the detection of bivalve beds and morphological changes on intertidal flats. *Estuarine, Coastal and Shelf Science* 171, 1–10. doi:10.1016/j.ecss.2016.01.025.

Garel, E., Ferreira, Ó., 2013. Fortnightly changes in water transport direction across the mouth of a narrow estuary. *Estuaries and Coasts* 36, 286–299. doi:10.1007/s12237-012-9566-z.

Ge, J., Shen, F., Guo, W., Chen, C., Ding, P., 2015. Estimation of critical shear stress for erosion in the Changjiang Estuary: A synergy research of observation, GOCI sensing and modeling. *Journal of Geophysical Research: Oceans* 120, 8439–8465. doi:10.1002/2015JC010992.

Gensac, E., Martinez, J.-M., Vantrepotte, V., Anthony, E. J., 2016. Seasonal and inter-annual dynamics of suspended sediment at the mouth of the Amazon River: The role of continental and oceanic forcing, and implications for coastal geomorphology and mud bank formation. *Continental Shelf Research* 118, 49–62. doi:10.1016/j.csr.2016.02.009.

Giddings, S. N., Fong, D. A., Monismith, S. G., 2011. Role of straining and advection in the intratidal evolution of stratification, vertical mixing, and longitudinal dispersion of a shallow, macrotidal, salt wedge estuary. *Journal of Geophysical Research: Oceans* 116, C03003. doi:10.1029/2010JC006482.

Goudie, A., 2013. Characterising the distribution and morphology of creeks and pans on salt marshes in England and Wales using Google Earth. *Estuarine, Coastal and Shelf Science* 129, 112–123. doi:10.1016/j.ecss.2013.05.015.

Green, M. O., 2011. Very small waves and associated sediment resuspension on an estuarine intertidal flat. *Estuarine, Coastal and Shelf Science* 93, 449–459. doi:10.1016/j.ecss.2011.05.021

Guastella, L. A., Smith, A. M., 2014. Coastal dynamics on a soft coastline from serendipitous webcams: KwaZulu-Natal, South Africa. *Estuarine, Coastal and Shelf Science* 150, 76–85. doi:10.1016/j.ecss.2013.12.009.

Huguenard, K. D., Bogucki, D. J., Ortiz-Suslow, D. G., Laxague, N. J. M., MacMahan, J. H., Özgökmen, T. M., Haus, B. K., Reniers, A. J. H. M., Hargrove, J., Soloviev, A.V., Graber H., 2016. On the nature of the frontal zone of the Choctawhatchee Bay plume in the Gulf of Mexico. *Journal of Geophysical Research: Oceans* 121, 1322–1345. doi:10.1002/2015JC010988.

Huguenard, K. D., Valle-Levinson, A., Li, M., Chant, R. J., Souza, A. J., 2015. Linkage between lateral circulation and near-surface vertical mixing in a coastal plain estuary. *Journal of Geophysical Research: Oceans* 120, 4048–4067. doi:10.1002/2014JC010679.

Huntley, D., Leeks, G., Walling, D., 2001. From rivers to coastal seas: The background and context of the land-ocean interaction study. In: (eds. Huntley, D., Leeks, G., Walling, D.) *Land Ocean Interaction – Measuring and Modelling Fluxes from River Basins to Coastal Seas*, London: IWA Publishing, 1–7. doi:10.2166/9781780402222.

Jennerjahn, T. C., Mitchell, S. B., 2013. Pressures, stresses, shocks and trends in estuarine ecosystems – An introduction and synthesis. *Estuarine, Coastal and Shelf Science* 130, 1–8. doi:10.1016/j.ecss.2013.07.008.

Kaitala, S., Kettunen, J., Seppälä, J., 2014. Introduction to Special Issue: 5th ferrybox workshop – Celebrating 20 years of the Alg@line. *Journal of Marine Systems* 140, 1–3. doi:10.1016/j.jmarsys.2014.10.001.

Kilcher, L. F., Nash, J. D., 2010. Structure and dynamics of the Columbia River tidal plume front. *Journal of Geophysical Research: Oceans* 115, C05S90. doi:10.1029/2009JC006066.

Korosov, A., Counillon, F., Johannessen, J. A., 2015. Monitoring the spreading of the Amazon freshwater plume by MODIS, SMOS, Aquarius, and TOPAZ. *Journal of Geophysical Research: Oceans* 120, 268–283. doi:10.1002/2014JC010155.

Li, C., 2011. Free surface motions: Tides, seiches, and subtidal variations. *Reference Module in Earth Systems and Environmental Sciences: Treatise on Estuarine and Coastal Science* 2, 91–122. doi:10.1016/B978-0-12-374711-2.00202-3.

Liu, G., Zhu, J., Wang, Y., Wu, H., Wu, J., 2011. Tripod measured residual currents and sediment flux: Impacts on the silting of the Deepwater Navigation Channel in the Changjiang Estuary. *Estuarine, Coastal and Shelf Science* 93, 192–201. doi:10.1016/j.ecss.2010.08.008.

MacCready, P., Banas, N. S., 2011. Residual circulation, mixing, and dispersion. *Reference Module in Earth Systems and Environmental Sciences: Treatise on Estuarine and Coastal Science* 2, 75–89. doi:10.1016/B978-0–12-374711-2.00205–9.

MacVean, L. J., Lacy, J. R., 2014. Interactions between waves, sediment, and turbulence on a shallow estuarine mudflat. *Journal of Geophysical Research: Oceans* 119, 1534–1553. doi:10.1002/2013JC009477.

Mamede, R., Rodrigues, A. M., Freitas, R., Quintino, V., 2015. Single-beam acoustic variability associated with seabed habitats. *Journal of Sea Research* 100, 152–159. doi:10.1016/j.seares.2015.04.007.

Mancero-Mosquera, I., Gačić, M., Mazzoldi, A., 2010. The effect of wind on the residual current velocities in the inlets of Venice lagoon. *Continental Shelf Research* 30, 915–923. doi:10.1016/j.csr.2010.02.011.

Manning, A. J., Schoellhamer, D. H., 2013. Factors controlling floc settling velocity along a longitudinal estuarine transect. *Marine Geology* 345, 266–280. doi:10.1016/j.margeo.2013.06.018.

Many, G., Bourrin, F., Durrieu de Madron, X., Pairaud, I., Gangloff, A., Doxaran, D., Odyd, A., Verney, R., Menniti, C., Le Berre, D., Jacquet, M., 2016. Particle assemblage characterization in the Rhone River ROFI. *Journal of Marine Systems* 157, 39–51. doi:10.1016/j.jmarsys.2015.12.010.

Mariotti, G., Fagherazzi, S., 2011. Asymmetric fluxes of water and sediments in a mesotidal mudflat channel. *Continental Shelf Research* 31, 23–36. doi:10.1016/j.csr.2010.10.014.

Masunaga, E., Yamazaki, H., 2014. A new tow-yo instrument to observe high-resolution coastal phenomena. *Journal of Marine Systems* 129, 425–436. doi:10.1016/j.jmarsys.2013.09.005.

Mitchell, S. B., Jennerjahn, T. C., Vizzini, S., Zhang, W., 2015. Changes to processes in estuaries and coastal waters due to intense multiple pressures – An introduction and synthesis. *Estuarine, Coastal and Shelf Science* 156, 1–6. doi:10.1016/j.ecss.2014.12.027.

Molleri, G. S. F., de M. Novo, E. M. L., Kampel, M., 2010. Space-time variability of the Amazon River plume based on satellite ocean color. *Continental Shelf Research* 30, 342–352. doi:10.1016/j.csr.2009.11.015.

Moreira, D., Simionato, C. G., Gohin, F., Cayocca, F., Tejedor, M. L. C., 2013. Suspended matter mean distribution and seasonal cycle in the Río de La Plata estuary and the adjacent shelf from ocean color satellite (MODIS) and in-situ observations. *Continental Shelf Research* 68, 51–66. doi:10.1016/j.csr.2013.08.015.

Mullarney, J. C., Henderson, S. M., 2011. Hydraulically controlled front trapping on a tidal flat. *Journal of Geophysical Research: Oceans* 116, C04023. doi:10.1029/2010JC006520.

Muscarella, P. A., Barton, N. P., Lipphardt, B. L., Veron, D. E., Wong, K. C., Kirwan, A. D., 2011. Surface currents and winds at the Delaware Bay mouth. *Continental Shelf Research* 31, 1282–1293. doi:10.1016/j.csr.2011.05.003.

Nagarajan, R., Jonathan, M. P., Roy, P. D., Muthusankar, G., Lakshumanan, C., 2015. Decadal evolution of a spit in the Baram river mouth in eastern Malaysia. *Continental Shelf Research* 105, 18–25. doi:10.1016/j.csr.2015.06.006.

Nauw, J. J., Merckelbach, L. M., Ridderinkhof, H., van Aken, H. M., 2014. Long-term ferry-based observations of the suspended sediment fluxes through the Marsdiep inlet using acoustic Doppler current profilers. *Journal of Sea Research* 87, 17–29. doi:10.1016/j.seares .2013.11.013.

Nepf, H. M., 2011. Flow over and through biota. *Reference Module in Earth Systems and Environmental Sciences: Treatise on Estuarine and Coastal Science* 2, 267–288. doi:10.1016/B978-0-12-374711-2.00213–8.

Nowacki, D. J., Ogston, A. S., Nittrouer, C. A., Fricke, A. T., Van Pham, D. T., 2015. Sediment dynamics in the lower Mekong River: Transition from tidal river to estuary. *Journal of Geophysical Research: Oceans* 120, 6363–6383. doi:10.1002/2015JC010754.

O'Callaghan, J., Stevens, C., 2011. Wind stresses on estuaries. *Reference Module in Earth Systems and Environmental Sciences: Treatise on Estuarine and Coastal Science* 2, 151–169. doi:10.1016/B978-0-12-374711-2.00211–4.

O'Donncha, F., Hartnett, M., Nash, S., Ren, L., Ragnoli, E., 2015. Characterizing observed circulation patterns within a bay using HF radar and numerical model simulations. *Journal of Marine Systems* 142, 96–110. doi:10.1016/j.jmarsys.2014.10.004.

Orton, P. M., Zappa, C. J., McGillis, W. R., 2010. Tidal and atmospheric influences on near-surface turbulence in an estuary. *Journal of Geophysical Research: Oceans* 115, C12029. doi:10.1029/2010JC006312.

Pan, J., Gu, Y., 2016. Cruise observation and numerical modeling of turbulent mixing in the Pearl River estuary in summer. *Continental Shelf Research* 120, 122–138. doi:10.1016/j.csr .2016.03.019.

Petus, C., Marieu, V., Novoa, S., Chust, G., Bruneau, N., Froidefond, J.-M., 2014. Monitoring spatio-temporal variability of the Adour River turbid plume (Bay of Biscay, France) with MODIS 250-m imagery. *Continental Shelf Research* 74, 35–49. doi:10.1016/j.csr.2013.11.011.

Piedracoba, S., Rosón, G., Varela, R. A., 2016. Origin and development of recurrent dipolar vorticity structures in the outer Ría de Vigo (NW Spain). *Continental Shelf Research* 118, 143–153. doi:10.1016/j.csr.2016.03.001.

Prandle, D., Lane, A., Souza, A. J., 2011. Coastal circulation. *Reference Module in Earth Systems and Environmental Sciences: Treatise on Estuarine and Coastal Science* 2, 237–266. doi:10.1016/B978-0-12-374711-2.00210–2.

Ralston, D. K., Geyer, W. R., Warner, J. C., 2012. Bathymetric controls on sediment transport in the Hudson River estuary: Lateral asymmetry and frontal trapping. *Journal of Geophysical Research: Oceans* 117, C10013. doi:10.1029/2012JC008124.

Ren, J., Wu, J., 2014. Sediment trapping by haloclines of a river plume in the Pearl River Estuary. *Continental Shelf Research* 82, 1–8. doi:10.1016/j.csr.2014.03.016.

Rodríguez-Calderón, C., Kuehl, S. A., 2013. Spatial and temporal patterns in erosion and deposition in the York River, Chesapeake Bay, VA. *Estuarine, Coastal and Shelf Science* 117, 148–158. doi:10.1016/j.ecss.2012.11.004.

Rogowski, P., Terrill, E., Chen, J., 2014. Observations of the frontal region of a buoyant river plume using an autonomous underwater vehicle. *Journal of Geophysical Research: Oceans* 119, 7549–7567. doi:10.1002/2014JC010392.

Rogowski, P., Terrill, E., Otero, M., Hazard, L., Middleton, W., 2012. Mapping ocean outfall plumes and their mixing using autonomous underwater vehicles. *Journal of Geophysical Research: Oceans* 117, C07016. doi:10.1029/2011JC007804.

Russell, P., Vennell, R., 2014. Distribution of vertical velocity inferred from secondary flow in a curved tidal channel. *Journal of Geophysical Research: Oceans* 119, 6010–6023. doi:10.1002/ 2014JC010003.

Salehi, M., Strom, K., 2012. Measurement of critical shear stress for mud mixtures in the San Jacinto estuary under different wave and current combinations. *Continental Shelf Research* 47, 78–92. doi:10.1016/j.csr.2012.07.004.

Scully, M. E., Geyer, W. R., 2012. The role of advection, straining, and mixing on the tidal variability of estuarine stratification. *Journal of Physical Oceanography* 42, 855–868. doi:10.1175/JPO-D-10-05010.1.

Shen, F., Zhou, Y., Li, J., He, Q., Verhoef, W., 2013. Remotely sensed variability of the suspended sediment concentration and its response to decreased river discharge in the Yangtze estuary and adjacent coast. *Continental Shelf Research* 69, 52–61. doi:10.1016/j.csr .2013.09.002.

Sommerfield, C. K., Wong, K.-C., 2011. Mechanisms of sediment flux and turbidity maintenance in the Delaware Estuary. *Journal of Geophysical Research: Oceans* 116, C01005. doi:10.1029/2010JC006462.

Song, D., Wang, X. H., Cao, Z., Guan, W., 2013. Suspended sediment transport in the Deepwater Navigation Channel, Yangtze River Estuary, China, in the dry season 2009: 1. Observations over spring and neap tidal cycles. *Journal of Geophysical Research: Oceans* 118, 5555–5567. doi:10.1002/jgrc.20410.

Stacey, M. T., Rippeth, T. P., Nash, J. D., 2011. Turbulence and stratification in estuaries and coastal seas. *Reference Module in Earth Systems and Environmental Sciences: Treatise on Estuarine and Coastal Science* 2, 9–35. doi:10.1016/B978-0-12-374711-2.00204-7.

Talke, S. A., Horner-Devine, A. R., Chickadel, C. C., Jessup, A. T., 2013. Turbulent kinetic energy and coherent structures in a tidal river. *Journal of Geophysical Research: Oceans* 118, 6965–6981. doi:10.1002/2012JC008103.

Todd, B. J., Shaw, J., Li, M. Z., Kostylev, V. E., Wu, Y., 2014. Distribution of subtidal sedimentary bedforms in a macrotidal setting: The Bay of Fundy, Atlantic Canada. *Continental Shelf Research* 83, 64–85. doi:10.1016/j.csr.2013.11.017.

Townend, I. H., 2012. The estimation of estuary dimensions using a simplified form model and the exogenous controls. *Earth Surface Processes and Landforms* 37, 1573–1583. http://dx.doi .org/10.1002/esp.3256.

Uncles, R. J., 2002. Estuarine physical processes research: Some recent studies and progress. *Estuarine, Coastal and Shelf Science* 55, 829–856. doi:10.1006/ecss.2002.1032.

Uncles, R. J., 2011. Small-scale surface fronts in estuaries. *Reference Module in Earth Systems and Environmental Sciences: Treatise on Estuarine and Coastal Science* 2, 53–74. doi:10.1016/B978-0-12-374711-2.00207-2.

Uncles, R. J., Stephens, J. A., Harris, C., 2015. Estuaries of southwest England: Salinity, suspended particulate matter, loss-on-ignition and morphology. *Progress in Oceanography* 137, 385–408. http://dx.doi.org/10.1016/j.pocean.2015.04.030.

Valle-Levinson, A., 2011. Large Estuaries (Effects of Rotation). *Reference Module in Earth Systems and Environmental Sciences: Treatise on Estuarine and Coastal Science* 2, 123–140. doi:10.1016/B978-0-12-374711-2.00208-4.

Valle-Levinson, A., Huguenard, K., Ross, L., Branyon, J., MacMahan, J., Reniers, A., 2015. Tidal and nontidal exchange at a subtropical inlet: Destin Inlet, Northwest Florida. *Estuarine, Coastal and Shelf Science* 155, doi:10.1016/j.ecss.2015.01.020.

Valle-Levinson, A., Schettini, C. A. F., 2016. Fortnightly switching of residual flow drivers in a tropical semiarid estuary. *Estuarine, Coastal and Shelf Science* 169, 46–55. doi:10.1016/j.ecss.2015.12.008.

Venier, C., Figueiredo da Silva, J., McLelland, S. J., Duck, R. W., Lanzoni, S., 2012. Experimental investigation of the impact of macroalgal mats on flow dynamics and sediment stability in shallow tidal areas. *Estuarine, Coastal and Shelf Science* 112, 52–60. doi:10.1016/j.ecss.2011.12.035.

Wang, Y. P., Voulgaris, G., Li, Y., Yang, Y., Gao, J., Chen, J., Gao, S., 2013. Sediment resuspension, flocculation, and settling in a macrotidal estuary. *Journal of Geophysical Research: Oceans* 118, 5591–5608. doi:10.1002/jgrc.20340.

Wargula, A., Raubenheimer, B., Elgar, S., 2014. Wave-driven along-channel subtidal flows in a well-mixed ocean inlet. *Journal of Geophysical Research: Oceans* 119, 2987–3001. doi:10.1002/2014JC009839.

Waterhouse, A. F., Tutak, B., Valle-Levinson, A., Sheng, Y. P., 2013. Influence of two tropical storms on the residual flow in a subtropical tidal inlet. *Estuaries and Coasts* 36, 1037–1053. doi:10.1007/s12237-013–9606-3.

Waterhouse, A. F., Valle-Levinson, A., 2010. Transverse structure of subtidal flow in a weakly stratified subtropical tidal inlet. *Continental Shelf Research* 30, 281–292. doi:10.1016/j.csr.2009.11.008.

Whitney, M. M., Codiga, D. L., 2011. Response of a large stratified estuary to wind events: Observations, simulations, and theory for Long Island Sound. *Journal of Physical Oceanography* 41. doi:10.1175/2011JPO4552.1.

Winterwerp, J. C., 2011. The physical analyses of muddy sedimentation processes. *Reference Module in Earth Systems and Environmental Sciences: Treatise on Estuarine and Coastal Science* 2, 311–360. doi:10.1016/B978-0–12-374711-2.00214-X.

Wolf, J., Brown, J. M., Bolaños, R., Hedges, T. S., 2011. Waves in coastal and estuarine waters. *Reference Module in Earth Systems and Environmental Sciences: Treatise on Estuarine and Coastal Science* 2, 171–212. doi:10.1016/B978-0–12-374711-2.00203–5.

Wright, L. D., 1995. *Morphodynamics of Inner Continental Shelves*. Marine Science Series (eds. M. J. Kennish and P. L. Lutz), Boca Raton, FL: CRC Press Inc.

Wu, J., Liu, J. T., Wang, X., 2012. Sediment trapping of turbidity maxima in the Changjiang Estuary. *Marine Geology* 303–306, 14–25. doi:10.1016/j.margeo.2012.02.011.

Yang, S. L., Shi, B. W., Bouma, T. J., Ysebaert, T., Luo, X. X., 2012. Wave attenuation at a salt marsh margin: A case study of an exposed coast on the Yangtze Estuary. *Estuaries and Coasts* 35, 169–182. doi:10.1007/s12237-011–9424-4.

Zhu, X.-H., Zhang, C., Wu, Q., Kaneko, A., Fan, X., Li, B., 2012. Measuring discharge in a river with tidal bores by use of the coastal acoustic tomography system. *Estuarine, Coastal and Shelf Science* 104–105, 54–65. doi:10.1016/j.ecss.2012.03.022.

Zippel, S., Thomson, J., 2015. Wave breaking and turbulence at a tidal inlet. *Journal of Geophysical Research: Oceans* 120, 1016–1031. doi:10.1002/2014JC010025.

2 Bathymetric and Tidal Measurements and Their Processing

V. J. Abbott

1 Introduction

Estuarine and coastal bathymetric measurements, as adjusted for tidal variations, or studies of the tides themselves, take place within a context of global activities. This is clear from the variations in water level that originate from oceanic responses to tide-raising forces (e.g., IOC, 2006) through to the determination of satellite-derived positions from Earth-centred, Earth-fixed reference frames. The latter ignore local undulations in the local Earth model and corrections to the associated mapping. Fieldwork can take place without reference to the global picture, but variations in mean sea level and the common use of the Global Positioning System (GPS), e.g., Lekkerkerk (2007), demand a clear understanding of the wider influences.

Mean sea level (MSL) is the surface chosen to approximate the model of the Earth itself, the geoid[1], this being a notional surface where the topography (the hills and valleys and ocean basins) is treated separately. At the coast, MSL is determined by time series tidal measurements, which are long enough to average-out meteorological variations. The United Kingdom's current fundamental gauge in southwest England has operated for one hundred years, although the value of MSL in use as Ordnance Datum (Newlyn), OD (N), was obtained after just six years, between 1915 and 1921. Satellite observations this century have identified the gauge's setting as being within a sea-surface topographic valley and affected by vertical oscillations associated with ocean tide loading. Coupled with long-term changes in MSL, which is currently standing 0.15 m above the value of OD (N), the simple use of OD (N) as a surrogate for MSL may be misleading.

The geoid is not the reference surface in common use. There are geoid models of increasing sophistication, with global values determined every 2.5' of arc and with some more detailed national studies. However, it is more common to use an ellipsoid to model the shape of the Earth. The ellipsoid of choice varies between a close fit to a portion of the Earth (a nation or a continent) to a global best fit. The best known of the latter is the World Geodetic System 1984 (WGS84) due to its association with GPS.

---

[1] The geoid is an equipotential gravitational surface and thus changes in the underlying geological structure and even of the density of water alter the link between MSL and the geoid. In Australia, there is a ±0.5m geoid/sea-level offset, north to south, due to the densities of the warmer northern and the colder southern waters (Geoscience Australia, 2016).

A geoid-WGS84 difference surface indicates variations of +85 m to −106 m. The offset within Australia is +72 m to −33 m (Geoscience Australia, 2016); in the UK, WGS84 lies 50–60 m below the geoid. Thus, GPS-derived heights are of little use without a local correction.

GPS plan positions are also of limited use without a local correction – plate tectonics result in an increasing width of the Atlantic Ocean and because WGS84 is attached to the contiguous United States, WGS84 positions in the UK change by 0.025 m per year. Over the last 30 years, a point in the UK will have apparently moved 0.75 m. In Australia the change is currently 0.1 m per year (Geoscience Australia, 2016).

This chapter sets out to describe useful technologies and methodologies, referring where necessary to a wider picture.

2   Vertical Tidal Movement[2]

The longest series of tidal observations has been observed in Brest, France, with records from 1807, and although there have been breaks, they are still being recorded. Some older records exist for four European cities (Amsterdam, Stockholm, Kronstadt and Liverpool), with measured values in the 1700s (PSMSL, 2016). The advent of cheap and reliable recording devices has greatly aided the spread of gauges, and the work of the Global Sea Level Observing System (known for historical reasons as GLOSS) offers training in tide gauge operation.

GLOSS data (GLOSS, 2016) feed into further international cooperation on long-term records at the Permanent Service for Mean Sea Level (PSMSL), based at the National Oceanography Centre, Liverpool, UK (formerly the Proudman Oceanographic Laboratory, Liverpool, UK). Their web page on relative sea level trends (PSMSL, 2016) indicates a variety of rises and falls, although the data are not adjusted for local land movements, which includes Earth tides, ocean tide loading and isostatic change. Whereas the first of these is negligible for practical tidal measurements, it is, as is the second, principally observable once the tidal signal is removed, e.g., using satellite observations, otherwise the associated rises and falls are incorporated into the tide gauge tidal values. Isostatic changes, due to, amongst other things, the land rebounding following the release of pressure from the last ice ages, may well be observable as a change in mean sea level. A consequence around the Baltic Sea is the loss of water depth within harbours. In the UK, Scotland is rising between 0.004 and 0.02 m per year as a consequence of isostatic change.

Separately, eustatic changes that are generated as a consequence of glacial and ice sheet melt will cause sea levels to rise, and this rise is increased due to the

---

[2] Water will circulate in an ellipse in the open ocean, whereas in estuaries the movement is predominantly linear and commonly in quadrature with the tidal height, reaching maxima on the mid-fall and mid-rise and minima at the high and low waters. However, the particular situation should be assessed at any place; for example, in the coastal zone, the tidal streams and vertical tide (heights) are often nearly in phase and form a progressive tidal wave (e.g., Pugh, 1987).

warming of global waters. Together (isostatic and eustatic), the change is currently estimated at +0.003 m year^{-1}.

Archived data (from about 2,000 stations) may be obtained through the PSMSL, and this may be of use in field surveys if work is undertaken close to the observing gauge. When tide data are not readily available, it may be tempting to use tidal predictions to reduce the need for soundings. However, meteorological influences have been averaged out due to the underpinning number of observations (perhaps over many years). For any fieldwork, the influence of weather on the day can have a significant effect, with a slow-moving high pressure tending to lower the tidal water-level surface and a northerly wind tending to reduce the water level along a south-facing coast. Thus it is generally best to establish a measuring device close to the survey area. More widely, an adjacent datum can be established using the simultaneous observations from a permanent gauge by means of a datum transfer technique. Such a datum transfer is more simply undertaken in waters that vary with straightforward semidiurnal changes, utilising records of four low waters and the intervening three high waters. These can be used to give a new datum value that is weighted to the central readings, thereby avoiding the extremes of passing meteorological changes. Such a technique is described in detail by Glen (1979) in the previous edition of this handbook, in *Hydrographer of the Navy* (1969) and, more generally, in IHO (2005). For diurnal regimes, the process is more complicated, requiring analysis for the principal four tide-raising constituents. Finally, making a link (by levelling) to two nearby points ensures that the device can be re-established into the same datum, should it be disturbed.

The datum for the tide gauge is crucial to support the reduction of soundings to a common reference level. This datum may be set at MSL (or for a bridge construction in the UK, say, its near-equivalent, Ordnance Datum (Newlyn)); or, especially to tie in with a navigational chart or to support safe navigation, a low water level - commonly Lowest Astronomical Tide (LAT). Whichever reference surface is chosen, measurements of water depth (or intertidal heights) can be taken at any state of the tide and reduced to the common surface. Although 'reduction' is used because most tidal heights are subtracted from measured depths, use of an MSL surface will result in as many tidal values being added to the raw depth (subtracting a negative number as the tide gauge will have zero in the centre and tides will be both positive and negative). Actually, this can occur even with a low water datum (such as LAT) because, on occasion, the water surface will fall below Chart Datum, requiring the tide reading to be added to the raw sounding (again, subtracting a negative number).

For a new survey:

- Establish a tide pole close to the survey site, and monitor the tidal curve.
- Search for an existing Chart Datum value. If there is an existing gauge, observations at the old and new sites will show whether the curve is of a similar shape, range and time (i.e., the times of high and low water). If so, a datum transfer may be possible. Because an established gauge is likely to be referenced to permanent marks ashore (possibly even a land levelling network), it may be possible to level to the new gauge from the national land levelling datum. If there is a

rigorous, high-quality, national GPS network (including rigorous transformation parameters), such height transfer may be possible using GPS-measured heights.

- Determine an appropriate datum value for the new work – making permanent offsets to it on land away from possible damage should the gauge be disturbed. Either: determine an offset from the pole's zero, or, more safely, adjust the zero of the pole to set it at the local datum.

Despite the range of technological approaches to measuring tidal heights, a tide pole remains a useful approach. It is more difficult to use at night, suffers from problems of visibility across a long shelving beach and suffers from bio-fouling over the long term. Nevertheless, in some places its establishment is straightforward; in some of these, the pole can be easily read; once fixed it probably will not move; and it can be cleaned. Even where an automatic transmitting gauge is used, a tide pole offers a quick check on the height value of the gauge.

Tide poles might be poorly designed, poorly placed and poorly used – just as an automatic gauge might be. A pole's numbers must be large enough to be read at the distance forced on the observer by the high tide. For much work, it can be sufficient to estimate centimetre precision from a decimetre graduation. The pole should be vertical, and there should be a place from which to read it – this may be from a boat if it is close to the survey area. The pole should not be impounded at low water (e.g., by a bar across an estuary), and it should be cleaned occasionally. It is better to mount the pole so that its zero is at the appropriate datum, but any simple arithmetic offsets can be easily applied; however, they can also be forgotten or applied the wrong way. Most of these issues apply to an automatic gauge too.

In waters closely connected to the Atlantic Ocean, the tidal variation is most usually semidiurnal; i.e., there are two low waters and two high waters in (just over) a day. The Atlantic is most responsive to the semidiurnal tide raising forces of the Moon and the Sun, but the ocean's reaction is not uniform, nor is the progression of the tidal wave into coastal waters and their estuaries. The range of the tide increases as the tidal wave propagates onto the continental shelves and into the shallow waters around the coasts. The particular tide, local to one place, may vary significantly from a simple sinusoidal curve and incorporate distortions of, say, a slow rise, an inconsistent rate of rise and a quick fall. Some places exhibit a double low water or a double high water (Pugh, 1987). The amplitude and times of high and low water may vary between adjacent estuaries.

Tide gauges can be classified by their power supplies (floating on the water surface, clockwork-driven drums with recording paper, compressed air, electricity) or be classified by their deployment above or below the water surface. Most usually, a platform or an area above the reach of the water is required, sometimes with a nearby vertical surface extending down below the lowest water level. To deploy a float, a vertical tube may be designed into the construction of a pier, or a sturdy cylinder may be attached to the outside. With a small diameter water inlet (one-tenth of the diameter of the vertical well) below the lowest level of the sea, the water movements inside the tube are calmed (the tube is commonly called a 'stilling well'), and the float's vertical travel within the tube can be measured. The facility must be maintained to protect against the effects of both biological growth in the inlet and tube and the settling of sediment at the bottom of the tube.

Other types of gauge sensors might be deployed in such a tube, benefitting from protection against the force of the waves and measuring the internal water level. If fresh water ingress occurs in the tube, the density differential would raise the level of the water – a float would give a high reading; a submerged pressure gauge would be unaffected.

Each technology should be assessed for anomalous effects. A pressure gauge will exhibit a higher reading if there is a horizontal water movement over the face of the sensor. Measurements of travel time with an acoustic above-water gauge will be affected by the temperature and humidity of the air. Typically, the latter has a bar fixed below the transducer to enable real-time calibration; the first reflection off the bar allows a real-time adjustment.

Another above-water gauge, utilising microwave energy, is small and portable with the bulkiest element being the cantilevered arm to extend the transducer out over the water surface. As with most electrically powered gauges, it averages many readings, thereby filtering-out short-term water surface movements. The battery power required is small and the storage capacity large, and it can use radio or telecommunications technology to transmit the data. These can be available over the mobile phone service (directly or through the Internet) or over Twitter.

In-water gauges include the pressure sensor, as mentioned, and bubbler gauges that use compressed air as the power source. Although there may be a clockwork mechanism to drive a rotating drum, the compressed air is used to balance out the potential ingress of water into a small bore tube. The back pressure caused by the fluctuations in water level is measured through an above-water quartz transducer, and through this a record is made of tidal change. As the air pressure must be set high enough to balance out the highest of tides (or the water will enter the mechanism), at low levels the air escapes – bubbling (slowly) – requiring the occasional replacement of the compressed air cylinder.

Tidal heights also may be measured by the high precision position and timing satellites within the Global Navigation Satellite Systems (GNSS; e.g., GNSS, 2016). A high-accuracy, real-time technique is required, and this is described later. Because the tide may only be a means of adjusting the soundings, the system may be on board the vessel and the measured values of tide and depth combined in near-real time.

3   Tidal Analysis

The tidal values most commonly reported to the public are predictions for ports and are published in national Tide Tables, in local booklets and on the local news. Predicted tides are useful to the fisherman, the navigator and the marine scientist for planning purposes, but are unreliable for precise values on any particular day.

The principal tide-raising forces are due to the gravitational attraction between the Earth, its waters, the Moon and the Sun. These are well established, and although the forces vary, they are regular and periodic. The Sun's mass is so much larger than the Moon's, yet its distance (and in the gravitational attraction relationship, force is inversely proportional to the distance squared) results in the Sun's effect being just 46 per cent of the Moon's. Changing distances between the Earth and Moon (by about 10 per cent) and between the Earth and Sun (by about 3 per cent) also result in larger lunar-induced than solar-induced consequences. The Moon's declination, being the sum of its inclination (about 5°) to the Sun's apparent path (the ecliptic) and the inclination of the Earth's rotation to the ecliptic (about 23°), varies from about –28° to +28° in 9.3 years, and thus 18.6 years for a full cycle (see more on this later).

Astronomical theory may be used to determine the consequences of the gravitational attraction of the Moon and Sun on the waters of the Earth, and in most cases, the dominant features are the semidiurnal (twice a day) tide-raising constituent of the mean Moon (designated M_2), semidiurnal tide-raising constituent of the mean Sun (S_2), and the diurnal constituents, K_1 and O_1 of the Moon and Sun, respectively (letters close to M in the alphabet denote lunar constituents and close to S, solar). The 'mean' Moon is supposed to travel around the equator at a constant radius. The wide range of orbital variations and their combinations result in the need for further terms, and 256 are easily isolated. These include quarter diurnal, fortnightly and annual. With 12 months of hourly observations, over 100 terms can be determined, and 114 with 4.5 years (TASK, 2016).

A real tide curve can be explained as the sum of all the curves originating from the tidal parts, and the results vary from the simple sinusoidal to those with a double low water, a double high, or distinct inequalities between consecutive high and consecutive low waters. To determine the constituents at a particular location, the tide must be monitored and the values analysed. This may be by semiharmonic or by Fourier analysis. Software is available for these computations (see, for example, TASK, 2016).

The length of observation time is an indication of the subtleties in deriving some of the tide-raising constituents. One can obtain a mean tide level from consecutive high and low waters, but the value is unlikely to be close to that accepted as MSL. In the UK, the (then) value of MSL to which the Ordnance Survey aligned their vertical Ordnance Datum (Newlyn) was obtained from six years of observations. It was correct at that time for that six-year period (1915–1921). MSL has variously risen and fallen since then (Pugh, 2004) but now, a hundred years later, it is about 0.15 m higher.

Some of the constituents have a really small effect, unlike the principal solar semidiurnal constituent (S_2) at 46 per cent of the Moon's (M_2) but, e.g., just 1 per cent for the annual solar constituent, S_a. The Sun's diurnal constituent (S_1) has a 'speed' of $15°$ h^{-1} ($360°/24$ h^{-1}) because we use the Sun as our timepiece. Speeds of other constituents vary from $0.04°$ h^{-1} for an annual change in the Sun's apparent movement (S_a) to $175°$ h^{-1} for $4M_2S_{12}$ (TASK, 2016). The phases of these constituents necessarily vary against each other over time, sometimes adding together, sometimes subtracting.

Longer observing times result in better predictions, but there are significant advantages in observing tides for at least 29 days (a lunar month, which includes two neap-spring cycles), a year (which includes seasonal effects) and 19 years, a period that includes both the precession of the lunar nodes (18.6 years, as earlier) and the Metonic cycle (repeating the Earth/Moon/Sun relative positions). No one has yet observed the tide over the precession of the Earth's axis (26,000 years).

The data gathered for the analysis can be assessed for the quality of the tide gauge and its maintenance, and independent checks can be made for water-level height (against, say, a tide pole or by tape-measuring to the water's surface) and for time. Precise time is readily available, but a time check at the gauge should not be forgotten. The tide pole must be established through rigorous levelling, precise erection and careful cleaning. For a seabed pressure gauge, the consequences of the changing atmospheric pressure should be incorporated. For a float-actuated gauge, the float must have free movement within the stilling well and be installed as designed for the associated mechanical assembly. For a bubbler tide gauge, the high-pressure air cylinder must be renewed periodically. For any of these, there may be adjustments to the records. There may be missing tide-gauge records due to periods of maintenance, equipment failure, storm damage and possibly vandalism. The software must be able to cope with these eventualities and the data-gaps patched.

The analysis of the data should be constrained to produce the number of tide-raising constituents in proportion to the time of observation; the output will identify the constituent, its amplitude, phase and speed. This, though, is just a stepping-stone because the constituents are used to predict (or to hindcast) values and curves. The output will permit the generation of MSL, an estimate of LAT and other intermediate and extreme surfaces. Scientific measurements may be related to MSL and Chart Datum, which is often set at, or close to, LAT.

Yet, for all the astronomical theory, the care, time and expense of the process, the predicted tide will rarely equal the observed tide; e.g., in a long-term observation programme and its analysis, the meteorological effects are averaged out. However, on a day-by-day basis the meteorology plays its part, supressing the tidal surface or allowing it to rise above the predicted level. A PC screen shot (Figure 2-1; Valeport, 2014) from modern tide-gauge data recorded 11 July 2014 in Millbay Docks, Plymouth, UK (located in Figure 2-2), processed and displayed through a modern web application 'Port-Log' (OceanWise, 2014), indicates an offset that is possibly due to a water-level 'surge'. Meteorological affects may produce, e.g., a downward movement of the sea surface directly due to the presence of a slow-moving high-pressure region. However, some effects are water induced, such as, e.g., the formation of a seiche within a confined water area – as seen in some estuaries, ports, docks, harbours and lakes (e.g., Proudman, 1953; Uncles et al., 2014).

4   Reference Frames and Positions

A point on Earth does not have unique values of latitude and longitude. Values of latitude are measured as an angular difference from the equator, both north and south, subtended at the centre of the Earth. Latitude has long been the simplest to determine, requiring tables of the astronomical bodies, an approximate measure of time and the angular height above the horizon of a star, or the Sun, or the Moon. Longitude

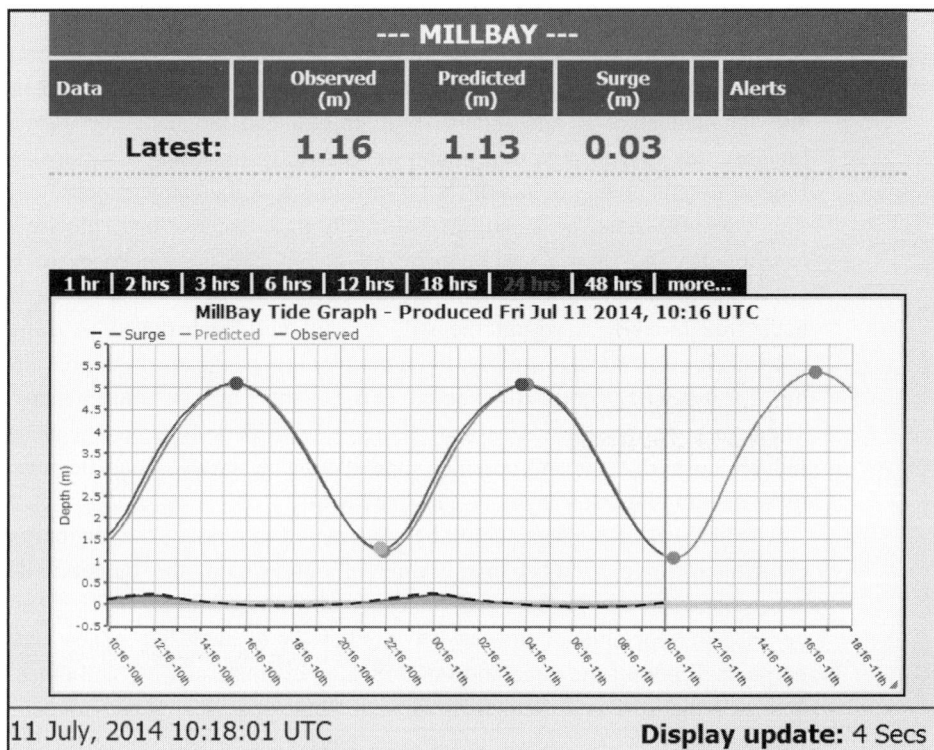

Figure 2-1 A PC screen grab, modified for greyscale presentation, of the web application 'Port-Log' (OceanWise, 2014) indicating predicted and observed tidal heights (Valeport, 2014) and the water-level 'surge' as the difference between them (courtesy OceanWise Ltd; reproduced with permission).

was more difficult, partly because there was no obvious reference meridian and partly because time was required more precisely, which was particularly difficult at sea until Harrison perfected his chronometers (see Sobel, 1996, for a readable if not entirely dispassionate account).

As surveying became more precise, the irregularities in the shape of the Earth became significant. Countries adopted models of the Earth relevant to the undulations in their vicinity. The variations between Britain and France were significant enough to suggest to each that the Earth might be longer north–south than west–east, or vice versa. Although theory and survey expeditions had settled the general principle by the mid-eighteenth century, the Earth's local variations continue to make local models relevant to local maps and charts. The current model for Britain was established by George Airy in 1830 (Airy went on to become the UK's Astronomer Royal).

There have been world-fitting models for a hundred years, and the natural undulations in the real world (the geoid) necessarily result in separations of any model (an ellipsoid, or by common usage, a spheroid) by about 100 m in places. The currently used, predominant model, WGS84 (more correctly, Geodetic Reference System 1980 – GRS80) exhibits this offset, though in places the offset is zero. WGS84 is the reference

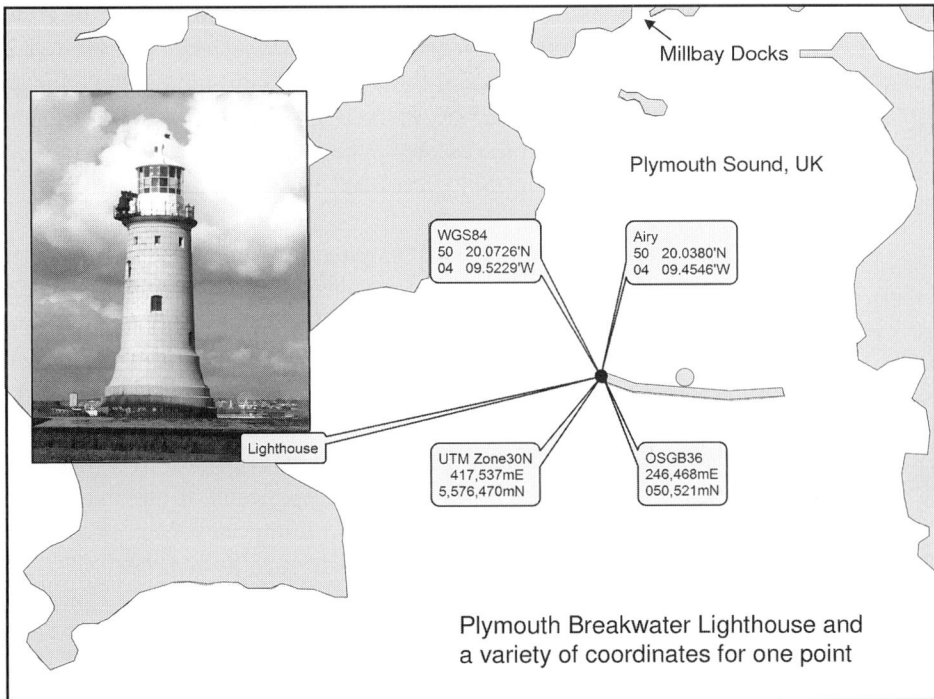

Figure 2-2 Coordinates in four systems for the location of the lighthouse on the western end of the outer breakwater in Plymouth Sound, UK. The coastline is from the USGS, Landsat 7.

model for the US satellite system GPS; other satellite systems (GLONASS, Galileo, Beidou) are independent and have their own models, observations and transformations to produce uniformity.

The Earth wobbles as it rotates, making the definition of the spin axis (north/south) and the implied equator a matter of observation and analysis. The offset between local, regional and international models produces a consequential offset in field surveying results. The definition of longitude (formerly measured from various European capital observatories for their local models), is measured from the United States for GPS-derived positions. Returning to Europe, the combined horizontal offsets (WGS84 to the UK's Airy spheroid) is of the order of 100 m. A field scientist may thus quote in latitude and longitude but needs to include the reference frame (e.g., Fig. 2–2). It may be that a given project uses GPS referenced to WGS84, but someone plotting these values on a map of a European nation may find a discrepancy in the locations. For maritime work, such offsets could place river measurements ashore and vice versa.

The degrees of latitude and longitude are the same size at the equator, but the latter cover a smaller and smaller space as the meridians converge to the poles; i.e., narrowing from approximately 100 km to zero. At about 52° north or south, 1" of arc (1/3600 of a degree) of latitude is approximately 30 m in length, and 1" of arc of longitude is approximately 18 m. Precision in values of latitude and longitude can be stated by using degrees and decimals; or degrees, minutes and decimals of a minute; or degrees, minutes, seconds and decimals of a second. Six decimal places of a degree, or one decimal place of a second, are required to quote to a spatial precision of 2 to 3 m.

Data provided by others should be inspected to ensure that the units are as quoted. A longitude value, e.g., –8.123456, is likely to be west of a zero meridian, but may be in any of the varieties expressed above, whereas –8.654321 must be in degrees and decimal degrees. Scanning through a data list may provide unambiguous indicators of the units.

Positions quoted in latitude and longitude are inconvenient for many purposes, whereas metric grids have the simple advantage of using Base 10, although feet (from one of the many definitions of 'foot' available) are still in use and may refer to feet and decimals of a foot rather than feet and inches. The national mapping organisation will have made the computational effort of converting from the spheroid and its reference lines, and national maps will be available with two or three orthogonal axes; these enable measurements in plan or height and conversions from a three-dimensional Global Navigation Satellite System to those applicable to the map. In Great Britain, this conversion is rigorously made between a European version of the WGS84 latitude and longitude (European Transformation Reference System 1989, ETRS89) using the Ordnance Survey transformations:

- Ordnance Survey Transformation Network (OSTN15) to derive plan coordinates (horizontal, two-dimensional OSGB36 eastings and northings), or
- Ordnance Survey Geoid Model (OSGM15) to derive height values (vertical, one-dimensional heights above Ordnance Datum (Newlyn)).

ETRS89 values are obtained when using one roving GPS receiver in conjunction with another at a base station tied into ETRS89.

The normal arrangement of a grid depends on the preprojection of the double curve of the Earth (north/south and west/east) onto a plane. The grid is set in a fixed relationship to the Earth, including its orientation. The National Grid of Great Britain is oriented along the line of the meridian 2° west of Greenwich. Another commonly used grid, the Universal Transverse Mercator (UTM), is locally oriented to meridians every 6° around the world, from 177° west of Greenwich. The zero of the former is southwest of the Isles of Scilly, resulting in positive coordinates over the whole of the country. The zero for the UTM zone (30°N, Central Meridian 3°W), which covers the bulk of the UK, has its northings counted from the equator; values vary by 5 million metres between the systems.

The variety inherent in all of these coordinate systems is illustrated for a lighthouse located in Plymouth Sound, southwest UK (Figure 2-2).

5 Satellite positioning

Satellite systems are at least three-dimensional (3-D), or 4-D over the time of tectonic movements, working on an Earth-centred, Earth-fixed, 3-D Cartesian coordinate system. For GPS, the positive 'x' axis crosses the equator on the meridian close to (but not on) the Greenwich Meridian (and minus x at 180° east or west, toward the International Date Line); positive 'y' at 90° east near India and minus y through the Americas; positive z through an agreed definition of the North spin axis and minus z through Antarctica. Each GPS position solution solves four unknowns and includes a host of uncertainties.

These GPS values are rarely seen, however, with coordinates more often expressed in degrees (and some subdivision) north or south and east or west. Height may well be in metres above the model of the Earth, and whereas a global model will coincide with the Earth in some places, in most it will not. Mean sea level may then not be zero on WGS84; as an example, near Cork, Ireland, it is +56 m, and near Chennai, India, it is −92 m. The horizontal reference frame may be printed on navigational charts to permit the direct use of the WGS84 latitude and longitude straight from a GPS receiver.

When using a map of the land, a simple position correction of WGS84 to a local model (the Airy spheroid in the UK) will move the latitude and longitude values to tie-in better with the land mapping system. If working at a medium scale (say, 1:5000) or smaller (i.e., toward 1:50000), then any uncertainty resulting from such a single transformation from WGS84 to Airy values (in the UK) will be unnoticeable. With larger scale (more detailed) work, a more rigorous transformation is required, which usually is available from the national mapping organisation and often built into standard surveying software.

The satellite positioning systems are best described as Global Navigation Satellite Systems (e.g., GNSS, 2016) due to their variety (GPS, GLONASS, Galileo, Beidou) and their use in multiconstellation receivers. Beidou (or Compass) satellites are due to make a significant addition soon. At the time of writing, GPS is the most developed system, the most stable and (with certain provisos) can be used alone.

GPS has 32 working satellites in orbit, up from the planned minimum of 21, plus 3 active spares. The satellites are reliable, and coverage on the ground is extensive. At sea, it is possible to receive up to 11 satellites at one time, but reception may be constrained by cliffs, buildings and overhanging trees in coastal and estuarine waters. Extra satellites from other constellations may be useful, but the obstructions will not change, and the window on the sky will remain limited. PDOP (Position Dilution of Precision) is a value indicating a poorer (>5) or better (tending toward <1) arrangement of the satellite constellation in the sky. This is continually changing as the satellites move in their orbits (although not changing very quickly). If one can upload the obstructions (e.g., the extent of a cliff face) into the planning software, a more realistic value of PDOP will be calculated. This planning element may indicate to the surveyor the need to operate at a particular time of day. It also acts as a commentary on the likely accuracy of the derived positions.

Most surveyors will work with the satellites' expected paths, which have been predicted and are broadcast from each satellite (the Broadcast Ephemeris). The results are acceptable for most work, including measurements afloat. However, the satellites' paths through space are monitored to determine their actual routes, and the Precise Ephemeris is available after a couple of days. From these data one can improve the quality of the computed positions.

Each GPS satellite transmits on 1.2 and 1.5 GHz, with a binary code superimposed onto these carrier signals. A receiver can read the code to establish the satellite number, its path through space and supplementary information. The code from four satellites can be used in four simultaneous equations to determine the 3-D position of the receiver in the Cartesian coordinate system, along with the offset of the receiver's clock from GPS time. It is also possible to strip off the code and use the carrier signal itself.

A (currently £100) receiver permits the determination of horizontal position to an accuracy level of about ±8 m. This is good enough for many studies. Such a receiver will have a countrywide conversion from WGS84 to a national framework, offering an overlay of the computed position onto the national mapping at an accuracy of about another ±5 m. Increases in the cost of the receiver provide increased channels to receive more satellites on both frequencies that enable the calculation of delays along the signal paths (specifically the ionospheric delay in the upper atmosphere). Using a second receiver as a monitor station permits variations in the lower atmosphere to be estimated; commonly there is a data link to send these and other offsets to the mobile receiver. Accuracies improve from ±8 m toward ±1 m with a differential link applied to the pseudo-ranges and to ±0.03 m with a Real Time Kinematic (RTK) link applied to the carrier signal. The rigorous national mapping organisation's transformation can improve the conversion from WGS84 to the national framework to ±0.2 m.

For some, a single-point position is required, perhaps to establish a local reference station for later RTK GPS work or to set up a tide gauge. In the UK, this postprocessed point can be determined with reference to the Ordnance Survey's OSNet, giving an ETRS89 position to ±0.03 m. Because the separation from the OS Base station is important (the closer, the more similar the atmosphere), a separation of 20 km or less requires observations over an hour and two hours for 40 km. Such guidelines are available from the RICS (2010) and others.

Further, there is a range of satellite-based augmentation systems (SBAS) covering various continents and countries (Europe/EGNOS, North America/WAAS, India/GAGAN and Japan/MSAS). Using precisely positioned ground monitor stations and a satellite uplink, the reliability of the original satellite positions can be monitored in real time and an improvement imputed into the position calculation.

6   Bathymetry

6.1   Technologies

The principal technology used to measure depth is the echo sounder (Chapter 3). This works by measuring the time between the emission and reception of a short pulse of acoustic energy. The transmission/reception is from the face of the transducer, which is necessarily below the surface of the water, and yet the most common temporary reference level is the water surface itself. Therefore, the draft of the transducer must be determined for each survey. Using the water surface as the initial reference permits

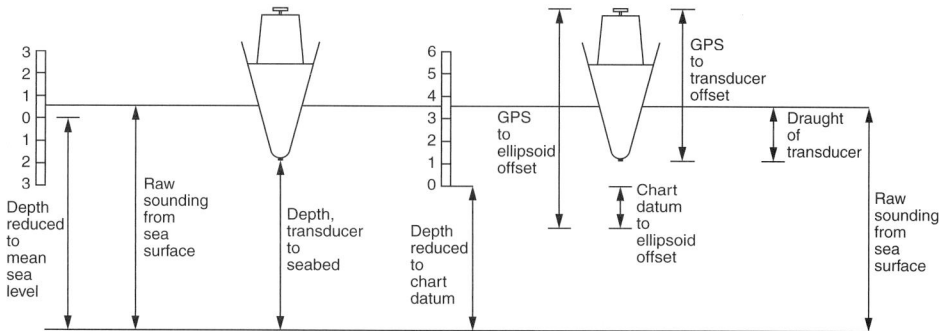

Figure 2-3 The correction of raw to reduced soundings, illustrating: the application of tides to MSL, the application of tides to Chart Datum and the use of GPS heights to Chart Datum.

the tidal height above Chart Datum (or above/below MSL) to be monitored at the tide pole and the value subtracted from the raw sounding (Figure 2-3).

Survey-grade echo sounders permit calibration for the draft of the transducer and for setting a speed of sound through the water. For sounding in waters less than 100-m deep, a 200 kHz, single-frequency, survey-grade echo sounder is appropriate. The sounder will not be set to exactly 200 kHz, but rather this represents a band that is different from those at 1 MHz, 600 kHz, 400 kHz or 30 kHz, for which the consequential propagation characteristics are

- Higher frequencies/higher absorption/higher precision and
- Lower frequencies/greater range/lower resolution.

A single-frequency echo sounder is cheaper and more portable than a dual-frequency instrument, although the latter may offer differentiation of an unconsolidated mud layer (using 200 kHz) from the consolidated or rocky lower layer (using 30 kHz). A survey-grade sounder can be adjusted for transmit power, time-variable gain, a lighter/darker print record (on a paper record), the draft of the transducer (affected by the draft of the vessel or the depth setting of a side-mounted pole) and the speed of sound through the water column. It is best to understand the technology and choose the appropriate settings, and there are some consequences of changing these settings, such as altering the point at which the returning pulse is recognised by the sounder and thus altering the calibration value.

The depth of the transducer can be determined directly by premarking a side-mounted transducer pole, and the value can be entered directly into the echo sounder. However, for a hull-mounted transducer the only sure way to determine the draft is via a bar check. The support wires must be measured and marked, the bar must be as long as the vessel is broad, and the bar must be suspended beneath the vessel at a known, shallow depth. Any errors due to an incorrect speed-of-sound setting will be nearly inconsequential over the short range to the bar, although it can be useful to repeat the whole calibration. In general terms, this first setting is to determine a 'zero error' (analogous to using a tape measure when the zero end part of the tape is missing). The second calibration point is a deep depth – perhaps deep enough to equate to the maximum depth within the survey area. However, it is difficult to achieve a reflection (a return) from the bar when much deeper than 10 m. Therefore, calibrating for waters of greater depth will require a cast from a velocimeter. This second setting is a 'scale error'. The speed of sound propagation varies through the water column – therefore, what value of speed to use? The sounder offers only one setting, and a choice must be made. Are the waters over the mud banks the most important or those in the channel for a deep draft vessel? Using a bar check, it is common to use the speed of sound returned from the deepest value (likely to be 10 or 12 m). Any offset at shallower depths must be disregarded. For deeper waters, individual readings returned from a velocimeter must be averaged.

Although one is likely to work with whatever echo sounder is available, some choices can be made at the time of purchase, or sometimes varied through the instrument's settings. A shorter pulse length will give a higher precision depth and more likely differentiate between one object above another in the water column; e.g., a fish shoal near to but separated from the seabed. However, a longer pulse length (containing more energy) will be needed for sounding through deeper waters.

A single beam echo sounder does not transmit a pencil-thin beam but a beam that spreads out; from a circular transducer the beam can be modelled as a simple cone (though in reality it is more complex). For a particular frequency, the beam spread angle can be approximated by the empirical formula, $B = 65\lambda/d$. For a 200-kHz sounder with a diameter of 0.05 m, and using a speed of sound of 1500 m s^{-1}, the wavelength λ computes as 0.0075 m, and the apex angle of the cone, B, comes out at just under 10°.

This beam spreading has some effect on the resolution. Over a rugged seabed the intricacies will not be clear, the deeper gullies will be masked by signal returns from peaks fore and aft and/or left and right of the boat's track. The minimum depth will be recorded, but the range of habitat may be masked. There is a similar problem with multibeam echo sounders (Chapter 3), where a typical beam width may be 2°×2° or even smaller – the resolution is improved but there is still a limitation.

The position of the sounding is usually stated as the position of the GPS. Therefore, any offsets of the GPS antenna from the echo sounder transducer must be applied. Whereas it may be possible to mount the transducer directly on top of a side-mounting echo sounder pole, any other configuration will require careful measurement. For swathe sounding (Chapter 3), the offsets are best measured using land surveying techniques or laser scanning.

Sometimes a chart may be produced that illustrates each sounding in metres and decimetres, with the decimal point used to illustrate the exact position of the depth. The latter, though, is a misunderstanding of the area of the seabed that is insonified by the sounding pulse and from within which the shortest return will give the least depth. The position of a sounding is more likely to be an area of seabed as covered by the numbers written to represent this depth.

These paragraphs lead to the question of the vertical accuracy of the soundings – working on the water is not like working on land. The International Hydrographic Organisation (IHO) gives formulae for computing the likely accuracy (IHO, 2008). In 30 m of water one might expect soundings to be accurate to ±0.8 m. A rigorous approach to the calculations is detailed in IHO (2005).

6.2   Bathymetry: Survey Platforms

Probably the most common vessel is a monohull, but catamarans do provide stability that is helpful in achieving useable data. Dillingham (2009) addresses the operation of survey vessels.

The sensor locations must be related to each other, principally in two dimensions (fore/aft; port/starboard) unless high-precision GPS is being used for the tidal correction to the vertical datum, in which case the heights of the sensors also are required. The horizontal relationships can be measured by tape, but this is not very satisfactory; Fibron tapes stretch under tension, and it is difficult to acquire a horizontal surface on which to lay the tape. For a little more investment in time, a 'Total Station' survey (Chapters 3 and 9) with the vessel on hard standing (perhaps on a slipway over low water) will produce better results, and these will match the requirements of a multibeam sensor. It is best to premark reference points on the vessel, such as a common reference point or the Centre of Gravity (see later), sensor locations, towing points and the outline of the vessel. On a steel vessel, the points may be premarked with a spot weld or punch point and labelled with a unique alphanumeric punch. A local network external to the vessel will allow a detailed survey using bearings and distances, and some of the marks will be observable from more than one station. The initial survey might start with arbitrary coordinates of one point (say 300 m E; 500 m N; and 50-m height), referenced to another point using an arbitrary direction for north (say), using the horizontal baseline distance to the second point to determine the new northing and the vertical offset for the height. The control network (a triangle or, if established before the vessel is in place, a braced quadrilateral) can be computed and all measurements tightly held within it. Multiple measurements to points will produce confidence in the overall quality.

The network must then be adjusted for three issues: a change in zero to the common reference point, a change in orientation to reflect the fore/aft alignment of the vessel and the change in trim of the vessel when afloat. The first network issue requires an adjustment, typically to the location of the Motion Reference Unit (MRU) that is definitely used in multibeam work but is unlikely to form part of single-beam echo sounding. If not the MRU, then a convenient point on the centreline can be chosen from which other measurements can be taken. Some methodologies call for the establishment of the Centre of Gravity, a point determined from design drawings and stability calculations but likely to be inconvenient because: (1), it may not rest against a surface but be within the space of the vessel; and (2), it will change as fuel, water and equipment are used and deployed – making the effort inefficient. The second network-issue adjustment requires a transformation to realign the coordinate framework. The last requires measurements from points around the vessel to the water's surface, although this trim setting may still not match that of the vessel when underway. For multibeam echo sounding there will be further adjustments made in the Patch Test (see Chapter 3) during calibration.

Towed sensors will have their point of tow determined in the foregoing, but the amount of cable deployed and the speed of the vessel through the water (with and against the current) will affect the actual sensor location. Further work is required to settle this. Finally, the vessel can be surveyed with a laser scanning survey. This can be very efficient in determining a whole vessel survey onto which photographic imagery can be morphed. Each point is determined in three dimensions, and with multiple setups the overall results will be of very high quality.

6.3   Bathymetry: Methodology

With the echo sounder calibrated for transducer draft and speed of sound, the survey may start – usually on a regular set of lines running into the coast. This is time consuming, and it takes as much time to stop one line, manoeuvre and get going on the next as it does to run short lines within an estuary. Single-beam echo sounding survey lines are conventionally run orthogonally to the contours; this is in contrast to sidescan sonar and multibeam echo sounding, which typically are run parallel to the contours. Nevertheless, with precise positioning and no offset between the antenna and the echo sounder transducer, lines run along the rising water line may permit access in a reasonable time. Tide values must be logged, although they can also be transmitted to the boat or to a shore station.

A convoluted coastline would not permit a simple set of parallel lines, and some compromise is necessary between consistent line spacing, optimal directions, and efficient coverage of the area. Line planning is a standard facility of commercial survey software, but can be operated in the field simply if lines can be north/south or east/west, where use can be made of consecutive grid values (easting or northings) or graticule values (latitude or longitude degrees and decimals or its variants). Transits may also be set on the shore by a helper with a second mark, indicating front and back points on each line. These can be set out by tape and optical squares, by RTK GPS or, if working at small scale, by stand-alone GPS.

6.4   Bathymetry: Uncertainties

The variables in the measurement process directly affect the outcome of the final product. For example, having an incorrect offset for the GPS antenna with respect to the echo sounder transducer may set the soundings aside from their true position by, say, half the width of the boat. Such an offset may or may not be important, and a judgement must be made. To quantify and model the uncertainties is a useful step. Hare et al. (2011) have written on this subject, identifying the variables associated with the vessel, the sensors, the environment, integration within the computer and the calibration process.

7   Planning and Surveying

The risk assessments associated with maritime land and hydrographic surveying should address the vessel, the weather, personal safety from mobilisation to demobilisation, navigation and land-based issues. A contracted vessel may well come equipped with a full risk assessment, broad enough to address individual operating procedures. A small dinghy or inflatable, convenient for trailering and launching close to the site, may never have been assessed. No matter how confident the operator or calm the water, it is crucial to wear life jackets, appropriately fastened.

Support is readily available to help assess issues of navigation (other vessels, underwater obstructions), operations (fire, groundings) and health (hypothermia, hyper-thermia, drowning). Activities can be absorbing, but 'Rules of the Road' must be adhered to. Many operations are likely to be in shallow waters. A reconnaissance at low water may reveal obstructions in the intertidal zone and, for those objects below the low water level, provide sufficient clearance at high water for a shallow draft vessel.

In the coastal zone, the fetch (the length of water over which the wind blows toward the survey site) may be small and the sea calmer than elsewhere. This is not always the case. Where the wind is significant there may be waves breaking on the beach and operational or safety issues with a shallow draft vessel. All operations should be undertaken with caution.

On land, mud flats should be approached cautiously, being wary of soft material (or, indeed, sand, which may have areas or channels unable to support an individual's weight). A life jacket may be appropriate, but it is better not to work alone, and it is useful to ensure communication where there might not be mobile phone coverage.

8   Final Remarks

It is always useful to take an initial overview of the data, considering whether there are straightforward anomalies. The data need to have been clearly annotated in the field, and where everything is digital, a daily log may clarify issues in later processing. Lines are not necessarily run consecutively, other vessels may interfere with or require a pause in operations, passing over another vessel's track may blank depth measurements due to aeration, or vessel wash may induce temporary roll.

The tides need to be processed, perhaps by reconstructing a tidal curve from sparse readings, or simply converting the logged readings into a file type readable by the software. The initial overview may show anomalous values (out of the tidal range or affected by the wash of a passing vessel, say). Values that are clearly wrong may be edited out.

The soundings must be viewed, and many can be removed as spurious (when obviously too shallow or too deep) or resulting from reflections within the water column. This is a matter of making a judgement because the guiding principle is that any one data point is as valid as the next. An isolated or apparently detached peak may come from a sunken ship's mast or running a line close to a cliff where the spreading sound pulse glances off an underwater rock.

With a pulse repetition interval of 10 Hz, many soundings are gathered along the survey line, and yet adjacent lines are comparatively widely spaced. There are too many soundings along the line to show them all, and some selection must be made, possibly to show the shallows, then the deeps and then to give sufficient information in between so that the seabed form is understood. The selection process is a standard feature within survey software and not within a Geographical Information System (GIS). That the lines are widely spaced, perhaps every 0.01 m at the plotted scale of the survey (i.e., for a survey scale of 1:2,500, every 25 m), will not matter if the topology is smoothly undulating and there are no hazards to navigation. The latter can be checked with a sidescan sonar instrument (see Chapter 3).

With soundings reduced for tidal height (above or below the datum value), the data can be plotted within the survey software or transferred as a simple *xyz* file to a GIS. If much other work is being done in a GIS, this is a useful place to combine the data. Contouring may not be quite as smooth, and it may not be possible to add standard charting features, but the growth in capabilities of GIS makes them a useful archiving and printing tool.

References

Dillingham, D., 2009. The safe and effective operation of inshore survey vessels. *The Hydrographic Journal* 131 & 132. The Hydrographic Society, Plymouth, UK.

Geoscience Australia, 2016. www.ga.gov.au/earth-monitoring/geodesy/geodetic-datums/geoid.html and connected pages [accessed March 2016].

Glen, N. C., 1979. Tidal measurement. In: (ed. Dyer, K. R.) *Estuarine Hydrography and Sedimentation*, Cambridge, UK: Cambridge University Press, 19–40.

GLOSS, 2016, www.gloss-sealevel.org [accessed March 2016].

GNSS, 2016. Global Navigation Satellite Systems. www.insidegnss.com [accessed March 2016].

Hare, R., Eakins, B., Amante, C., 2011. Modelling bathymetric uncertainty. *International Hydrographic Review* November 2011, 31–42. [online] available at: ushydro.thsoa.org/hy11/0428A_07.pdf [accessed March 2016].

Hydrographer of the Navy, 1969. Admiralty manual of hydrographic surveying, Volume 2: Tides and Tidal Streams, *Taunton.*

IHO, 2005. Manual on hydrography, C-13 (1st edition; corrections to February 2011), IHO, Monaco. [online] available at www.iho.int/iho_pubs/IHO_Download.htm [accessed March 2016].

IHO, 2008. S44, IHO standards for hydrographic surveys (5th edition). [online] available at: www.iho.int/iho_pubs/IHO_Download.htm [accessed March 2016].

IOC, 2006. *Manual on Sea Level Measurement and Interpretation*, Vol. 4, JCOM Technical Report 31, Paris: Intergovernmental Oceanographic Commission of UNESCO. [online] available at: www.psmsl.org/train_and_info/training/manuals/manual_14_final_21_09_06.pdf [accessed March 2016].

Lekkerkerk, H.-J., 2007. *GPS Handbook for Professional GPS Users*. Lelystad, the Netherlands: Pilot Survey Services.

OceanWise, 2014. OceanWise Ltd, Dovedale House, Alton, Hampshire GU34 1NB, UK. www.oceanwise.eu [accessed March 2016].

Proudman, J., 1953. *Dynamical Oceanography*. London: Methuen,

PSMSL, 2016. Permanent Service for Mean Sea Level, Liverpool L3 5DA, UK. www.psmsl.org and connected pages [accessed March 2016].

Pugh, D. T., 1987. *Tides, Surges and Mean Sea-Level*. Chichester: John Wiley.

Pugh, D. T., 2004. *Changing Sea Levels: Effects of Tides, Weather and Climate*, Cambridge, UK: Cambridge University Press.

RICS, 2010. *Guidelines for the Use of GNSS in Land Surveying and Mapping*, 2nd Edition. London, UK: RICS.

Sobel, D., 1996. *Longitude: The True Story of a Lone Genius Who Solved the Greatest Scientific Problem of His Time*. London, UK: Fourth Estate.

TASK, 2016. *Tidal Analysis Software Kit*. Marine Data Products Team, UK National Oceanography Centre, Southampton, UK. http://noc.ac.uk/using-science/products/tidal-harmonic-analysis [accessed March 2016].

Uncles, R. J., Stephens, J. A., Harris, C., 2014. Infragravity currents in a small ría: Estuary-amplified coastal edge waves? *Estuarine, Coastal and Shelf Science* 150, 242–251.

Valeport, 2014. Valeport Ltd, Totnes, TQ9 5EW, UK. www.valeport.co.uk [accessed March 2016]. Millbay Docks tide data are displayed at: https://valeport.port-log.net/live/Display.php? Site=33&Observed=24&Predicted=8 [accessed March 2016].

3 Acoustic Seabed Survey Methods, Analysis and Applications

G. E. Jones, V. J. Abbott, A. J. Manning, M. Jakt

1 Introduction

Bathymetric sounding by single-beam echo sounders (SBES) gathers data in profiles that have a high density of soundings along each line but nothing between them. To overcome this shortfall, current technology 'looks' between the adjacent sounding lines (with sidescan sonar) and, to increase the coverage of soundings across the seabed, makes wide area (swathe or swath – the terms are used interchangeably in this chapter) precise measurements of depth. Low-frequency acoustics are also used to sense below the seabed (albeit in profiles) for the underlying sediments. Applications of these techniques range through visualisation of the seabed for the navigator, understanding habitats better, identifying features of geological interest for improved understanding and providing support that facilitates construction works on the seabed.

Sidescan sonar (sideways looking sonar) was developed to image the seabed to identify the possible presence of seabed features between SBES lines and to give confidence to those using charts for navigation or engineering purposes. For some, however, this qualitative approach to seabed assessment was insufficient, and research focussed on deriving precise measurements of depth across a swathe of seabed, rather than a portrayal of the seabed form. This was achieved through the development of swathe sounders, which, in their early days, were primarily focussed on the acquisition of a high density of depth soundings; thus, the need for imaging between sounding lines could be removed. However, early systems did not routinely measure signal amplitude, a feature of sidescan, and this is now an important output variable. Nevertheless, in deeper waters the separation between the shipborne sensor and the seabed causes the loss of some of the differentiating features, such as shadowing, which aids sidescan sonar record interpretation. Accordingly, both towed sidescan sonar and shipborne swathe systems may still be deployed.

The deployment of lasers on aircraft (Lidar) has offered another method to measure the depth, and at greater speeds than surface vessels, whilst keeping the operators safe when working very shallow or hazardous waters. Conversely, the deployment of acoustic systems on remotely operated vehicles (ROVs) offers higher resolution in deeper waters by taking the sensors closer to the seabed. Acoustic frequencies are chosen for their precision, range or penetration. Lower frequencies can penetrate the seabed to image the sediments and rocks beneath, and profiles can be measured and areas of strata modelled.

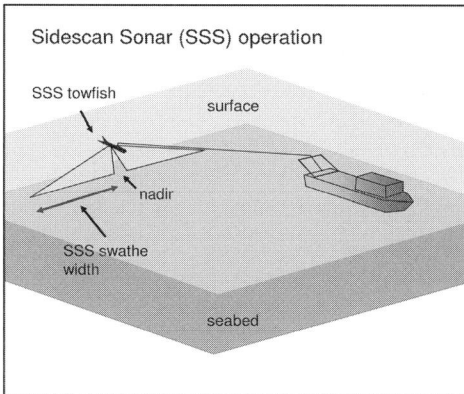

Figure 3-1 A schematic of a sidescan sonar (SSS) in operation. The torpedo-shaped instrument (a towfish) is towed from a vessel's rear frame and emits sound pulses into the water from and to each side of the vessel.

2   Sidescan Sonar

Sidescan sonar (SSS) has become a standard and essential tool for seabed imaging, geological mapping and habitat mapping (Figure 3-1); it developed from early 12-kHz systems (in 1963), through progressively higher frequencies (100 kHz, 500 kHz) to today's 900-kHz and 1600-kHz systems. Until the mid-1980s, commercial sidescan images were produced as paper records on a line-scan recorder, with each acquired line of data being added to a 'waterfall' view (e.g., Figure 3-2). Computerisation of the display systems and of the sidescan towfish led to improved imagery and digital data storage. Digital sonars have replaced the single or switchable analogue systems and have the ability to show both, say, 100-kHz and 500-kHz return signals simultaneously. Additionally, the ability to control the signal offers a higher resolution, focussed beam, synthetic aperture and chirp (frequency-pulsed) sonar systems.

Current sonars include the Klein System 4900, offering dual frequencies of 455/900 kHz, and the EdgeTech 4125, which offers dual frequencies of 600/1600 kHz. Other developments are multibeam sidescan sonars, 'gap-filler' sidescan sonars and Sonardyne's Solstice, which uses 'Multipath Suppression Array Technology'. Operating at between 725 and 775 kHz, the Solstice offers a 200-m swathe that utilises two 80° x 0.15° beams and 0.05–0.2 m resolution across the full swathe, together with integrated swathe bathymetry.

2.1   Principle of Operation

The principle of operation of sidescan sonar (Figure 3-1) is similar to that of an echo sounder, although a pair of rectangular transducers is utilised. The transducers are inclined down from the horizontal and consist of an array of piezo-electrostrictive, lead zirconite titanium (PZT) ceramic elements, set against an acoustically matched epoxy

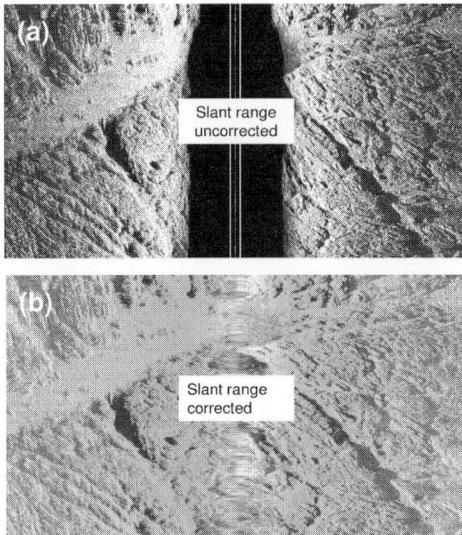

Figure 3-2 A real-world example of a sidescan sonar 'waterfall' record, uncorrected (a) and corrected (b) for slant range. Both images are from a sidescan survey undertaken in Heybrook Bay, near the mouth and to the east of Plymouth Sound, UK (see Figure 2-2 of Chapter 2), during December 2014.

resin sounding face. Applying electrical energy to the acoustic elements causes them to vibrate, which creates a pressure wave that is propagated into the water. On reflection from the seabed, a portion of the transmitted energy is reflected back to the transducer, which vibrates the transducer elements that convert the pressure changes into electrical energy. Sampling of the returned pressure wave, over a period of time equivalent to a range from the sonar, offers an interpretation of the variable amplitudes of the reflected signal as a function of the acoustic properties of the seabed. Repeated transmissions as the vessel and sonar move forward along a line of travel, and recorded on a waterfall display (Figure 3-2), build up a sidescan sonar trace, or image of the seabed.

The piezo-electrostrictive effect may be illustrated best by reviewing the qualities of quartz or other crystals and ceramics that have similar characteristics. These respond to an alternating electric current with a change in the dimension of the crystal, such that it will resonate with a frequency of vibration that is inversely proportional to its thickness. These transducers generally are designed to vibrate for a range of frequencies from the order of 5 kHz up to 1600 kHz, dependent on the functional design and purpose of the sonar. Typically, 5–10 kHz systems are used for long-range oceanic surveys (30 km, say), 50–100 kHz for medium-range geophysical purposes (1500 m), whilst 200-, 500- and 600-kHz systems are used for engineering surveys with decreasing ranges from 300 to 100 m. Sonars with frequencies 700/900/1600 kHz are designated as search and recover systems as well as search and characterisation systems, with operational ranges reducing from 100 to 20 m. Effectively, the higher the frequency of the sonar system, the greater the signal attenuation within the water column and the shorter the attainable range.

Each transducer will have marginally different acoustic characteristics and there is therefore a need to design systems with a matched pair of transducers, each transducer offering a comparable frequency, power output and acoustic pulse-form, so as to offer comparable transmit and record sensitivity to each side of the sensor.

In coastal and estuarine areas, the use of higher frequencies is of greater use, and the selection of a dual-frequency system containing a primary frequency of, say, 400 kHz or 600 kHz, with a secondary frequency at either a higher or low frequency, would probably suit the coastal engineer best.

2.2 Directivity and Beamwidth

Transducers are designed to send out beams of sound of various shapes and with a level of directivity such that accurate measurements and differentiation between seabed features can be achieved. There is value in using acoustic beams that are narrow in the direction of movement, which would be coincidental with the build-up of adjacent swathes of data, and broad in the across-track direction, to allow for reflections from close to the nadir beneath the transducer (Figure 3-1) and out to the desired range. Coverage from each side may not overlap because cross-interference would occur between transducers.

In the first instance, directivity is directly related to the wavelength and therefore the frequency of the projector, and also to the size of the transducer, which normally is rectangular. Directivity is a measure of the transducer's beam width, which is defined as the angle between the half-power points (the loss of half of the maximum power transmission along the acoustic beam's central axis). The beam width for a rectangular transducer is given by the expression:

$$B° = \frac{50\lambda}{L} \qquad (1)$$

in which $B°$ is the beam width, λ is the wavelength of the projected sound frequency and L is the dimension of the transmitting surface perpendicular to the plane of the beam being considered. The units are unusual in Eq. 1, requiring SI values on the right-hand side but the numerical answer being in degrees on the left.

Given a constant frequency pulse at 600 kHz, and a velocity of sound in water of 1500 ms^{-1}, the wavelength of the transmission would be 0.0025 m (1500 ms^{-1} divided by the frequency of 600000 s^{-1}). Hence, for a rectangular transducer of dimensions 0.4 m by 0.0025 m, the beam width would be:

$$along - track = \frac{50 \times 0.0025}{0.4} = 0.31° \qquad (2)$$

$$across - track = \frac{50 \times 0.0025}{0.0025} = 50.00° \qquad (3)$$

Directivity also can be improved by placing reflectors around the transducer, or by phasing the transmission of a more complex array of acoustic elements. This last is particular relevant for systems utilising Chirp technologies that result in long pulse lengths and focussed beam systems.

2.3   Resolution

Resolution can be determined out to the side (across track) and down the survey line (along track). Across track, two closely spaced objects can be differentiated if they are a distance apart greater than half a pulse length. For a pulse duration of 100 µs, the pulse length would be 0.15 m (at a nominal sound speed of 1500 ms^{-1}) giving a resolution of 0.075 m.

When using a Chirp sonar, the range resolution is defined by a function of the speed of sound divided by twice the bandwidth. Thus, for a system with a 50-kHz bandwidth, the range resolution (1500/(2x50000)) would be 0.015 m; however, for a lower signal loss the energy requirements are significantly higher.

Along-track resolution is again a function of the ability of a system to differentiate between two closely spaced objects. Thus, it is a function of the along-track beam angle and distance from the sonar towfish; e.g., the resolution at a range of 50 m would be 0.27 m for a 0.31° along-track beam angle, but would deteriorate to 0.54 m at a range of 100 m.

2.4   Sonar Records and Interpretation

As the acoustic signal is propagated through the water, it is spread across the surface of the expanding wave front and further reduced by attenuation and absorption within the water column. The result is that the longer the signal path, or range, the lower the signal level that reaches the seabed, with resultant lower levels of reflectance. Within the 'raw' sonar record this will be characterised by a weaker return signal and lowered contrast levels between differing sediment types and morphological changes. Whilst initial gain levels can be adjusted to lighten and darken the overall record, or compensate for signal loss, recording systems offer a function, time variable gain (TVG). TVG allows for a time-relative amplification of the signal level, with greater distances of travel receiving increased levels of amplification. Most acquisition/processing software will offer a number of options, which may include

- Automatic application of time variable gain (which may result in either an average 'grey scale' being set across the sonar record), or
- The option to manually select an algorithm based on spherical or cylindrical signal spreading for signal amplification, or
- The manual selection of amplification levels

The ultimate objective will be to obtain an average and aesthetical grey scale, with an appropriate level of contrast between differing sediment bodies and seabed features.

Once an appropriate waterfall image is obtained, the operator can then choose to view the acquired data in 'normal' slant range mode, where the waterfall progressively builds up adding new swathes to the image as the vessel progresses, and where the port and starboard scans portray the water column and seabed to each side of the vessel (Figure 3-2a) or, if using seabed detection algorithms, view a range-corrected portrayal of the seabed, where the water column has been removed (Figure 3-2b). An advantage of acquiring data that displays the water column is the safeguard that an operator can visually observe the towfish height above the seabed. This should be kept at a ratio of 1:10 with respect to the range in use (e.g., a flying height of 15 m for a 150 m range), down to a minimum where the fish is at a safe flying height above the seabed, typically about 10 m. The 1:10 ratio offers good contrast, both between the acoustic returns from varying sediment types, and for useful shadows cast behind rocks and obstacles to highlight the objects' characteristics.

Signal reflectance from the seabed itself will also fluctuate as a function of sediment-related acoustic impedance (reflecting porosity, water content and sediment density), or reflectivity, the grazing angle of incidence, seabed slope, and roughness. Most of the signal will be reflected away from the transducer, with only a small amount being refracted into the sediment and only a small portion being returned to the receiving transducer. The reflectivity (R) of the seabed for a normal angle of incidence is given by:

$$R = \frac{(Z_2 - Z_1)}{(Z_2 + Z_1)} = \frac{A_R}{A_1} \tag{4}$$

where:

A_R is the amplitude of the reflected signal
A_1 is the amplitude of the incident signal
Z_1 is the acoustic impedance of water
Z_2 is the acoustic impedance of the seabed material

In general, on a flat seabed, the reflectivity is directly dependent on sediment type, with coarser materials normally reflecting a stronger return and hence displaying increased intensity on the sonar record. However, it is significant that the acoustic impedance also has an inverse relationship with porosity, with sediments of higher porosity possessing lower impedance and hence less reflectivity, whereas low porosity materials have a higher impedance and reflectivity. The result is that differing sedimentary materials might have similar levels of reflectivity, so that ground-truth sediment data would be needed to identify the true reflectivity-sediment relationship (Chapter 6).

Reflectivity also is affected by the angle of incidence, such that not only are lower returns of reflectivity obtained from more distant materials, but also topographic slopes that are inclined away from the sonar receiver will result in lower reflectivity (even zero return). A topographic crest will mask the seabed beyond, creating a shadow zone within which no features can be detected; conversely, a seabed that slopes toward the receiver will show higher reflectivity than its surroundings.

In cases of low acoustic impedance within the sediment, an acoustic signal might also succeed in penetrating the immediate seabed to produce reflected energy from underlying stratification, although penetration is more likely to occur where the incident angles are higher. This may prove to be problematic in estuaries, where there might often be fluid muds. In these circumstances it is often useful to observe both high and low frequencies within a dual-frequency sonar system, where the higher frequency may be able to detect the surface of the fluid mud, whilst the lower frequency may reflect from the consolidated estuary bed. Again, ground truthing of the sediment beds may be needed to qualify and quantify the sonar returns and the sedimentary materials involved.

2.5   Signal to Noise

As with all acoustic systems, the focus is on the abundance of valuable signals over undesired noise, which might arise from a range of ambient and self-generated sources. Sometimes the data may exhibit fuzzy returns, features of ill-defined quality, or of a distinct pattern. These may be the result of several factors including, but not limited to, features detected within the water column, such as the residual hydrodynamic cavitation caused by the wake from another vessel, reflections from buoys, pontoons, the hulls of moored vessels, or even the swim bladders of shoals of fish. These last, and cavitation, often appear as darker impressions on the record, with fish frequently appearing as circular or elliptical blurs. Within stratified waters, turbulence along the interface of differing bodies of water as well as surface roughness in shallow waters might also give an interference pattern – rather like looking through moving water.

Reflections from the interface between different, perhaps highly stratified, layers of water may prevent signal reception where the signal strength is in excess of that reflected from the seabed. In such cases, it will be necessary to tow the sonar beneath any thermocline or halocline. Reflections from the surface can occur either with a rough or a calm surface. In the former case, particularly in shallow water, side lobes of the principal acoustic pulse may be reflected from the sea surface, overlaying the seabed returns. In the latter case, a calm sea surface may act as a perfect reflector, reflecting the energy toward the receiver – in and out of phase with the seabed return – such as to first cancel and then enhance the amplitude. These interference fringes (Lloyd's mirror effect) run parallel to the ship's track and may be observed to follow changes in bathymetry; this has led to the development of interferometric swathe sounding (see later).

Other sources of noise could be self-generated electrical noise, hydrodynamic noise, as well as that from the vessel's own echo sounder. Systems have to be tested for noise produced from within the towed body, within the processing system and by the vessel. Similarly, a towed sidescan sonar instrument will be in a torpedo-like housing (Figure 3-1), whereas the tow cable attachment, the cable in the water, or the vessel itself may produce noise as they pass through the water. The survey vessel or nearby vessels may produce acoustic noise on a similar or nearby frequency, possibly at a subharmonic within the bandwidth of the sonar.

On the face of the transducer, a loss of the sonar signal (quenching) may occur either by towing too fast or by transducer rolling, where the noise generated by the flow of water across the transducer may be greater than the signal generated or recorded. Conversely, hydrodynamic cavitation may occur if the emitted signal power is too high in proportion to the back pressure available at the depth of tow of the sensor.

2.6   Survey Accuracy

The accuracy and completeness of a sidescan sonar survey depends on a wide variety of factors, including vessel speed; pulse repetition rate; height of towfish or transducer; effects of roll, pitch and yaw; correct setting-up of equipment; accurate navigation and the quality of positioning.

With the transducers designed to receive returns from some distance away (from the principal, central transmit and receive beams), the return from nearly beneath the towfish depends on the minor transmissions (sidelobes) to either side of a sidescan sonar's near-linear transmission. Even so, there can be a blank directly below the fish, the width of which is a function of the angle of declination of the central beam of the transducer from the horizontal. The first return from the seabed gives an approximation of the towfish flying height; thus, it is possible to correct the slant range from the transducer to the seabed using Pythagoras' Theorem and the towfish height (albeit there will remain an area of distortion around the nadir, Figure 3-1 and Figure 3-2). True (horizontal) range is given by Eq. 5:

$$R_t = \left(R_s{}^2 - H_t{}^2 \right)^{1/2} \tag{5}$$

where R_s is the slant range and H_t the height of the transducer above the seabed.

Good-quality records are achieved best by navigating straight survey lines and overlapping the nadir by adjacent lines. For example, if it is proposed to acquire data using a 75-m range at a towfish height of 10 m, then a line-spacing of 60 m would be suitable. This gives more than 100 per cent overlap, so that the area beneath the nadir is covered and any seabed feature can be seen from two aspects, thereby countering any variation in signal return that the feature might reflect as a function of its orientation and the nature of the surrounding seabed.

The quality of the sonar record may also be affected by the along-track resolution, dependent on the pulse repetition rate (PRR, which is also termed the PRI – pulse repetition interval) and forward vessel speed. For example, given a velocity of sound in water of 1500 ms^{-1}, a 400-kHz system, and a PRR of 10 Hz, an along-track resolution of 0.5°, and a 75-m range setting, then for an optimum coverage at 25 m the acoustic beam footprint would be:

$$2 \times range \times \tan\left(beamwidth/2 \right) = 2 \times 25 \times \tan\left(0.5°/2 \right) = 0.22m \tag{6}$$

For 100 per cent insonification, the maximum vessel speed must be less than beam footprint sweep time (0.22 m, Eq. 6, divided by 0.1 s) and, hence, no more than 2.2 ms^{-1} or 4.3 knots. Naturally, a slower vessel speed would achieve a higher resolution, but the vessel may then lose steerage way.

With a flat seabed it is possible to compute the approximate height of objects (H_0) standing above the seabed because they cast an acoustic shadow. This is achieved through trigonometry and the properties of similar triangles (Eq. 7); however, with a sloping seabed the unknown angle of the slope significantly degrades the calculation:

$$H_0 = \frac{L_s \times H_t}{L_s + R_s} \tag{7}$$

in which R_s is the slant range to the object, H_t the height of the transducer above the seabed and L_s the slant length of the shadow.

The derivation of an object's position is approximate, and the precision degrades with an increasing separation of the towed sensor and the survey vessel. The towfish position may be computed as a function of tow distance (layback) and a reciprocal of either the vessel heading or the course 'made-good'; the offset (port or starboard) distance to an object is applied to the tracked position of the vessel and its towing point. Nevertheless, the vessel will operate under variable wind, wave, tidal stream and other current conditions, so the towfish may well move forward at a variable angular offset to the vessel and on a variable heading. A best-fit position for a feature can be determined by averaging the computed position as derived from two adjacent lines of opposite headings; this can be enhanced if the object is boxed-in by four perpendicular lines. Alternatively, survey operations can be undertaken using acoustic range-and-bearing positioning systems to precisely position the towfish, although this would increase survey costs and be difficult to achieve in shallow water.

2.7   Application

Sidescan sonar has been the main survey tool for examination of the geology, condition and make-up of the seabed and for detection of artefacts on the seabed. In the past it was used to identify variations in the seabed between bathymetric survey lines, which have been now largely been overtaken by swathe sounding systems. However, swathe systems have been limited in their ability to gather quality amplitude data – the consequence of reflections from different seabed types. Accordingly, many operators and their clients continue to utilise sidescan sonar.

With object detection protocols, sidescan sonar is widely used by archaeologists to identify artefacts that may be distinguished from the geology, or may clearly be identified as man-made; nature, for instance, does not normally produce features that are regularly shaped. It is also used by engineers to discover obstructions on the seabed that might impede dredging operations, such as cables, or to monitor the condition of seabed infrastructure; e.g., the scour around foundations. The imagery may be used to support the planning of armour emplacements around seabed defences or as an aid to monitor the condition of pipelines; acoustic surveys along a pipeline's length can provide engineers with information on the support beneath the pipeline, thereby warning them if the pipelines are unsupported, scoured or suspended in the water column. Sidescan surveys also provide engineers with imagery that shows whether pipelines and cables are buried, or lie exposed on the seabed, where their integrity may be endangered by bottom-trawling. Historically, such activities were undertaken using sonars of frequency 200–500 kHz, but operators are increasingly using higher frequencies for higher resolution. Indeed, dual-frequency systems incorporating the higher frequencies of 600, 900 and 1600 kHz are being touted as search-and-recovery systems, with users such as the police and border agencies utilising them to search for bodies or suspicious packages on the seabed or attached to the hull of a vessel. Nevertheless, the mapping and quantification of scour around dockside facilities and bridge supports is increasingly being taken over by swathe bathymetric systems.

3   Swathe Bathymetric Systems

Swathe systems have arisen through three lines of research and development: (i) using multiple echo sounder transducers, (ii) interferometric bathymetric swathe sounding, which is a development based on sidescan sonar and observations of the Lloyd's mirror effect, and (iii) Multiple Beamforming Echo Sounder (MBES).

Most swathe bathymetric systems have backscatter detection capabilities, although they remain limited by: (a) their mounting on the vessel, with a consequential greater height above the seabed in comparison with a towed sensor; (b) swathe widths that are limited by the beam angle in use; and (c) resolution limited by the number of beams within the swathe system (or some limited multiple). In any case, the resolution is lower than a true sidescan sonar system. Whilst backscatter to a transducer that is hull-mounted may be more readily geo-referenced, a less than optimum seabed-to-sensor separation may reduce the data contrast and the ability to see shadow details. Swathe widths, which may need to be changed as bathymetric depth varies, cannot compete with those achievable with a sidescan sonar system.

3.1   Multiple Transducer System

Multiple transducer echo sounders (Figure 3-3), which came into use during the 1970s and 1980s, are the simplest of the three swathe bathymetric sounding options. Offered by a range of manufacturers, an instrument might comprise an array of up to 50 transducers mounted on booms extending on either side of the survey vessel. Placed typically 1 m

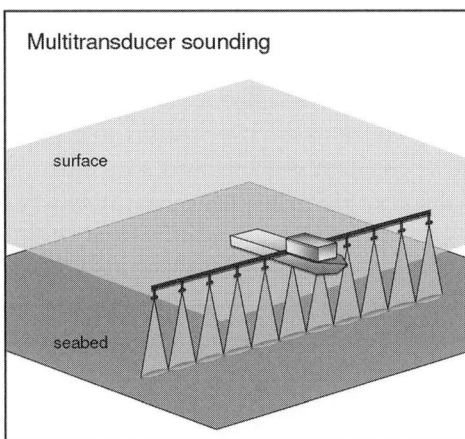

Multitransducer sounding

surface

seabed

Figure 3-3 An illustration of a multitransducer echo sounder in operation, showing the two arms, guyed in reality, with transducers on each and attached to the hull of the vessel.

apart, each transducer transmits and receives and has its own depth-digitizer board, the sum of which is multiplexed prior to onward transmittal to the sounding computer. Each of the transducers within the array is individually positioned in three dimensions in terms of its offset from the vessel's Common Reference Point, modified by the vessel's heading, a boom inclination angle and the attitude of the vessel (i.e., roll, pitch and yaw).

Whilst utilised by a range of organisations such as the US Army Corp of Engineers (USACE, 2002), the Canadian and Swedish Hydrographic Services and ports such as Hamburg and Rotterdam, and perhaps ideal for shallow areas and restricted waterways, the uptake of multiple transducer systems has been constrained by manoeuvrability, stability and complexity of operation. Some demand remains, as evidenced by the German Federal Institute of Hydrology deployments (Wirth and Brüggeman, 2011). Automatic Surface Vehicles may be useful in this context.

3.2   Interferometric Systems

Sidescan sonar uses sound that is projected at low incident angles to the seabed to obtain reflections from objects standing clear of the seabed. Banding occurs in calm, shallow waters, due to the Lloyd's mirror effect, and the bands follow the bathymetry. Interferometric Bathymetric Swathe Sounders are an extension of high-resolution digital sidescan systems that use this phenomenon, where the effect is reconstructed with multiple transducer staves (i.e., the transmitter and/or receiver transducer sections) at half-wavelength separation. Amplitude levels are measured for the sidescan effects and depths are derived.

Whereas a typical sidescan transducer has one transmit and one receive stave on each side of the fish, an Interferometric Sounder will have one transmit and multiple receive staves. This not only allows the system to acquire high-resolution sidescan imagery through phase comparison techniques between the additional receiver staves, but also to simultaneously collect coregistered bathymetric data.

Early systems were developed between 1979 and 1985 utilising a Science and Engineering Research Council (SERC, UK) funded research project at the UK's University of Bath. The aim was to produce a bathymetric system that could collect depth data on either side of a vessel's track to the same quality as a single-beam echo sounder located under the vessel. The system was mounted on a towed platform with one transducer and two receivers that operated at 300 kHz and possessed a 1° beam and a 60° swathe (Cloet and Edwards, 1986). As a towed system, the array had to accommodate heading and attitude sensors to orientate the swathe of soundings. Since those early days, improvements in electronics, faster computers and better algorithms, as well as an increased number of receiver elements within the transducers and the adaptation to vessel mounting, have greatly improved the performance of these systems.

3.2.1   Principles of Operation

The GeoSwath Plus Wide Swath Bathymetry System (Kongsberg, 2016a) comprises two transducers, each housing one transmit and four receiver staves of piezoelectric elements (Figure 3-4). The bottom stave transmits the sonar signal and the topmost

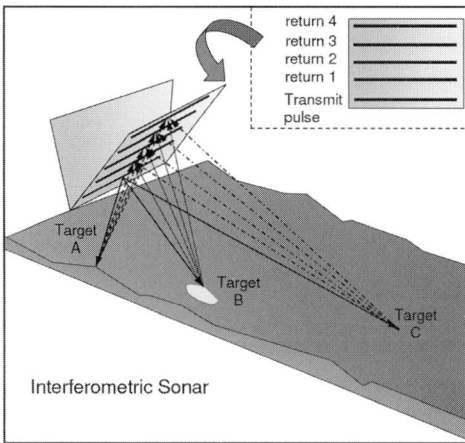

Figure 3-4 The elements of a wide-swathe bathymetric system transducer (the staves), showing the transmitted and received signals. In this illustration there is one transmit stave, located at the bottom of the transducer, and four receiver staves in line above it. Only one swathe is shown, as generated by one of two transducer arrays mounted in a 'V' configuration.

measures the amplitude of the returning signal to give the sidescan imagery. All four receivers are used to derive the relative, between-stave phase delays, which in turn give the angle of return of the sonar signal (Figure 3-5). The relative delays allow the angle of return of the sonar signal to be measured to a fraction of a degree; measurements are taken at very short intervals, and it is this rapid and accurate phase measurement that gives the GeoSwath its very high data density (Hiller and Hogarth, 2006).

Interferometric sonar is particularly useful in shallow waters because it can obtain sidescan sonar-style backscatter at distances out to 20 times the water depth. Near vertical structures, the above-horizontal swathe (up to a 250° view angle, i.e. 35° above the horizontal on each side) gives imagery of, say, jetty supports and the mud bank between them. The precise measurement of depth is more restricted, with bathymetry obtained to about seven times the water depth. The angular resolution is about 0.03°, and the range resolution is 0.03 m (Hiller and Lewis, 2004).

There has been an interferometric-sonar issue with poor coverage at nadir (i.e., directly under the transducer) with less-effective results obtained from an acute angle of incidence with the seafloor; nevertheless, surveying with 100 per cent overlap overcomes this problem, and the size of the nadir gap has been reduced to a few centimetres with newer versions. Another gap in data can occur directly in front of a vertical feature.

The distinct advantage of the high-resolution dataset can be lost by overfiltering. Where the full dataset has been retained, even propeller blades on a downed aircraft engine have been identified. Interfacing with third-party software reduces the data density to a much lower level – similar to that of a Multi Beam Echo Sounder; whereas an MBES may have 256–512 data points per ping, a GeoSwath Plus system may acquire 2000–3000 data points per ping. However, unless some form of data reduction is used, much MBES processing software cannot cope with this data density.

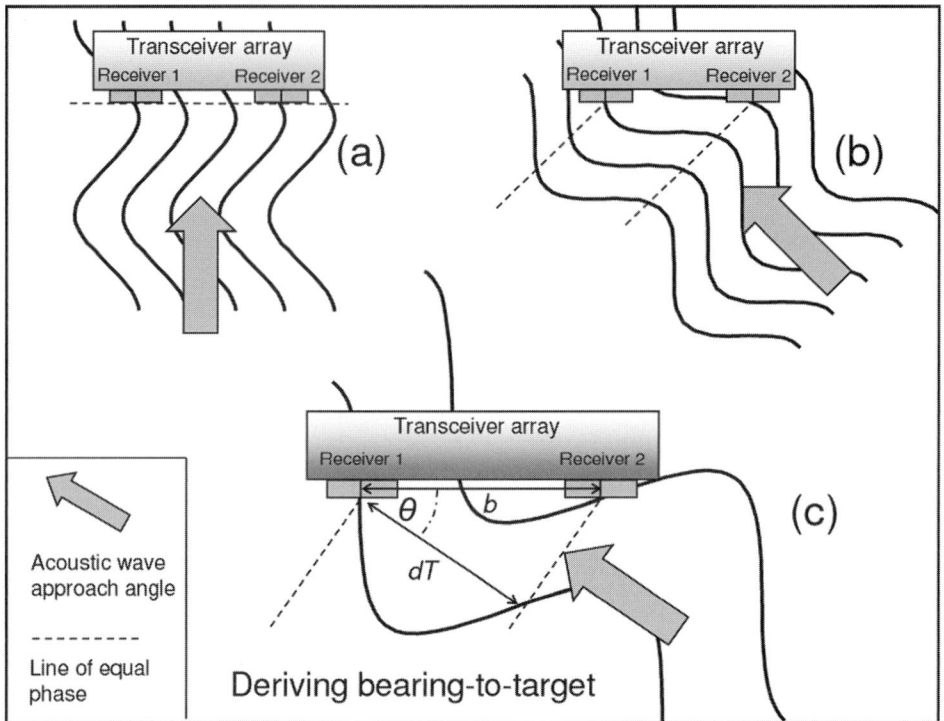

Figure 3-5 The derivation of angle (bearing-to-target) from differential (phase difference) signal detection. (a), In this illustration, acoustic waves approach the transceiver array head-on, and there is no phase difference between receiver elements; (b), in this case, acoustic waves approach at 45°, and there is a measureable difference in phase between receiver elements; (c), in general, if the acoustic waves, of frequency f, approach at an angle θ, and with a sound speed of c, then θ can be derived from knowledge of the baseline length between receiver elements, b, and the measured phase difference between receivers, dT: $\theta = \arccos(dT \cdot c/(2\pi \cdot f \cdot b))$. Note that dT represents a phase difference in (c), and not a distance.

There are several advantages to using interferometric sonar (Williams, 2015):

- Six to seven times the water depth for commercial bathymetric applications and up to 20 times for backscatter mapping (Swathe Services, 2005)
- Very high angular resolution of 0.03° and a range resolution of 0.03 m (Hiller and Lewis, 2004)
- Up to 240° view angle, ideal for mapping vertical features (Kongsberg, 2016a)
- Coregistered geo-referenced sidescan sonar backscatter (Swathe Services, 2005)
- IHO (International Hydrographic Organisation) SP-44, Special Order compliance (Kongsberg, 2016a)

3.3   Beam-Forming Echo Sounders

It is possible to trace the development of multibeam echo sounders back to 1964 and 12-kHz deep sea systems. The Danish company Reson in particular has developed

455-kHz systems and, across the industry, current developments include over 500 beams per ping; multiple pinging at the same time across more than one frequency; multiple signal detections within a single beam; attitude stabilisation for roll, pitch and yaw; variable frequency selection from 200 kHz to 700 kHz; software-driven swathe sector rotation; angular selection from between 10° and 165°; individual beams to the order of a nominal 0.5° cross-section; and very high resolution backscatter outputs.

3.3.1 Principles of Operation

In most multibeam echo sounders, transmit and receive arrays are constructed with piezoelectric elements placed at regular intervals. By applying voltages to the elements at different time intervals, creative and destructive interference effects produce a main central lobe and a series of decreasing side lobes within each sonar beam. The length of the array (aperture) determines the number of lobes and thus the beam width of the main lobe. The longer the array, the narrower the main lobe; this inverse relationship is evident in a radar antenna and in a sidescan sonar transducer.

Deploying two long transducers at right angles allows the transmission of one long, thin beam and the reception of one long, thin beam; the combination focuses on just a small area of the transmission, and a small area of the seabed is sensed[1]. Refining the narrowness of the transmit beam and the receive section increases the number of beams and the resolution of the system; however, beam spread results in a broader footprint at greater depths and increasingly toward each edge of the beam.

The backscattered amplitude data can be used to produce a plot of acoustic intensity. This is akin to the imagery recovered from sidescan sonar. In its simplest form, the number and spacing of the backscatter returns are a function of the number of beams; e.g., if an MBES has 256 beams then one would have a swathe displaying 256 signals. However, advanced computer algorithms and engineering are allowing for within-beam sampling and beam splitting, as well as beam correction for angle and range, such that a higher resolution of the seabed reflectance footprints can be displayed.

A further development is the acquisition of water-column data. Previously, priority was given to detection of the seabed, and systems used bottom tracking algorithms. Once the seabed had been detected it was assumed that a depth change between subsequent pings (up to some maximum) would fall within a 'bottom track window of change'. Nevertheless, concerns that shoal points or obstructions could be missed, particularly over wrecks, resulted in the recording of the full vertical beam profile. Recording these data quickly generates terabytes of data but does allow for the three-dimensional portrayal of targets within the water column, which might be the mast of a wreck, a volume of fish, or differing bodies of water, dependent on frequency and signal sensitivity.

3.3.2 Resolution

From preceding modes of sounding operation, it follows that the highest resolution achievable with an MBES is from directly below the multibeam transducer (at nadir).

---

[1] This approach was applied to radio astronomy by Bernard Mills, 1920–2011.

The footprint size can be calculated at any given angle by considering the angular resolution of the system, the frequency of the sonar, the operational water depth, the swathe sector angle and the number of beams.

3.3.3 MBES Calibration

The multibeam setup requires the careful measurement of offsets (Chapter 2) between the sensors (GNSS[2] antennae, Motion Reference Unit and MBES transducer) and, typically, a Common Reference Point, a tow point and any other transducers or sensors. It is best if the transducer deployment can be exactly duplicated each survey. The offset measurements may need to be in 3-D if using GNSS height; otherwise, at least the draught of the transducer below the water level will be required. A temperature and salinity profile with measurements at known depths will be used to determine point velocity measurements using, say, Medwin's Equation, Eq. 8 (Medwin, 1975):

$$c = 1449.2 + 4.6T - 0.055T^2 + 0.00029T^3 + (1.34 - 0.010T)(S - 35) + 0.016D \quad (8)$$

Where c is the speed of propagation of an acoustic wave, S is salinity (dimensionless, but commonly expressed as psu), T water temperature ($°C$) and D (<1000 m in Eq. 8) the depth of measurement of a number of sample points. More directly, sound velocity can be measured using a sound velocimeter (e.g., AML, 2016; Valeport, 2016); these instruments are used to establish a water-column sound velocity profile to enable calculations of ray paths to the seabed.

Finally, the interrelationship of the settings (particularly between the assumed heading of the vessel and measurements orthogonal to this) need to be checked against readings from the seabed (a patch test). Residual time, roll, pitch and yaw offsets can be determined. Time or latency between sensor measurements should be overcome by using the 1 pulse per second GNSS output. The calibration, or patch test, can then concentrate on roll, pitch and yaw by running a series of lines over a patch of seabed that contains both a flat area and an area with well-defined features. The process varies depending on whether there is one transducer head or two, each transducer has to measure one section of seabed.

4 Lidar

Lasers have applications in land and hydrographic surveying. They are deployable ashore on tripods or on beach buggies, afloat with geo-registered swathe surveys, as well as airborne (Chapter 11) to measure the landform or, over clear waters, the bathymetry (Figure 3-6). With survey platforms on the move (i.e., buggies, boats and planes), it is crucial to position and orientate the platform at a high update rate. Lasers can operate in pulsed mode, timing the signal from its emission to reception (this being

---

[2] GNSS: Global Navigation Satellite Systems (i.e., GPS and the other satellite positioning systems such as GLONASS, Galileo and Beidou).

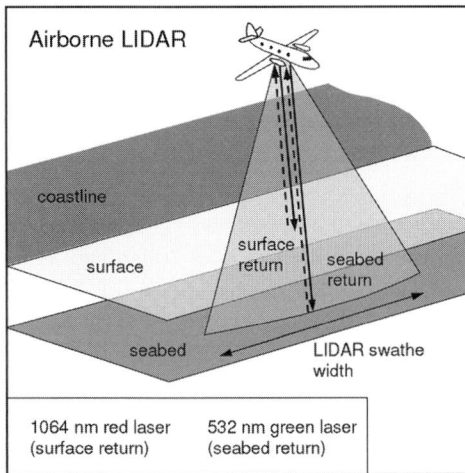

Figure 3-6 Illustration of an aircraft-deployed bathymetric Lidar, showing signal transmissions to, and reflections from, the water surface (red light) and the seabed (blue/green) light.

from a variety of surfaces), or use a modulated continuous wave by measuring the phase difference between the outgoing and returning signal.

4.1   Resolution

Lasers are monochromatic, are very intense, have low divergence and are coherent. They can be unsafe, and all are classified with one of seven alphanumeric values between 1 and 4 (Public Health England, 2014); the safety level may be associated with the mechanics of its construction, and any disassembly may make the instrument more dangerous. Typically, one should not look into the laser light and especially not with any magnifying optics.

Resolution may be assessed from the spacing between adjacent laser shots. With a terrestrial laser, the rate of firing, spin speed of the mirrors and the turning on the horizontal axis all play their part. The consequential resolution is distance-dependent due to angular divergence. For airborne lasers, the shot-point interval across the swathe is coupled with the aircraft's height, forward speed and variations due to turbulence to determine resolution in any one small area.

4.2   Recording Sensitivity

The amplitude of the received signal varies with the transmit power, the ambient light level, the roughness of the surface being illuminated, the angle of incidence to the surface, meteorological conditions and the object's reflectivity – a truly black surface may not reflect the laser light at all.

The gap between scan points requires a surface-fitting module to make, say, a vessel appear in its true form, otherwise one can look through these points and see the far surface of the vessel that had been obtained from a different scan. Such a feature is not a requirement for determining offsets on a vessel and is irrelevant to cliff monitoring.

Airborne Lidars might generate their laser light on a wavelength of 1064 nm (red). If the frequency is doubled, so that the wavelength is halved, the resulting second output is at 532 nm (blue-green). Each laser beam produces an ellipse on the ground or water surface (red) or on the seabed (blue-green) of a size related to the flying height; a typical size being a red 0.2 m by 0.4 m ellipse at a flying height of 200 m and a green 2 m circle on the seabed. The latter increased size is also a consequence of the widening of the coherent beam on its transition from air to water.

In bathymetric Lidar, the red-light return yields the surface data that are reflected from the first few centimetres of the water column, the blue-green reflects from the seabed, and their difference is related to the water depth (Figure 3-6). Depth uncertainties are increased by an undulating sea surface (swell) and the offset between the flying height (probably a GNSS-derived ellipsoidal height) and Chart Datum (a local value of low water, however so defined). White or dark seabeds will result in different intensity returns, and some thought should be given to the surface covering of any cube used to test the Lidar (see later).

4.3   Survey Accuracy

For terrestrial laser scanners, instrumental accuracy typically is approximately ± 0.003 m at 50 m and ± 0.006 m at 100 m, with a maximum range to 120 m (although different power levels are available). The angular precision typically might be ± 8 seconds of arc in the horizontal and vertical planes.

Local control accuracy will depend on the siting of the targets, which may be just significant features on the object being surveyed, or may more usefully be targets designed to be recognised by the software and manufactured to rotate in two axes around a central point. These targets, deployed around the survey area, need to be turned to face the scanner at each new setup whilst avoiding problems of misidentification or changing perspective from different directions.

With the instrument properly levelled (i.e., within the range of the compensator to determine the local vertical), each scan must pick up at least two targets that are visible in a second scan to be able to coregister the two scans. Typically, more targets are deployed, which eases the redeployment of the instrument whilst maintaining lines of sight to the targets. With insufficient targets to hand, each may be scanned and then moved to a new location and scanned once more, but the scanner's position must not alter. Target identification can take place before and/or after the main scan.

For a cliff face survey, the targets may be set up on the beach or along the cliff top; indeed, if there is a protecting fence, simple brackets can be attached to fence posts and the targets redeployed to repeatable positions. A terrestrial laser scanner afloat, say for a vessel's dimensional control survey, may have the vertical compensator turned off; in that case, three targets must be in view – from one scan to another – to ensure coregistration between scans.

Terrestrial surveys can be given real-world coordinates by surveying the targets into a national reference frame or, say, the latitude and longitude system as modelled by GPS[3] through WGS84[4]. High-precision GNSS is accurate to about ± 0.03 m, and it may be better to have control points sited remotely and reduce the consequences of this inherent inaccuracy by establishing a longer baseline. Tying-in the GNSS-derived control to the local survey through measurements with a Total Station (Chapters 2 and 9) would provide a higher degree of precision in orientation. Similarly, heights provided by conventional spirit levelling give the best precision. Nevertheless, these are not needed in many local surveys, and a local scan may be appropriate.

Marine laser scanners use GNSS as their real-time local reference frame. Because the vessel is likely to be taking swathe soundings at the same time, the two data sets can be coregistered. It is crucial to have a vessel motion sensor feeding into both systems. The positional quality will be derived from the GNSS input, being poor with a single receiver (± 8 m in plan and ± 12 m in height) and improved with a relative technique; most commonly, Real-Time or Post-Processed Kinematic give the highest-quality results (± 0.03 m in plan and ± 0.05 m in height). Such approaches can also free the survey from the need for tidal information, provided there is a local connection between the spheroid and Chart Datum; in the UK, such a system is the Vertical Offshore Reference Frame, VORF (Ziebart et al., 2007). The vessel sensor survey must measure the vertical offsets.

These parameters apply to aircraft-based systems too. Checking the offsets in an aircraft may require traversing through to the optics and relating the orientation of the scanner to the aircraft frame. The 3-D coordinates of the aircraft nose wheel may be checked against a known point on the airport apron. An initial flight may be taken over a simple flat surface, even a car park with white line markings; for bathymetric surveys, a 0.5-m cube or a 2-m cube may be deployed to ensure appropriate detection (IHO, 2008), although the IHO standards are designed for charting to facilitate safe navigation.

4.4 Applications

Terrestrial lasers can be tripod mounted and views taken around an object, such as a hulk, from two or three directions to obtain a full 3-D image; numerous sites may be required to avoid shadowing. With coregistered photography, four setups may store 6 GB of data. This technique is useful to assess the relationship between the GNSS antennae on a survey vessel and the multibeam head; it has the advantage of providing an image of the vessel too, which is a convincing backdrop to a survey report. Nevertheless, measurements derived from a Total Station (Chapters 2 and 9) are adequate for that purpose.

---

[3] GPS: the US Global Positioning System.
[4] WGS84: World Geodetic System as defined for GPS; other GNSS use other reference frames; local interpretations of WGS84 will result from using a local, relative Base Station in another system, such as ETRS89 (or its successors) in Europe.

Observing stations along a beach can be used to measure a cliff line, indeed to monitor it for change, and marine scanning can 'tie together' the bathymetry and the coastline with a highly detailed dataset. The marine survey may take place at high and low water to ensure overlap, and the coastal features might be a beach, sea defences, historical fortifications, jetties, underneath jetties and port equipment.

Airborne scanning with Lidar might include all the terrestrial features, but from a near-vertical perspective. Processing can establish ground level by removing trees and scrub that hide the ground surface. The ability of bathymetric Lidar to penetrate clear water and give a depth to the seabed is very useful in dangerous or inaccessible areas, although wave action against rocks may generate aerated water and absorb the laser-generated light. Muddy coastal waters will also absorb the light, and there may be little seabed coverage.

5   Subbottom Profilers

When sound energy travels through a fluid or a gas, it is usually referred to as an acoustic wave, whereas when travelling through a solid, such as soil, sand or rocks, it is usually referred to as a seismic wave and possesses seismic energy. Subbottom profiling or marine reflection seismic survey may be thought of as similar to single-beam echo sounding, except that the acoustic pulse reflects from the subsurface geological formations (Figure 3-7). Rather than record returns solely from the water–seabed interface, the lower frequencies penetrate to, and return from, subseabed layers (Figure 3-7a and b) or buried objects. Thus, as with echo sounding, the two-way travel time of the reflection is multiplied by the velocities of propagation, which may vary, say, from seawater at 1500 ms^{-1} to granite at 6000 ms^{-1} (Lekkerkerk and Theijs, 2011).

A profiler emits acoustic or seismic energy and may be mounted on a vessel or towed. The transmitted energy is then reflected from boundaries between various layers that possess different bulk densities and sound velocities (Lekkerkerk and Theijs, 2011). The reflected signal is received either by the same tuned transducer that generated the outgoing signal (as with a simple pinger, by some Chirp systems or by a parametric system), or by a ship-towed hydrophone (or array of hydrophones). In earlier systems, the receiver converted the analogue reflected signal, which was plotted on paper; more frequently today, the signal is digitized, displayed, and logged with high-speed computers, where it can be portrayed as a two-dimensional profile or, with advanced processing, a three-dimensional volume.

5.1   Principle of Operation

Marine seismics use a range of acoustic sources, differing in their energy intensity and frequency. At the higher end of the spectrum they are often of piezostrictive design, very similar to that of echo sounders. However, lower frequencies and greater power are required for deeper subseabed penetration, so that mechanical sound sources are used. The interplay of frequency, power and acoustic source contributes both to

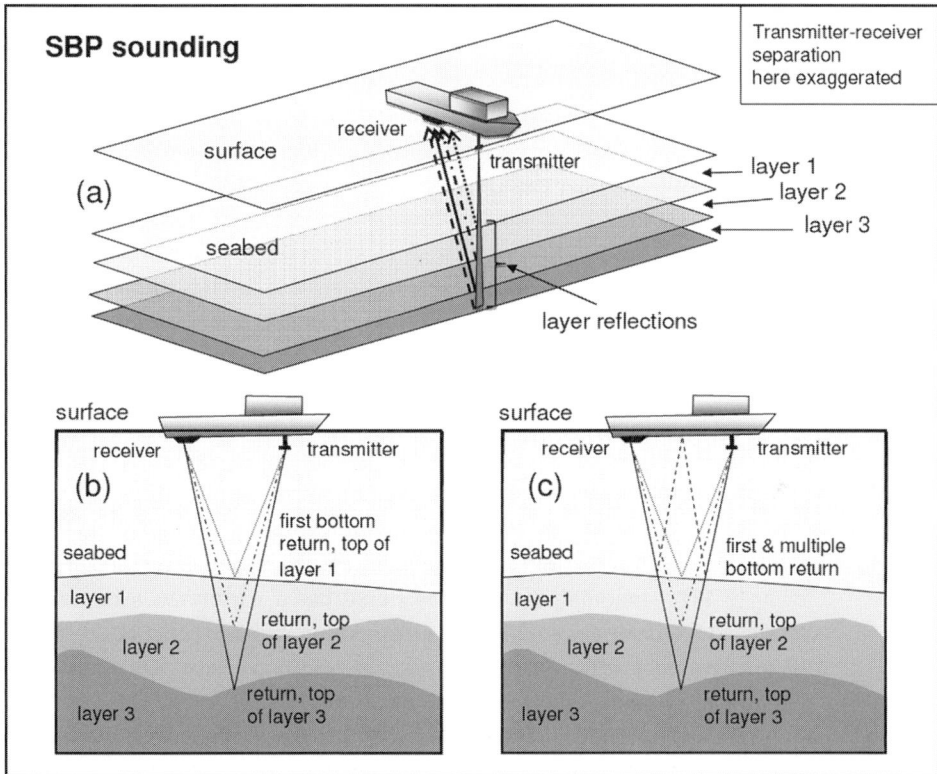

Figure 3-7 Elements of a marine seismic reflection survey, with an illustration of subbottom profiling (SBP) that shows reflections from the seabed and from subbottom layers 1, 2 and 3, (a, b); and reflections from the seabed and from subbottom layers 1, 2 and 3, together with a multiple return from the seabed, (c).

subseabed penetration and resolution (common marine seismic sources and their characteristics are discussed later).

Marine reflection seismic surveys introduce issues that do not occur with echo sounders, where interest is confined to the first return. This is primarily caused by the water surface being an almost perfect reflector of seismic waves. Accordingly, acoustic signals become trapped and reflected within the water column, becoming detectable at the receiver as multiples of the water depth. Such multiple reflections can obscure the true seismic record from subseabed depths greater than the water depth. This coherent noise is more significant in shallow water (Simpkin and Davis, 1993) and in single-channel systems, with the result that it becomes increasingly difficult to detect subbottom changes at depths greater than one water-depth.

5.2   Sound Sources

The subseabed penetration by a seismic wave depends on the energy and frequency of the wave; generally, the lower the frequency the greater the penetration. Increasing the energy also increases the penetration. Further discussion on the theory of sound frequency may be found in Urick (1983) or Lekkerkerk and Theijs (2011). Energy sources are divided into two types: resonant (piezoelectric) and impulsive. By convention, signals produced by resonant sources are referred to as sonar pulses and those by impulsive sources as seismic wavelets (Verbeek and McGee, 1995). The following subsections introduce the variety of sound sources used within marine seismic surveying.

5.2.1   Parametric Echo Sounder

The parametric echo sounder uses two, simultaneously emitted signals possessing closely similar frequencies, f_1 and f_2. Summing the two waves leads to a low 'beat' (or difference) frequency $(f_1 - f_2)$ that modulates the higher frequency f_1 and f_2 waves. These higher-frequency waves are used to measure depth to the seabed. The lower-frequency $(f_1 - f_2)$ wave penetrates the seabed but with the spatial resolution of the higher frequency waves (Lekkerkerk and Theijs, 2011). For example, the Innomar SES-2000 Standard Parametric Sub-Bottom Profiler generates two primary frequencies within the unit's 94–110 kHz range, resulting in a difference-frequency within the range of 5–15 kHz (Innomar, 2009). This signal has a narrow acoustic beam, as if it were produced by the original higher frequencies, but with the penetration ability of the lower 'beat' frequency. The system has a small transducer size and a resolution of 0.05 m within and up to a subbottom depth of 50 m.

As with all systems, the depth of penetration is dependent on the bulk density of the sediment layers; within compacted sediments or rock, the quoted depths of penetration may be severely diminished. A parametric sounder may attain tens of metres in soft sediments but much less in sand or rock.

5.2.2   Pinger

The pinger is much like a low-frequency echo sounder, in which the transmitted pulse travels onward to penetrate the seabed. Typically, it is low energy (about 5 Joules) with tuneable sources and operates between 2 and 12 kHz (Kongsberg, 2016b). Like parametric sounders, pingers have limited penetration, but have a resolution of 0.1–0.2 m (Kearey et al., 2002). Again, they usually use small piezoelectric transducers, which also act as the receivers, such that separate hydrophones are not required (Kearey et al., 2002). An example of a commonly used pinger system is the Kongsberg GeoPulse Sub-Bottom Profiler (Kongsberg, 2016b).

5.2.3   Chirp

Chirp profilers are essentially pingers that use a digitally produced, linear frequency-modulated signal, rather than a sonar pulse on a particular frequency. The sound pulse is swept through a wide-frequency band, thus containing high and low frequencies, which improves penetration compared with the pinger but maintains the resolution (Lekkerkerk and Theijs, 2011). Penetration is dependent on sediment types and on the frequencies selected within the modulated transmitted signal, but may vary from 3 m in coarse sand to 200 m in soft, fine sediments. Resolution is between 0.04 and 0.4 m, depending on the frequency range (US Geological Survey, 2007). Examples of chirp systems are the dual-frequency CHIRP subbottom profiler (Falmouth, 2016) and the Kongsberg GeoChirp 3D (Kongsberg, 2016c).

5.2.4 Boomer

Boomers are based on mechanical sounding devices, which generate seismic wavelets by discharging large capacitors through a spring-loaded coil, forcing an aluminium plate to move rapidly downward. This produces a compressional wave in the water (Kearey et al., 2002). Boomers are a broad-band noise source producing seismic wavelets with a range of 300 Hz to 3 kHz (US Geological Survey, 2007) and with 100 to 1500 Joules of energy (Lekkerkerk and Theijs, 2011) and penetration to a few hundred metres at a resolution as small as 0.5 m (Kearey et al., 2002). There are a number of suppliers of boomer systems, with the more common systems being those produced by Applied Acoustics (Willoughby et al., 2012) and C-Products.

Whilst piezoelectric devices utilise transducers capable of both transmitting the acoustic pulse and receiving the returning signal, boomers and other mechanical transmitters do not have the necessary internal recording ability and must therefore use a separate recording hydrophone or hydrophone array.

5.2.5 Sparker

Sparkers are another example of a mechanical transmitter, generating seismic wavelets by discharging large capacitors directly into the water through an array of electrodes towed behind the vessel. Operating voltages are between 3.5 and 4 kV and peak current may exceed 200 A. The discharge leads to the formation of a plasma bubble that implodes and generates the seismic wavelet (Kearey et al., 2002). Sparker energies range from 100 Joules up to 200 kJoules (McQuillin et al., 1984) and more often are used for deeper subseabed penetration in conjunction with multichannel recording techniques. Conventional sparkers are affected by lateral variations of electrical conductivity in water and are subjected to progressive deterioration of their electrodes, which causes the electrodes to wear out and require replacement (Bellefleur et al., 2006). Sparkers generate broadband seismic wavelets with frequencies from 50 Hz to 4 kHz and can penetrate several hundred metres of sediment with a resolution of a few metres (US Geological Survey, 2007). More recently, developments in electrode technology have diminished electrode erosion rates, and a sparker can maintain its pulse-shape generation throughout its lifetime (Geo Marine Survey Systems BV, 2010). Conventional sparkers are offered by Applied Acoustics and by Geo Marine Survey Systems.

5.2.6 Air Gun

The air gun is a pneumatic sound source. The system consists of an onboard air compressor and storage tanks, an onboard electronic firing circuit controlled by the seismic recording system, and one or more air guns towed astern of the survey vessel. On command from the seismic recording system, the air gun releases a specified volume of air into the water. The release of air produces hydrodynamic cavitation, which results in a steep-fronted shock wave followed by several oscillations that result from the repeated collapse and expansion of the air bubble (US Geological Survey, 2007). These oscillations have the unwanted effect of lengthening the seismic wavelet. To suppress this effect it is possible to detonate near the water's surface, so that the gas bubble escapes into the air. Although the oscillation is removed, much energy is wasted, and the down-going seismic wavelet is weakened (Kearey et al., 2002). The most effective method to eliminate bubble oscillations is to use an array of air guns with a variety of air-chamber capacities that are fired in sequence. The pressure signal recorded from the array at a great distance (i.e., the far field) is equal to the sum of the sound pressures of the individual guns. This is tuned to destructively interfere and reduce the effect of the bubble oscillation (Dobrin and Savit, 1988). Air guns range in size from 1 to 2000 inch^3 (Dobrin and Savit, 1988) or $1.6 \times 10^{-5} - 0.033 m^3$; gun pressures may be variable and a wide range is used, although 2000 psi ($1.4 \times 10^7 Pa$) is most common (Sheriff and Geldart, 1995).

Air guns usually generate frequencies from 100 Hz to approximately 1200 Hz, achieving a resolution of 10–15 m at a penetration of about 2000 m (US Geological Survey, 2007); higher pressures contribute to a somewhat higher frequency range of the seismic pulse (Dobrin and Savit, 1988). Air guns tend to be used with multichannel recording techniques and are predominantly used within the offshore industry. However, small airguns and airgun arrays at high air pressure are deemed to be safer than sparkers and may be used for selected nearshore survey work.

5.2.7   Water Gun

The water gun is a variation of the air gun in which the compressed air is not released but rather used to ram a piston forward, thereby ejecting a volume of water through a number of ports. Cavitation occurs when the piston is arrested and the seismic wavelet is generated by implosion of the water through the collapse of the cavity. Thus, water guns have no bubble oscillation and produce a cleaner signal pulse than air guns (McQuillin et al., 1984). The water gun produces seismic wavelets with frequencies from 20 Hz to 1500 Hz; a 15 cubic inch (0.00025 m^3) water gun will produce a pulse that may penetrate 300 m with about 10-m resolution (US Geological Survey, 2007).

Similar to the air gun, high air pressure and small chamber size produce a higher-frequency signal with high resolution and shallow penetration, whereas low air pressure and large chamber size yield a low-frequency signal with low resolution and deep penetration (US Geological Survey, 2007).

5.3   Receiving and Recording

Marine seismic acquisition systems utilise either a transducer capable of both transmission and reception, or are primarily divided into single- and multichannel systems, which comprise an acoustic source and a hydrophone streamer, the latter containing one or many hydrophones.

5.3.1 Single Channel

Single-channel systems are more commonly referred to as subbottom profilers and are usually used to investigate the geology from just a few metres to a few hundred metres beneath the seabed. They are used mainly for engineering and environmental purposes, and are seismic-reflection surveying reduced to its bare essentials: a marine acoustic or seismic source, a hydrophone streamer, an oceano-graphic thermal printer or digital recording system, and a positioning and navigation system (Kearey et al., 2002). The seismic wavelet rays travel from the acoustic source to the seabed/subsurface layers and are reflected back to the array of hydro-phones. The signal received by each element of the array is summed and averaged as an analogue signal, which is then recorded as a single reflection or trace, either digitally or on paper.

Hydrophones typically are pressure sensitive piezo-electric detectors, which convert changes in amplitude of the surrounding pressure field into voltages. The signal-to-noise ratio is reduced when an array of hydrophones is used to record data. Suitably spaced elements reduce random noise by a factor of $\sqrt{N}$, where N is the number of active elements (see section 5.3.2 and Mosher and Simpkin, 1999). This is due to summation of the signal N times while random noise cancels. As the hydrophone streamer is towed horizontally, the acoustic field from the seabed sediments will be in phase, and signals from other directions, as well as multiple (sea) surface reflectors, will tend to cancel. Other critical factors in the use of hydrophones are hydrophone element spacing, tow depth, array length with respect to water depth and source-receiver geometry (Mosher and Simpkin, 1999). One may refer to Mosher and Simpkin (1999) for a more in-depth discussion of the issues involved. Suffice to briefly mention that the closer the spacing of the hydrophone elements, then the higher the frequency that can be detected. In addition, one may have to adjust the tow depth, array length and source-receiver geometry with respect to the wavelength of the frequency of interest to optimise the recording of the signal return.

The signal may be recorded by either an analogue or a digital system. An oceano-graphic thermal printer can record the return signal, synchronise the firing of the seismic pulse and perform some basic analogue signal processing, such as gain control, time variable gain (TVG) and bandpass filtering (EPC Lab. Inc., 2016). The data may also be acquired digitally, additional signal processing performed, and the data exported to seismic interpretation software.

5.3.2 Multichannel

Multichannel systems within the offshore industry typically consist of high-energy, low-resolution seismic sources such as sparkers, air guns or water guns and, for a 2-D survey, a 2.4 km, 48- or 96-channel streamer (Sheriff and Geldart, 1995), or a 3 km, 100-channel streamer (Lavergne, 1989), or other combinations (Jones, 1999). The vessel may tow up to 26 of these hydrophone streamers for 3-D surveys (Petroleum Geo-Services (PGS), 2010).

Multichannel hydrophone elements are the same as those for single-channel hydro-phones, but arranged in groups of 3 to 50, with each group designated and digitized as a channel. Each streamer may have depressor paravanes and depth controllers to place it approximately 10-m below the water surface. Multichannel systems acquire all data digitally; the data are then processed digitally and exported to interpretation software. The processing of multichannel data requires a lot more effort, resources, expertise and cost than single-channel processing. Multichannel utilises the Common Depth Point (CDP) or Common Mid Point stacking method (first invented by Harry Mayne in 1950) to combine the signals from all the channels into a single seismic trace (Mayne, 1962). Also known as horizontal stacking, it increases the signal-to-noise ratio (S/N) by a factor of $\sqrt{N}$, where N is the number of CDP traces summed. Each channel records a single CDP trace and is termed a 'fold' (Kearey et al., 2002). The CDP method allows greater penetration and higher resolution of the subseabed reflectors, so that the limiting factor becomes power rather than depth.

More recently, nearshore operators have experimented with 20/24 channel streamers, of about 100-m length, in an effort to overcome the problem of multiple reflections and to improve depth resolution (very high resolution, VHR, systems).

5.3.3   VHR Multichannel Systems

A variety of systems have been trialled in southern Kiel Bay since at least 1992 by workers at the University of Ghent (Versteeg et al., 1992). Bellefleur et al. (2006) compared single- and multichannel data simultaneously acquired using a low-energy source in the St Lawrence River Estuary. They used a 2–8 kJ sparker with a single-channel, 23-element hydrophone streamer over 12 m and a 48-channel streamer, with each hydrophone separated by 5 m.

Another novel implementation of a multichannel system for subbottom pro-filing, undertaken by the UK's National Oceanography Centre (NOC; Gutowski et al., 2008), used a chirp energy source with frequencies from 1.5 to 13 kHz and mounted its 60 groups of four hydrophone elements on a rigid sled. Due to the high frequency used, and to prevent spatial aliasing, each hydrophone element was placed 0.0625 m apart and each hydrophone group 0.25 m apart. Each hydrophone group was summed over four hydrophone elements and recorded as a channel.

Positioning requirements are very critical, requiring horizontal accuracies of ± 0.04 m. The NOC system used an RTK GPS positioning system with a GPS-based attitude system (heading, pitch and roll) with four antennae mounted on 0.9 m poles at the four corners of the array. The system was able to achieve a static horizontal and vertical positioning accuracy of 0.007 m and 0.0182 m, respectively. The system was able to image a buried, collapsed coffer dam of dimensions 4.6-m wide and 16.5-m long in about 12 m of soft sediments. Thus, this system has been shown to provide true 3-D imaging of about the top 20 m of the subseabed with decimetre resolution in all three spatial dimensions.

More recently, Geomatrix have undertaken trials of an HMS-620 Bubble Pulser (boomer) and a 24-channel, 100 m, MicroEel streamer, utilising variable channel spacing in an effort to improve the signal-to-noise ratio and hence the subbottom resolution.

5.4 Applications

Single-channel, subbottom profiling systems are primarily used in site investigations for engineering or environmental purposes. These may include the construction of bridges, tunnels, harbours, jetties, quay walls, pipelines for sewerage outfalls and dredging for access channels. Other applications include the laying of pipelines, communications cables, hydrocarbon production platform footings and well heads, as well as the location of sites for drilling rigs (Lekkerkerk and Theijs, 2011).

Multichannel systems are usually used for prospecting hydrocarbons (Sheriff and Geldart, 1995; US Geological Survey, 2007), although small multichannel systems have been built and used as subbottom profilers. Their advantage arises because of their higher resolution and their use in aiding the removal of multiple reflections, such as might be caused by signal reverberation and/or bouncing between reflective layers. These systems are sometimes called Very High Resolution (VHR) 3-D systems (Versteeg et al., 1992), which refers to the use of higher-frequency seismic sources that produce high resolution data and the use of many channels to build a 3-D image of the subseabed.

6 Survey Planning

Initial investigations should start with a Desk Study prior to survey work. Google Earth can offer insights for this, even to the spotting of hulks that are lying within sediment. Existing charts, tidal predictions, values and locations of Chart Datum are all valuable. An estimation of the likely contour directions will help estimate the likely survey line direction – across the contours for SBES and with the contours for SSS and MBES. A Desk Study ought also to include shipping movements and the weather forecast – which should be monitored for change. Whereas Chart Datum will likely need to be determined anew when there is a separation between the survey site and an existing tide gauge greater than 15 km, a simple datum transfer may be applied in between as long as the shape and timing of the tidal curve is within acceptable limits (tidal range at both places similar and within 0.3 m and tidal time differences < 20 min).

There should be sufficient time allowed for the mobilisation of a survey vessel or aircraft of opportunity, where there may be issues of obtaining 'clean' power (stable electrical frequency and no voltage spikes), of having sufficient space, and having the ability to tie-down instruments securely as a precaution against poor weather, of mountings for antennae and transducers, and launching and towing points. It should be borne in mind that subbottom profilers can be very heavy, and some use very large voltages. For instrumentation, the carrying of spare batteries is much better than a potentially long journey and a day's work lost.

Health and safety concerns are very important for all boat work but must also address working (especially lone working) on beaches, which may be soft or suffer from rock falls, which may have problems with access/egress due to tides, and may have poor mobile-phone coverage. In any laser use, the in-built engineering controls should be maintained, and there should be a clear administrative system; sources of support for this topic are the courses run by Public Health England (2015).

6.1   Sidescan Sonar

The sidescan sonar layback is a mixture of the fixed distance between the GNSS antennae and the tow point, which with a movable 'H' frame may have a forward, inboard position and an outer position once the sidescan sonar fish is deployed, and the variable distance that is dependent on the length of cable paid out and the speed of the vessel; increasing the speed raises the fish height, which is particularly useful when turning from one line to another, as otherwise the fish would drop and perhaps hit the seabed. This layback should be determined by lines run in the opposite directions, although even this will be affected by vessel speed over the ground and vessel speed through the water.

6.2   Swathe Bathymetry

Survey mobilisation needs somewhere to undertake a patch test. A suitable area would include flat ground and a significant feature (a rock outcrop, perhaps). Running over the feature in the same direction at slow and fast speeds highlights any latency in the sensing and processing. Running two lines in opposite directions isolates any pitch error. After adjusting for pitch, two lines run in opposite directions over a flat surface highlight a roll error. Finally, a yaw (heading) error can be determined from two lines in the same direction run to each side of the seabed target. There may be a requirement to prove the system to a client by detecting a 0.5-m cube in waters less than 20 m and a 2-m cube in waters shallower than 40 m.

7   Acoustics Techniques for the Assessment of Habitats

The application of acoustic techniques in benthic assessment results from their ability to provide relatively rapid coverage of large seabed areas compared with conventional photographic and direct sampling methods. Acoustic methods have a great advantage in deep and or turbid waters, where optically based techniques become extremely limited.

Acoustic techniques have been used extensively in seismic research to map large-scale seabed and subseabed structures, although these typically utilised low-frequency acoustic systems due to the requirement to acquire data at considerable depths below the seabed. For marine habitat classification, higher-frequency acoustic systems operating in the range of tens of kHz or above are usually more relevant.

Acoustic monitoring of the seabed has commonly been used primarily as a sediment classification technique. A review of this in terms of the properties of sediment grain size, density, porosity, compressional and shear sound speeds and absorptions and surface roughness has been conducted by Sternlicht (1999). Detailed information both about the seabed and the first few metres below the seabed (i.e., the stratigraphy) are highly desirable for pipeline (e.g., hydrocarbon industry) and cable-laying (e.g., offshore windfarms) operations. Acoustic reflection and scattering from the seabed itself and from biota extending above the seabed are central to benthic assessment, whilst acoustic returns from biota below the seabed surface are not easily distinguished in most acoustic signals (Penrose et al., 2005). An extensive bibliography of literature relevant to seabed classification by acoustic techniques has been compiled by Hamilton (2005).

7.1 Acoustic Proxies and Surrogates for Habitats

Acoustic techniques, as with a number of other methods for benthic assessment, gather data on surrogate measures of habitat. This has led to the concepts of seabed acoustic 'roughness' and 'hardness' (Penrose et al., 2005). The term *acoustic hardness* can be used to describe the acoustic impedance of the substrate type and thus provides a measure of the impedance contrast offered to an acoustic wave by the water-seabed interface. The physical roughness of a surface influences the amount of sound backscattered to a transducer, so that the measured acoustic roughness (i.e., backscatter) is often used as a proxy for physical sediment roughness. Penrose et al. (2005) explain that for sedimentary seabeds, 'hardness' in particular can sometimes be linked to sediment density and compressional sound speed, which in turn links to other sediment variables.

The acoustic descriptors of hardness and roughness do not necessarily describe the geometrical roughness properties of the seabed, or even their actual physical hardness. Sedimentary roughness is a function of currents, tides and waves that also create bedforms. Bedforms have a profound impact on sediment entrainment and transport, due to the changing flow and sediment dynamics as the bedforms develop. However, high backscatter and reflectivity from just a small cluster of shells (or similar minerals) within particular sedimentary beds can significantly degrade the use of acoustic hardness and roughness parameters for inference of geometrical seabed roughness. Additionally, high densities of macroalgae and seagrasses can significantly affect acoustic returns and mask signals from the seabed itself.

Acoustic techniques provide several surrogate descriptors of bed habitats, and these descriptors are usually linked to more direct habitat parameters by a variety of techniques, including photography and spot sampling of sediments and biota. Such is the diversity and complexity of nearshore seabed habitats that it has led Snelgrove and Butman (1994) to conclude that:

...the complexity of soft-sediment communities may defy any simple paradigm relating to a single factor, and we propose a shift in focus towards understanding relationships between organism distributions and the dynamic sedimentary and hydrodynamic environment...

In reality, the situation is even more complicated. Hamilton et al. (1999) point out that because harder surfaces, such as rock, tend to have a greater roughness and a more random orientation of seabed facets than other sediments, this causes widely varying return pulse shapes and energies, which can have an average signal strength resembling that of mud, especially if adequate data averaging techniques are not employed.

Experimental studies by Brekhovskikh and Lysanov (1982) have demonstrated that sound reflection is determined by sediment variables only at comparatively low frequencies. At frequencies above a few kHz, the relief of the seabed plays a dominant role. Importantly, they found that the reflection from a very rough rocky bottom may appear to be less than that from muddy sediment. Similarly, losses due to roughness effects can cause sands with ripples, sandwaves, holes and scours to appear to some acoustic measures to have the same properties as mud. Suitable averaging of echoes can overcome much of this variability; however, acoustic bottom classification results are sometimes ambiguous, a point that must always be remembered (Penrose et al., 2005).

7.2   Data Collection Considerations

As with all oceanographic and marine scientific surveys, the high expense of survey vessel charter and marine scientific instrumentation means that a cost-effective survey is of paramount importance. Acoustic techniques are therefore highly attractive because they offer an option whereby large spatial coverage can be utilised. However, many acoustic systems, such as single-beam sensors, can only acquire data at a fairly slow vessel transit speed. A similar consideration must be made in variable water depths, where survey track widths can be a direct function of the sonar beam width; this has been assessed by Siwabessy (2001). Adequate interpolation between tracks is necessary if interpolation for full 2-D spatial representation of a particular survey seabed location is required. This often requires some assumptions regarding the spatial variability of the seabed habitat.

The horizontal 'bin' size of the insonified seabed (which is a function of depth and transducer angle) is another consideration when using acoustic interrogation of the seabed. The bin is typically regarded as the insonified area created by the central beam; however, in sufficiently shallow waters the patch can become smaller than the horizontal roughness length scale (including that due to bedforms). The converse may occur in deeper waters, with the insonified area exceeding the roughness horizontal length scale. These resultant grazing-angle effects have been examined by Dugelay et al. (2000).

7.3   Acoustic Systems for Habitat Mapping

This section outlines how different acoustic systems can be used to monitor and characterise seabed habitats.

7.3.1   Single-Beam Echo Sounder (SBES) Systems for Habitat Mapping

The single-beam echo sounder is an active sonar device that uses the interval from emission to the first return signal to yield the depth below the vessel. These instruments typically operate within the 30 to 200 kHz range. Dual-frequency echo sounders have been very successful in identifying fluid mud layers, with the higher-frequency pulse inferring the height of the lutocline above the seabed; they are often used in dredging-related surveys and assist with the determination of navigable depth.

It is possible to derive parameters relating to the nature of the seafloor from an analysis of the subsequent backscatter from an SBES. However, it is only possible to obtain a single set of parameters for each acoustic transmission from the SBES. Therefore, it is quite common to conduct backscatter analysis on the averages of returns from a number of 'pings' to derive a value for the hardness and roughness of the seabed. The values acquired are taken as applicable to a single position below the vessel; the surveys therefore provide a line of discrete points along the vessel's track for which a number of variables such as depth, hardness and roughness have been derived for each point (Penrose et al., 2005). The spacing of these points is a function of the vessel's speed, water depth and the number of pings used to calculate the required variables.

SBES, when used for acoustic seabed classification, tends to utilise a wider beam than standard echo sounding (beamwidth typically 12°–55°) to obtain information on seabed acoustic hardness (acoustic reflection coefficient) and acoustic roughness (as a backscatter coefficient). An inversion approach could enable seabed geo-acoustic parameters to be estimated from normal incidence data. However, the reliability of such inversion techniques is still open to question for real-world applications, especially when the seabed is composed of complex mixtures and distributions of sediments and vegetation, and there may also be acoustic return 'hotspots' present (e.g., clusters of shells).

7.3.1.1   *Return Signal Averaging*

The return echo can vary to a fair degree within a relatively small time interval, even when the seabed is not changing significantly. This can be a result of random survey vessel movements and natural variations within the water column. Therefore, it is good practice to average over a series of pings (typically ten). However, Hamilton et al. (1999) report that if the seabed is rocky and has a high roughness, averaging does not improve ping stability and can even suggest that rock be interpreted as mud. For such cases, averaging the third-highest values is a better alternative; this is analogous to the calculation of significant wave height.

7.3.1.2   *Reference Depth*

Even if the seabed type remains the same, the power and shape of the returned signal can change significantly with depth (Caughey and Kirlin, 1996; Clarke and Hamilton, 1999). For example, the return signal becomes more compressed in shallow waters, with the converse occurring in deeper waters. This is due to the fact that acoustic signals are sampled at equal time intervals and not equal angles (Caughey et al., 1994), thus more samples are collected at one particular angle to another for a deeper sample when compared with a shallower sample. Therefore, it is good practice to transform the SBES signal data to a reference depth (e.g., an average survey depth), thus enabling normalisation of the acoustic waves to a reference depth and to a corresponding standard group of incidence angles (i.e., in contrast to linearly spaced times). This process is outlined by Caughey et al. (1994).

7.3.1.3   *Calibration*

Empirical algorithms need to be developed to relate the acoustic parameters to specific bed sediment properties. A direct approach is one in which explicit correlations are determined to provide a fully quantitative calibration based on vegetation indices or grain-size bounds from the survey site.

For the assessment of seabed habitats, indirect calibration approaches are usually more applicable. For example, indirect calibration could be along the lines of initially applying image processing to the acoustic data in a geographical space, and then using ground-truth data to assign geographical classes to the acoustic data (Greenstreet et al., 1997; Fox et al., 1998). Ground-truth data might be derived from, e.g., video approaches or grab samples, depending on the nature of the habitat survey. This calibration can then be followed by a further iteration in which the calibrated data are compared with the original acoustics data, in order to check for potential errors. Geographical Information Systems can be used as a tool to compare patterns between ground-truth data and the calibrated acoustic data.

It is important to note that calibration and classification are generally functions of the bottom sampling strategy and methodology employed. Classifications can vary from site to site, even using the same acoustic system, due to the empirical and subtly variable nature of acoustic systems, together with system noise and ambient environmental variability.

7.3.1.4   *Slope Effects on Acoustic Systems*

The crossing of bubble wakes, and similar issues, can lead to large apparent, but false, depth changes. A similar affect can occur due to bottom slopes, especially for second-echo methods. Such errors must be removed during the data-processing stage. Furthermore, acoustic bottom classification systems insonify different areas (i.e., different sizes of the acoustic footprint) at different depths, so that depth changes may alter the results even for the same bottom type (Rukavina, 1997). It is not possible to distinguish between different bed sediment types within the acoustic footprint. Also, the signal-to-noise ratio decreases with increasing depth, so that a wide range of depths within a seabed area may cause poor classifications. A wide range of depths can also affect the reference-depth corrections.

7.3.2   **RoxAnn**

The RoxAnn (Marine Micro Systems) was the first acoustic system to use the first and second acoustic bottom returns to perform bottom sediment classification. The operating principle is such that the first acoustic seabed return is reflected directly from the seabed and the second is reflected twice from the seabed and once from the sea's surface (or the vessel's hull; e.g., Figure 3-7c). Theory suggests that the double seabed interaction of the second echo causes it to be strongly affected by the acoustic bottom hardness, roughness effects becoming secondary (Chivers et al., 1990). However, Hamilton et al. (1999) report that over rougher seabed types, e.g., those with ripple bedforms, the energy lost to the second echo by backscatter can lead to lower-than-expected values of

RoxAnn acoustical hardness parameters for a specific sediment type. Therefore, careful calibration against known sediment samples is necessary to make inferences about bottom type from acoustic data.

The RoxAnn system uses echo-integration methodology to derive values for an electronically gated tail part of the first return echo (E_1) and the whole of the first multiple return echo (E_2). E_2 is primarily a function of the gross reflectivity of the sediment and therefore hardness, whereas E_1 is influenced by the small to meso-scale backscatter from the seabed and is used to describe the roughness of the bottom. Various acoustically different seabed types can be discriminated by plotting E_1 against E_2 (Chivers et al., 1990; Heald and Pace, 1996). In principle, E_1 and E_2 are related to acoustic roughness and hardness, respectively, although each contains components of both.

RoxAnn seabed squares (Burns et al., 1989) have been used to assess seabed classifications. If the roughness index (E_1) is plotted against the hardness index (E_2), each square represents one particular seabed type and this is determined arbitrarily based on ground-truth data (Chivers et al., 1990). Although this appears to be a rigorous process, a number of problems in using the RoxAnn squares approach have been identified. Voulgaris and Collins (1990), Greenstreet et al. (1997) and Hamilton et al. (1999) all found that the E_1 and E_2 parameters were not independent but linearly related, such that the data form an elongated, roughly elliptical envelope, inclined to the E_1 and E_2 axes. Inconsistency in the allocation system of the RoxAnn squares to boundaries between different seabed types was observed by Greenstreet et al. (1997). As an alternative to the RoxAnn squares approach, Schlagintweit (1993) suggested that an unsupervised classification method would be an improvement; i.e., first let the acoustic system select the natural groupings and then utilise ground-truth data.

7.3.3 Alternative 1st and 2nd Echo Technique

CSIRO Marine Research developed its own system that used the 1st and 2nd echo technique. Using a SIMRAD EK 500 scientific echo sounder operating at three frequencies (12, 38 and 120 kHz), acoustic data was recorded using the ECHO software (Waring et al., 1994; Kloser et al., 1998). The ECHO software provides several algorithms to derive E_1 and E_2 parameters, including a constant depth and a constant angular algorithm.

Using the data collected by the ECHO software, Siwabessy (2001) developed an alternative approach to the use of echo sounder returns for bottom classification. In grouping bottom types, multivariate analysis (Principal Component Analysis and Cluster Analysis, PCA and CA) was used to reduce the dimensionality of the roughness indices from three to one. The same was done for the hardness indices. The 'k-means' technique (e.g., MacQueen, 1967) was applied to cluster the resulting ($E_1 - E_2$) pairs formed for each set of six parameters to see if this would produce separable seabed types. Four separable seabed types were produced, namely soft-smooth, soft-rough, hard-smooth and hard-rough. Penrose et al. (2005) have commented that there are some possible drawbacks to this technique, in that what is being done may merely be density clustering of a continuous data cloud and may have no actual physical meaning.

A seabed type, which is sampled less often than another type, may be lost (absorbed into another type), even if it is dramatically different from other types (unless it appears as an extreme outlier). Greenstreet et al. (1997) noted this problem for their bottom grab samples; these were evenly spaced throughout their study area, rather than having the same number for each bottom type. Hamilton (2001) concludes that there is a similar problem with the analysis of seabed acoustic data. A seabed classification using the PCA and CA approach by Althaus et al. (reported by Penrose et al., 2005) initially seemed to be robust in that the acoustic classification was consistent with independent ground-truth data. However, further investigation into the relationship between the derived bottom type and the acoustically assessed total biomass of benthic mobile biota, showed that there was no apparent trend linking the two parameters (Penrose et al., 2005). This seems to be an unresolved issue in acoustic bottom classification.

7.3.4   Comparisons

Siwabessy (2001) showed that the use of multiple frequencies increased the discrimination ability of echo sounder methods (e.g., discriminating between sand, rock and algae) because different scales of seafloor roughness and different seabed volume contributions were being examined simultaneously. The ECHOplus instrument (Bates and Whitehead 2001) is basically a digital version of the original RoxAnn, but with the advantage of a capability to undertake simultaneous, dual-frequency analysis.

When comparing RoxAnn and another system, QTC View, Hamilton et al. (1999) noted that the former's data were only consistent when the survey vessel maintained a constant track speed. Although QTC View appeared to outperform RoxAnn during the Hamilton et al. (1999) survey in the vicinity of the Great Barrier Reef lagoon, Australia, it was possible to directly map RoxAnn space to QTC classes. In contrast, a benthic habitat survey of four sites off the southwest coast of England (three of area 48 km^2 and one of 336 km^2; Brown et al., 2001) showed that there was minimal difference in performance between the QTC View and RoxAnn systems (Foster-Smith et al., 2001).

7.4   Swathe Sonar for Habitat Mapping

Swathe technologies differ fundamentally from single-beam sounders and subbottom profilers, both in footprint shape and in the provision of spatially referenced information about the variability of recorded parameters from within the footprint. Traditionally, sidescan sonars yield only backscatter data and little information about depth, providing a wide, often almost 'photo-realistic' image of the seabed. The operating fundamentals of sidescan systems were outlined earlier in this chapter. Additional overviews of the use of sidescan sonars are provided by Bennell (2001) and Kenny et al. (2003).

The primary product of multibeam sonars is bathymetry, and although backscatter is also recorded, the imagery acquired is generally accepted to be of a lower quality than that recorded from sidescan sonars. Thus sidescan sonars and multibeams are often used in unison to gain complementary data sets. More recently, there is a greater convergence with the data provided by both the interferometric version of sidescan sonars, which can provide depth information, and their multibeam counterparts.

8 Final Remarks

The currently deployed acoustic and electromagnetic energy sensors discussed here have been developed to assess the out-of-sight seabed. Having developed from the need to ensure safe navigation, their uses have grown with an understanding of the human effects on the environment and our multiplying uses of the coastal zone, the shelf seas and seabed. Although the search for, and exploitation of, hydrocarbons funded these developments, environmental surveys benefit directly with tools that enable an improved understanding of the morphology and variety of the seabed.

Some of these tools are expensive, and working on the water is intrinsically expensive too. Nevertheless, appropriate good practice can ensure best results and also safeguard against misleading results. Safe working at sea may be supported best by an experienced skipper; best practice with the data gathering and processing with these technologies may be supported best with the professionals experienced in using them. Nevertheless, following the guidelines can produce new and exciting data never seen before; we may have mapped the Moon and Mars to a resolution of 100 m, but we have a long way to go for our seas.

References

AML, 2016. AML Oceanographic, 2071 Malaview Avenue, Sidney, B.C. V8L 5X6, CANADA. www.amloceanographic.com [accessed September 2016].

Bates, C. R., Whitehead, E. J., 2001. ECHOplus measurements in Hopavågen Bay, Norway. The Oceanography Society, Biennial Scientific Meeting, April 2001 [online]. Available at: www.st-andrews.ac.uk/rasse/library/pdfs/Miami_final.pdf [accessed March 2016].

Bellefleur, G., Duchesne, M. J., Hunter, H., Long, B. F., Lavoie, D., 2006. Comparison of single- and multichannel high-resolution seismic data for shallow stratigraphy mapping in the St. Lawrence River estuary, Quebec. Geological Survey of Canada [online]. Available at: http://publications.gc.ca/collections/Collection/M44-2006-D2E.pdf [accessed March 2016].

Bennell, J. D., 2001. Procedural Guideline No. 1–5 Mosaicing of sidescan sonar images to map seabed features. In: Davies, J. (ed.), *Marine Monitoring Handbook*, Peterborough, UK: Joint Nature Conservation Committee, 4–11 [online]. Available at: http://jncc.defra.gov.uk/PDF/mmh_PG1-5.pdf [accessed March 2016].

Brekhovskikh, L., Lysanov, Y., 1982. *Fundamentals of Ocean Acoustics*. Springer Series in Electronics and Photonics 8, Berlin: Springer-Verlag, 240pp.

Brown, C. J., Hewer, A. J., Meadows, W. J., Limpenny, D. S., Cooper, K. M., Rees, H. L., Vivian, C. M. G., 2001. Mapping of gravel biotopes and an examination of the factors controlling the distribution, type and diversity of their biological communities. CEFAS Lowestoft Sci. Ser. Tech. Rep. 114 [online]. Available at: www.cefas.co.uk/publications/techrep/tech114.pdf [accessed March 2016].

Burns, D. R., Queen, C. B., Sisk, H., Mullarkey, W., Chivers, R. C., 1989. Rapid and convenient acoustic seabed discrimination. *Proceedings of the Institute of Acoustics* 11, 169–178.

Caughey, D. A., Kirlin, R. L., 1996. Blind deconvolution of echosounder envelopes. In: *Proceedings of the IEEE International Conference on Acoustics, Speech and Signal*

Processing, May 7–10, 1996, Atlanta, GA: Institute of Electrical and Electronics Engineers Signal Processing Society, Vol. 6, 3149–3152.

Caughey, D., Prager, B., Klymak, J., 1994. Sea Bottom Classification from Echo Sounding Data. Contractor's Report 94–56 prepared for Defence Research Establishment Pacific, Canada. Document number SC93-019-FR-001, 120 Quester Tangent Corporation, Marine Technology Centre, 99–9865 West Saanich Road, Sidney, British Columbia, V8L 3S1, Canada.

Chivers, R. C., Emerson, N., Burns, D. R., 1990. New acoustic processing for underway surveying. *Hydrographic Journal* 56, 9–17.

Clarke, P. A., Hamilton, L. J., 1999. Analysis of Echo Sounder Returns for Acoustic Bottom Classification. DSTO General Document DSTO-GD-0215 [online]. Available at: http://dspace .dsto.defence.gov.au/dspace/handle/1947/3739 [accessed March 2016].

Cloet, R. L., Edwards, C. R., 1986. The Bathymetric Swathe Sounding System. *The Hydrographic Journal* 40, 9–17.

Dobrin, M. B., Savit, C. H., 1988. *Introduction to Geophysical Prospecting*. New York: McGraw-Hill Book Co.

Dugelay, S., Pace, N. G., Heald, G. J., Brothers, R. J., 2000. Statistical analysis of high frequency acoustic scatter: What makes a statistical distribution? *Proceedings of the 5th European Conference on Underwater Acoustics*, 10–13 July 2000, Vol. 1, 269–274.

EPC Lab. Inc., 2016. EPC Labs Product Information Data Sheets [online]. Available at: www.eiva.com/Files/Billeder/RentalProductSheets/1086.pdf [Accessed March 20161].

Falmouth, 2016. Dual frequency CHIRP sub-bottom profiler [online]. Available at: www .falmouth.com/images/HMS-622_Chirp_Subbotom_10april2015.pdf [Accessed March 2016].

Foster-Smith, R., Brown, C., Meadows, W., White, W., Limpenny, D., 2001. Ensuring continuity in the development of broad-scale mapping methodology – direct comparison of RoxAnn and QTC View technologies. QTC technologies. SeaMap/CEFAS Report, 113pp.

Fox, D., Amend, M., Merems, A., Miller, B., Golden, J., 1998. Nearshore Rocky Reef Assessment. Newport, Oregon. Oregon Department of Fish and Wildlife, Marine Program.

Geo Marine Survey Systems BV, 2010. Geo Source 200–400 Marine Multi-Tip Sparker System – Technical Specification [online]. Available at: www.geomarinesurveysystems.com/products/ seismic/geo-source/ [Accessed March 2016].

Greenstreet, S. P. R., Tuck, I. D., Grewar, G. N., Armstrong, E., Reid, D. G., Wright, P. J., 1997. An assessment of the acoustic survey technique, RoxAnn, as a means of mapping seabed habitat. *ICES Journal of Marine Science* 54, 939–959.

Gutowski, M., Bull, J. M., Dix, J. K., Henstock, T. J., Hogarth, P., Hillier, T., Leighton, T. G., White, P. R., 2008. Three dimensional high-resolution acoustic imaging of the sub-seabed. *Applied Acoustics* 69, 412–421.

Hamilton, L. J., 2001. Acoustic seabed classification systems. Report No. DSTO-TN-0401, Defence Science and Technology Organisation, Aeronautical and Maritime Research Lab., Victoria, Australia [online]. Available at: http://dspace.dsto.defence.gov.au/dspace/bitstream/ 1947/3596/1/DSTO-TN-0401%20PR.pdf [Accessed March 2016].

Hamilton, L. J., 2005. A bibliography of acoustic seabed classification. Cooperative Research Centre for Coastal Zone and Waterway Management, Technical Report No. 27 [online]. Available at: www.ozcoasts.gov.au/pdf/CRC/27-Hamilton_Acoustic_Biblio.pdf [Accessed March 2016].

Hamilton, L. J., Mulhearn, P. J., Poeckert, R., 1999. Comparison of RoxAnn and QTC View acoustic bottom classification system performance for the Cairns area, Great Barrier Reef, Australia. *Continental Shelf Research* 19, 1577–1597.

Heald, G. J., Pace, N. G., 1996. An analysis of the 1st and 2nd backscatter for seabed classification. *Proceedings of the 3rd European Conference on Underwater Acoustics*, 24–28 June 1996, vol. II, 649–654.

Hiller, T., Hogarth, P. J., 2006. Application of phase measuring bathymetric sonars to nautical charting and environmental mapping. In: *Evolutions in Hydrography*, 6th–9th November 2006, Provincial House, Antwerp, Belgium: Proceedings of the 15th International Congress of the International Federation of Hydrographic Societies. Special Publication (Hydrographic Society) 55, 83–86.

Hiller, T., Lewis, K., 2004. Getting the most out of high resolution wide swath sonar data. In: *Diversity*, 2nd–4th November 2004, Salthill, Galway, Ireland. Proceedings of the 14th International Symposium of the International Federation of Hydrographic Societies. Special Publication (Hydrographic Society) 53, Paper 8.

IHO, 2008. International Hydrographic Organisation Standards for Hydrographic Surveys, Special Publication 44. www.iho.int/iho_pubs/standard/S-44_5E.pdf [accessed March 2016].

Jones, E. J. W., 1999. *Marine Geophysics*. Chichester, England: John Wiley and Sons Ltd.

Kearey, P., Brooks, M., Hill, I., 2002. *An Introduction to Geophysical Exploration* (3rd Edition). Oxford: Wiley-Blackwell.

Kenny, A. J., Cato, I., Desprez, M., Fader, G., Schüttenhelm, R. T. E., Side, J., 2003. An overview of seabed mapping technologies in the context of marine habitat classification. *ICES Journal of Marine Science* 60, 411–418.

Kloser, R. J., Sakov, P. V., Waring, J. R., Ryan, T. E., Gordon, S. R., 1998. Development of software for use in multi-frequency acoustic biomass assessments and ecological studies. CSIRO Report to FRDC, project T93/237.

Kongsberg, 2016a. GeoSwath Plus - shallow water multibeam echosounder [online]. Available at: www.km.kongsberg.com/ks/web/nokbg0240.nsf/AllWeb/AA3FAF5EEADC6020C125762B0 050914C?OpenDocument [accessed March 2016].

Kongsberg, 2016b. GeoPulse GeoAcoustics Pinger Sub-bottom Profiler [online]. Available at: www.km.kongsberg.com/ks/web/nokbg0240.nsf/AllWeb/CBED70CC9CA43D2DC12574C80 04A37AD?OpenDocument [Accessed March 2016].

Kongsberg, 2016c. GeoChirp 3D – high resolution sub-bottom profiler [online]. Available at: www.km.kongsberg.com/ks/web/nokbg0240.nsf/AllWeb/E0788FD1BE34357FC1257C9300 4ABB49?OpenDocument [accessed March 2016].

Lavergne, M., 1989. Seismic methods (Méthodes sismiques) Institut Français du Pétrole Publications. Translated from the French by Nissim Marshall. Published in London by Graham and Trotman; Paris by Technip; Norwell, MA, USA, by Kluwer. Sold and distributed in the USA and Canada by Kluwer Inc.

Lekkerkerk, H.-J., Theijs, M.-J., 2011. *Handbook of Offshore Surveying*. Volumes 1–3, Voorschoten, The Netherlands: Skilltrade BV.

MacQueen, J. B., 1967. Some methods for classification and analysis of multivariate observations. *Proceedings of the 5th Berkeley Symposium on Mathematical Statistics and Probability*, Berkeley: University of California Press, 281–297.

Mayne, W. H., 1962. Common reflection point horizontal data stacking techniques. *Geophysics* 27, 927–938.

McQuillin, R., Bacon, M., Barclay, W., 1984. *An Introduction to Seismic Interpretation: Reflection Seismics in Petroleum Exploration*. Houston, TX: Gulf Publishing Co.

Medwin, H., 1975. Speed of sound in water: A simple equation for realistic parameters. *The Journal of the Acoustical Society of America* 58, 1318–1319.

Mosher, D. C., Simpkin, P. G., 1999. Status and trends on marine high-resolution seismic reflection profiling: data acquisition. *Geoscience Canada* 26, 174–188.

Public Health England, 2014. www.gov.uk/government/publications/laser-radiation-safety-advice/laser-radiation-safety-advice/ [accessed March 2016].

Public Health England, 2015. www.phe-protectionservices.org.uk/lpa/ [accessed March 2016].

Penrose, J. D., Siwabessy, P. J. W., Gavrilov, A., Parnum, I., Hamilton, L. J., Bickers, A., Brooke, B., Ryan, D. A., Kennedy, P., 2005. Acoustic techniques for seabed classification. Cooperative Research Centre for Coastal Zone Estuary and Waterway Management, *Technical Report* 32.

Petroleum Geo-Services (PGS), 2010. PGS: Media Gallery and Ramform-W-Class [online]. Available at: http://www.pgs.com/ [Accessed March 2016].

Rukavina, N. A., 1997. Substrate mapping in the Great Lakes nearshore zone with a RoxAnn acoustic sea-bed classification system. In: *Proceedings of the Canadian Coastal Conference 1997*, Guelph, Ontario, 338–349.

Schlagintweit, G. E. O., 1993. Real-time acoustic bottom classification: a field evaluation of RoxAnn. In: *Proceedings of the IEEE Conference – Oceans '93, Engineering in Harmony with Ocean*, 18–21 October, 1993, Victoria, British Columbia, Canada, 214–219.

Sheriff, R. E., Geldart, L. P., 1995. *Exploration Seismology*, Cambridge, UK: Cambridge University Press.

Simpkin, P. G., Davis, A., 1993. For seismic profiling in very shallow water, a novel receiver. *Sea Technology* 34, 21–28.

Siwabessy, P. J. W., 2001. An investigation of the relationship between seabed type and benthic and bentho-pelagic biota using acoustic techniques. PhD Thesis. Curtin University of Technology, Perth, Australia.

Snelgrove, P. V. R., Butman, C. A., 1994. Animal-sediment relationships revisited: cause versus effect. *Oceanography and Marine Biology: an Annual Review* 32, 111–177.

Sternlicht, D. D., 1999. High frequency acoustic remote sensing of seafloor characteristics. PhD Thesis, Marine Physical Laboratory, Scripps Institution of Oceanography, La Jolla, California, USA.

Swathe Services, 2005, Personal Communication. Internal Document – GeoSwath and GeoTexture. Swathe Services – UK Division, 1 Calenick House, Truro Technology Park, Heron Way, Newham, Truro, Cornwall TR1 2XN. http://swathe-services.com/.

Urick, R. J., 1983. *Principles of Underwater Sound*. New York: McGraw-Hill.

USACE, 2002. Multiple transducer channel sweep systems for navigation projects. In: *USACE Engineering Manuals: Engineering and Design – Hydrographic Surveying*, EM 1110–2-1003, Chapter 10 [online]. Available at: www.publications.usace.army.mil/USACEPublications/EngineerManuals.aspx?udt_43544_param_page=4 [Accessed March 2016].

US Geological Survey, 2007. USGS Woods Hole Coastal and Marine Science Center, WHSC Seismic Profiling systems, Data Acquisition [online]. http://woodshole.er.usgs.gov/operations/sfmapping/equipment.htm [Accessed March 2016].

Valeport, 2016. Valeport Ltd, Totnes, TQ9 5EW, UK. www.valeport.co.uk [accessed September 2016].

Verbeek, N. H., McGee, T. M., 1995. Characteristics of high-resolution marine reflection profiling sources. *Journal of Applied Geophysics* 33, 251–269.

Versteeg, W., Verschuren, M., Henriet, J.-P., De Batist, M., 1992. High-resolution 3D and pseudo-3D seismic investigations in shallow water environments. In: Weydert, M. (ed.), *European Conference on Underwater Acoustics*. London: Elsevier Applied Science, 497–500.

Voulgaris, G., Collins, M. B., 1990. USP RoxAnn ground discrimination system: a preliminary evaluation. Admiralty Research Establishment, Portland, UK, UTH Tech. Memo. 36/90. RE005314. University of Southampton, Dept. of Oceanography, Marine Consultancy Services, Tech. Rep. No. SUDO/TEC/90/5C, 75pp.

Waring, J. R., Kloser, R. J., Pauly, T., 1994. ECHO – Managing fisheries acoustic data. Proceedings of the International Conference on Underwater Acoustics, University of New South Wales, Sydney, Australia, Dec. 1994, 22–24.

Williams, J., 2015. Personal Communication. Swathe Services, 1 Calenick House, Truro Technology Park, Heron Way, Newham, Truro, Cornwall TR1 2XN, UK. Email: james.williams@swathe-services.com

Willoughby, G., Macdonald, N., Darling, A., Hiller, T., 2012. Applying novel sub-bottom boomer technology to the submerged Wellington Fault, Applied Acoustics: Underwater Technology, Technical Paper, Shallow Survey Conference, Wellington, New Zealand, 31pp. Available online [accessed March 2016]: www.conference.co.nz/files/docs/shallow%20survey/presentations/08.pdf.

Wirth, H., Brüggemann, T., 2011. The development of a multiple transducer multi-beam echo sounder system for very shallow waters, TS05J – Hydrography in Practice, Paper No. 5200, FIG Working Week 2011 – Bridging the Gap Between Cultures, Marrakech, Morocco, 18–22 May 2011 [online]. Available at: www.fig.net/resources/proceedings/fig_proceedings/fig2011/papers/ts05j/ts05j_wirth_bruggemann_5200.pdf [accessed March 2016].

Ziebart, M., Iliffe, J., Turner, J., Oliveira, J., Adams, R., 2007. VORF – The UK Vertical Offshore Reference Frame: Enabling Real-time Hydrographic Surveying, In: *Proceedings of the 20th International Technical Meeting of the Satellite Division of The Institute of Navigation (ION GNSS 2007)*, Fort Worth, TX, September 2007, 1943–1949.

4 Temperature, Salinity, Density and Current Measurements and Analysis

A. J. Souza

1 Introduction

Why is it important to measure temperature and salinity (and thus density) in coastal waters? One reason is that a good description of the density field in estuaries and the coastal ocean enables a clear picture of the residual circulation to be formed; e.g., Hill et al. (1994) describe jet-like flows around the Western Irish Sea front, and Burchard and Hetland (2010) describe how the estuarine circulation depends on the horizontal Richardson number, which in turn is dependent on the horizontal density gradients. The classic Hansen and Rattray (1965) scheme classifies estuaries in terms of their dependence on haline stratification, as described by the ratio of the surface-to-bottom salinity difference to the vertically averaged salinity. Another reason why it is important to measure density is that mixing processes are affected by stratification, and this can have important impacts on the health of an estuary or a coastal area; e.g., areas of Long Island Sound can be affected by hypoxia due to the presence of stratification that suppresses gas exchange between the atmosphere and the bottom layer (O'Donnell et al., 2008).

Just as important as the accurate measurement of density is the accurate measurement of currents. For example, the same estuarine classification scheme of Hansen and Rattray (1965) requires estimates to be made of the surface and bottom currents. Calculations of the estuarine and horizontal Richardson numbers need good estimates to be made of the horizontal density gradients and the bottom turbulent velocity, while calculations of the gradient Richardson number need good descriptions of the vertical density distribution and associated velocity. In addition, provided currents can be measured sufficiently rapidly and accurately, then it is possible to obtain estimates of the different components of the turbulent kinetic energy (TKE) balance and thus understand how mixing occurs in estuaries and coastal waters.

The aim of this chapter is to present both a historical review and a description of current state-of-the-art techniques that are used to measure temperature, salinity and velocity. The chapter will look at in situ and remote sensing techniques and at those advances in reliability and accuracy of velocity measurements that have allowed the estimation of turbulence and turbulent quantities.

2   Temperature and Salinity (Density) Measurements

One of the most important measurements in oceanography is that of density, which traditionally has been achieved by separately measuring the in situ temperature and salinity of seawater. This was once carried out using water bottles and reversing thermometers. A sample of water and the local temperature were obtained at a certain depth and then the water sample was analysed to estimate its conductivity, from which both salinity and density could be calculated (Sverdrup et al., 1942).

Temperature is much easier to measure than salinity (via conductivity), so profiling instruments for measuring temperature have existed for some time. The bathythermograph (BT) was a mechanical instrument that used pressure-driven bellows to move a metal- or smoke-coated glass slide under a stylus driven by a liquid-filled bourdon tube that was sensitive to temperature; although this system is now obsolete, it was very reliable (Baker, 1981). The BT was the precursor to the expendable bathythermograph (XBT), which is used on moving vessels for the rapid assessment of ocean temperatures and has been used by the world's navies. The XBT uses a thermistor to sense the temperature and relies on a known rate of fall to determine the depth. The advent of reliable thermistors is one of the most important advances in oceanography; another important development is that of reliable, stable, conductivity cells.

The precursor of the conductivity-temperature-depth instruments (CTDs) used today was probably that developed by Jacobsen (1948), which was then further developed by Hamon and Brown (Hamon, 1955; Hamon and Brown, 1958). This first instrument was designed to operate in the upper 1000 m of the ocean; it had a range of 0–30°C with an accuracy of ±0.15°C and a salinity range of approximately 13 parts per thousand (ppt – the salinity-defining ratio commonly used at that time), with an accuracy of ±0.05 ppt. A conductivity cell was used with platinum electrodes. One of the novel features of the Hamon and Brown design was the use of a single-cored armoured cable as the only connection between the underwater unit and the ship. At the same time, Hamon had the idea of making a portable, bridge-based instrument for use in estuaries (Hamon, 1956). Brown developed a conductive cell (Brown and Hamon, 1961) that was then incorporated into the continuous profiling system. Known as the STD; it was designed and sold in the early 1960s by the Bissett-Berman Corporation. Further developments, using more stable electrodes, resulted in the first CTD, built by Neil Brown (Brown, 1974). This instrument then became the standard instrumentation for all oceanographic cruises until it was superseded by modern Seabird Electronics instruments.

Although measurements of temperature and salinity are usually carried out using CTDs, temperature alone is also measured using thermistors. Thermistors are relatively cheap and vary enormously, depending on the required accuracy, response time and cost. Conductivity, and hence salinity, is a more expensive measurement but salinity is an important core measurement in oceanography and climate science (Schmidt, 2008). Following the recent establishment of the international network of almost three thousand Argo floats and the launch of the Aquarius and SMOS (Soil Moisture and Ocean

Salinity) satellites, the salinity and temperature of the upper ocean is being monitored at a finer spatial and temporal scale than ever before (ACT, 2007a).

The present suite of commercial instruments estimates salinity by measuring conductivity. The basic difference among these conductivity systems lies in whether they use an inductive or a conductive sensor and whether the sensor measures in an internal or an external configuration. Further information on individual commercial instruments is provided by ACT (2007a). The advantage of self-recording systems is that the instruments can be moored and used to obtain long, high-resolution time series of density, or mounted on vessels of opportunity, such as ferries (e.g., Balfour et al., 2013). Other developments include communication technologies, which have evolved so that it is now possible to obtain real-time estimates of density and advances in robotics that have facilitated the mounting of sensors in automated underwater vehicles (AUVs) and gliders (Chapters 1 and 10).

The main problems with measuring density over long periods, either in moorings, ships of opportunity or gliders, are instrumental drift and the fouling of sensors. There are two kinds of fouling: biofouling, when organisms settle and grow on the instrument, and sediment fouling, when sediments cause clogging. Both factors are a problem in estuaries and coastal waters and must be considered when designing observational programs (see also Chapter 10). Three devices have been compared in this regard: the Teledyne Citadel NXIC CTD (recent Citadel CTD instruments are shown in Teledyne RD Instruments, 2016), the Sea-Bird SBE-16 (discontinued, Sea-Bird Scientific, 2017), and the JFE-Alec Electronics temperature and conductivity sensor with mechanical wiper (CTW) instrument (Horiuchi et al., 2010). The Sea-Bird and JFE-Alec systems use electrodes, whilst the Teledyne RDI uses a nonexternal inductive cell. The comparison showed that Teledyne NXCI CTD measurements appeared to be more robust than those obtained from the SBE-16, which has a smaller cell that became rapidly fouled with sediment; greater robustness was found in the JFE-Alec instrument, due to its physical wiper, even though its accuracy is less than that of the Sea-Bird or Teledyne instruments (Balfour et al., 2013).

Other instruments that use mechanical wipers include the YSI EXO2 produced by Xylem. The advantage of a (plunger) wiping system can easily be seen from a comparison of images of an instrument before and after its deployment (Figure 4-1, reproduced from ACT, 2009): the external part of the instrument is largely biofouled after deployment (compare the instrument casings shown in Figure 4-1a and b), whereas the actual conductivity cell remains clean (compare Figure 4-1c and d).

3   Velocity Measurements

Circulation (current speeds and paths) in many coastal waters are dominated by periodic tidal currents, which can reach values of the order of 1 ms^{-1}; nevertheless, the residual circulation, with slower currents driven by baroclinic processes or frictional and non-linear processes, can be more important when considering net fluxes (Bolanos et al., 2013). These currents are also important when studying the transport of nutrients,

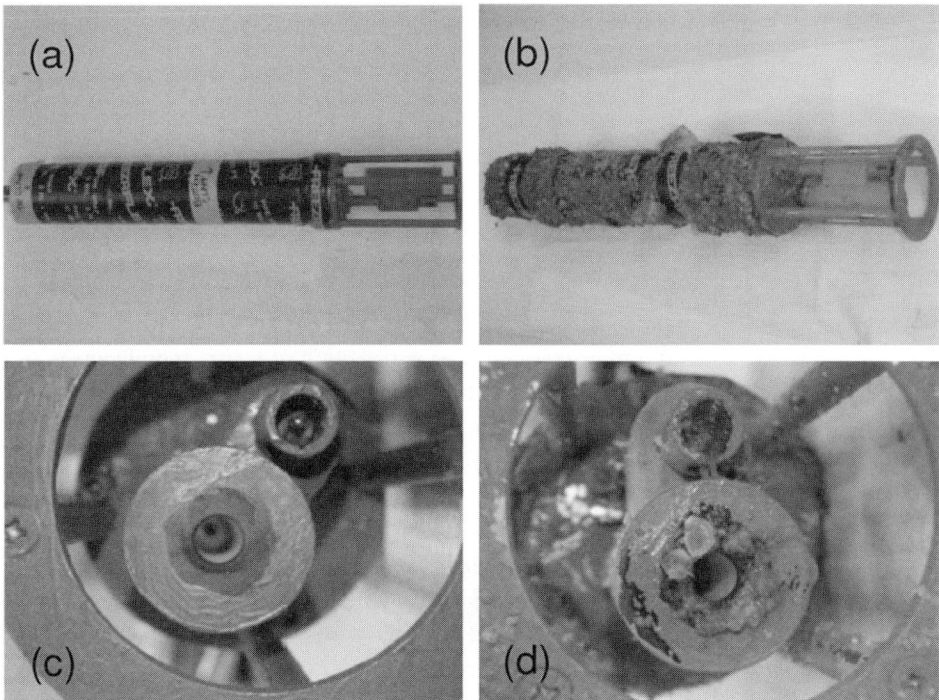

Figure 4-1 Images of the JFE Alec CTD, showing the benefit of a mechanical wipe to the integrity of the conductivity cell. (a), A lengthwise view of a typical instrument casing before deployment; (b), the same view after deployment; (c), an endwise view, looking toward the sensors before deployment; (d), the same view after deployment (reproduced with modifications from ACT, 2009; with permission).

contaminants and other variables of ecological importance. Residual currents tend to be of the order of a few cms^{-1}, with occasional maxima upward of 0.1 ms^{-1}, so there is some need for a high level of accuracy when measuring them (Souza et al., 1997).

Currents can be measured in Lagrangian mode, using drifters, bottles, etc., following their paths over a given time interval, or in an Eulerian manner – measuring the flow at a fixed point over a given time interval. Lagrangian observations are relatively simple to carry out and have evolved from the use of drifter buoys to radar tracking to the current satellite-tracking devices used today. Eulerian measurements progressed from the use of rotor and vane instruments (Baker, 2007) to electromagnetic and thence to acoustic instruments. This chapter will focus on the use of acoustic Doppler instruments because these are the most commonly used current meters at present (Chapter 1).

3.1   Acoustic Doppler Current Profiler (ADCP)

Acoustic Doppler current meters (velocimeters and profilers) are used to measure currents in the ocean. The idea, as its name indicates, is to use Doppler shifts to calculate velocities. The Doppler shift is the observed change of sound pitch that results from relative motion. An example of the Doppler effect is the changing sound made by a vehicle as it moves past an observer; e.g., an aeroplane engine appears to emit a higher pitch as it approaches and a lower pitch as it recedes. The change in pitch is proportional to the speed at which the vehicle is moving (the same technique is applied in speed cameras and speed guns used by police forces). If the instrumentation can measure how much the pitch changes, then it is also possible to measure how fast the vehicle moves (RDI, 1996). The speed of sound is:

$$C = F \times \lambda \tag{1}$$

where F is the frequency and λ the wavelength of sound. The Doppler shift, F_D, is the difference between the frequencies when there is no relative movement and when either the target or the source, or both, are moving relative to each other:

$$F_D = F\left(\frac{V}{C}\right) \tag{2}$$

In which V is the relative velocity between source and receiver. In the case of all acoustic Doppler profilers and current meters (but not velocimeters) the transducer works both as a transmitter and a receiver, so that Eq. 2 changes to:

$$F_D = 2F\left(\frac{V}{C}\right) \tag{3}$$

Therefore, if the original frequency and velocity of sound are known (Eq. 1), then it is possible to measure the frequency change (Eq. 2 and 3) and infer the along-sound (beam) velocity. Because it is only possible to measure the along-beam velocity, instruments have three or four transducers, so that the three velocity components can be measured. The problem with this technique is that each along-beam velocity is measured in a different part of the surrounding waters, due to beam spreading, so it is necessary to assume that the velocity field is the same within those waters (more information can be found in RDI, 1996).

3.1.1   Broadband and Coherent Techniques

Up to now the Doppler effect has been described in terms of the frequency considered, but to understand how broadband and phase coherent techniques work, it is necessary to discuss signal changes in the time domain. If a pulse of sound is sent to a stationary particle, the echo from this pulse of sound will have the same phase difference between it and the transmitted signal as all subsequent pulses and echoes (Figure 4-2a). If the particle moves away from the transmitter it will take longer for the sound to travel back and forth (Figure 4-2b). This change in travel time caused by the extra distance travelled is called the propagation delay, or time dilation. Echoes from a particle always look the same when the particle does not move; i.e., there is no propagation delay (Figure 4-2a). Echoes have the same relative phase, which means a zero phase change. In the case of superimposing a second echo on a particle that is moving away from the transducer, the second echo will be delayed with respect to the first. As an example illustration,

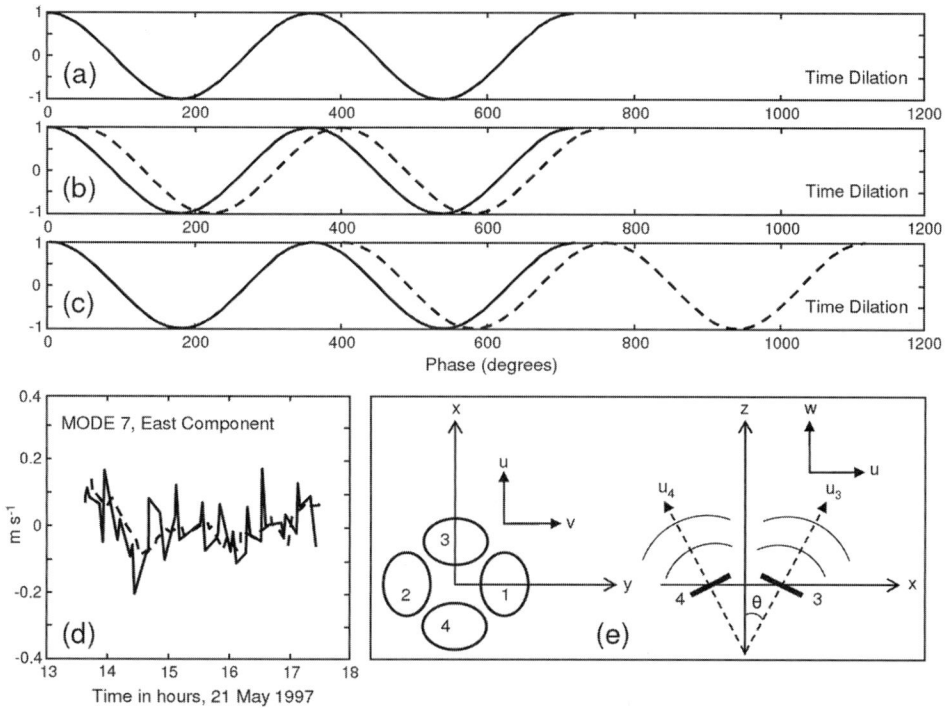

Figure 4-2 This diagram illustrates some acoustic wave properties and the geometry of the ADCP. Time dilation diagram: (a) echo of a nonmoving particle, (b) echo of a moving particle with a 40° phase, and (c) echo of a moving particle with a 400° phase (a–c reproduced with minor modifications from Souza et al., 2011; with permission from Elsevier). (d) Performance comparison of RDI ADCPs: broadband versus narrowband, showing velocity time series for broadband (dashed) and narrowband (continuous) ADCPs (© 1997 IEEE. Reprinted, with permission, from Wilson et al., 1997). (e) Geometry of the ADCP in Janus configuration, using, in this diagram, a left-handed coordinate system (x, y, z) that is not specified in terms of Earth coordinates, and with in-water current velocity components (u, v, w) and corresponding along-beam velocity components (u_1, u_2, u_3, u_4) for the four sensor heads.

the delayed echo (shown as a dashed line in Figure 4-2b) has a phase delay of 40°. A propagation delay corresponds to a change in distance, so if the delay is measured, and the speed of sound is known, it is possible to estimate how much the particle has moved. Because the time lag between pulses and echoes is known, the particle velocities can be calculated.

The main problem with measuring the phase is that it can only be measured between 0° and 360°; once it reaches 360° it starts again at 0°. Therefore, in the case described here (shown in Figure 4-2a–c), a change of 40° (Figure 4-2b) or a change of 400° (Figure 4-2c) appear the same, which is an effect known as phase ambiguity. To solve this problem it is necessary to determine how many times the phase has passed through 360°; this is called ambiguity resolution.

Broadband and pulse-coherent instruments use the autocorrelation method to process complicated real-world echoes to obtain velocity. In pulse-coherent processing, two pulses are transmitted into the water. As explained, the change in phase between the pulse pair is measured so that each pulse pair produces a single velocity estimate and thus the pair is defined as a 'ping'. The time between the two pulses determines the maximum velocity detectable and also determines the maximum range of the system. There is no other acoustic method that can produce such high-precision velocity data (with accuracies of the order of 0.001 ms^{-1}) for such small cells and rapid sampling (Lohrmann et al., 1995); this technique is used by ADV (Acoustic Doppler Velocimeters) instruments and is available in certain Doppler Profilers from Teledyne RDI, Nortek and Sontek.

Broadband works by transmitting a series of coded pulses, all in sequence inside a single long pulse, from which many echoes from many scatterers are obtained, all combined into a single echo. The propagation delay is extracted by computing the autocorrelation at the time lag separating the coded pulses. The success of this computation requires that the different echoes from the coded pulses (all buried inside the same echo) be correlated with one another (RDI, 1996). The advantage of broadband and pulse-coherent modes is that the accuracy of the data is improved enormously in comparison with the narrowband instrument (Figure 4-2d).

3.1.2   Three-Dimensional Velocities

As mentioned earlier, each transducer can only measure an along-transducer velocity, so that a minimum of three beams are needed to measure the three components of velocity; here, for simplicity, a four-beam Janus configuration is assumed (Figure 4-2e) to calculate both Earth coordinate velocities and Reynolds stresses. Because the beams measure the characteristics of different water columns, due to beam spread, it is necessary to assume horizontal homogeneity, which is reasonable over small spatial scales for most of the time in shelf seas and estuaries, but care must be taken when deploying the instruments in areas of high horizontal shear. An explanation is now provided of how to solve the three-dimensional (3-D) velocities in a four-beam array. This configuration has been chosen because the trigonometry is trivial and because it is the only configuration that can be used to calculate Reynolds stresses using the variance method (Lu and Lueck, 1999; Stacey et al., 1999).

The method is based on the fact that an ADCP has two pairs of opposing acoustic beams, and that each beam measures a velocity that is actually a weighted sum of the local horizontal and vertical velocities (Eq. 4 and 5). The velocities for each beam (Figure 4-2e) are given by the trigonometric relationships:

$$u_1 = v \sin(\theta) + w \cos(\theta); u_2 = -v \sin(\theta) + w \cos(\theta) \tag{4}$$

$$u_3 = u \sin(\theta) + w \cos(\theta); u_4 = -u \sin(\theta) + w \cos(\theta) \tag{5}$$

In which θ is the angle of the acoustic beam relative to the vertical (20° in the 4-beam case) and u, v and w are the two horizontal and one vertical velocity components (Stacey et al., 1999). From Eq. 4 and 5 it follows that:

$$u = \frac{u_3 - u_4}{2\sin(\theta)}; v = \frac{u_1 - u_2}{2\sin(\theta)}; w = \frac{u_1 + u_2}{2\cos(\theta)} = \frac{u_3 + u_4}{2\cos(\theta)} \tag{6}$$

One of the advantages of using ADCPs is that these instruments can be mounted on moving platforms, where the ship's velocity is removed using either the bottom tracking of the same ADCP in shallow waters or the calculated ship's velocity from the GPS. The tidal velocities can be removed using either a simple tidal analysis (Lwiza et al., 1991) or more complex polynomial functions (Candela et al., 1992; Carrillo et al., 2005). This approach can also be effectively used in ships of opportunity, such as the Texel ferry in the Netherlands (Buijsman and Ridderinhof, 2008).

4   Turbulence Estimates

The influence of turbulence on the dynamics of currents and waves and their interaction with near-bed processes has been acknowledged to be of great importance, although the turbulence itself remains poorly understood. The development of turbulence models is supported by the development of measurement techniques, such as the microstructure profiler and acoustic instruments, which can provide a direct comparison with simulated energy production and dissipation rates (ε).

4.1   Estimation of Turbulence Production and Dissipation Using ADCPs

The use of ADCPs has become common in recent years in the quest to measure and calculate turbulence parameters (Stacey et al., 1999; Rippeth et al., 2002; Souza et al., 2004). It is now possible to obtain reliable and continuous estimates of Reynolds stresses and shear and to easily derive shear production (P) and eddy viscosity (N_z). Estimates of eddy viscosity are obtained using the variance method.

The ADCP variance method for calculating Reynolds stresses is relatively simple and inexpensive because bottom-mounted ADCPs can be left for long-period deployments, in contrast to the use of shear profilers, which require a ship to be present. The Reynolds stresses are calculated using the variance method, first explained by Lohrmann et al. (1990) and applied by Stacey et al. (1999), Rippeth et al. (2002) and Souza et al. (2004). The method has gained popularity after Stacey et al. (1999) made great efforts to study the theoretical errors of the method, while other workers spent some time achieving a clear validation by comparing ADCP Reynolds stresses with estimates obtained from other instrumentation (Howarth and Souza, 2005; Souza and Howarth, 2005), such that it is now possible to have some confidence in TKE production estimates.

The advantage of the ADCP is its simplicity and versatility. The instrument is simple to set up and use and allows the user to obtain full water-column estimates of P, ε and suspended particulate matter (SPM); if properly configured, it also provides directional wave spectra using either a PUV method or similar, where the wave height is measured from acoustic measurements of sea level (Chapter 5). Estimates of SPM concentrations depend on the user's knowledge of particle or floc sizes, or on calibrations derived from gravimetric samples.

More recently, Wiles et al. (2006) used the structure function method to estimate TKE dissipation, with promising results, although the dissipation values appear to be slightly underestimated using this method. There is an issue with the variance technique in that it may fail when the turbulence length scale is smaller than the bin size. This arises because the average of the square of the turbulent velocity (which is the required quantity) differs from the square of the average value (which is the quantity obtained from the technique).

4.1.1   The Variance Method

To calculate the Reynolds stresses it is necessary to separate the velocities into their mean and fluctuating quantities. Writing Eq. 6 for the fluctuating quantities, then the ensemble means of the velocity products are (Stacey et al., 1999):

$$\langle u'w' \rangle = \frac{\left\langle (u_3')^2 \right\rangle - \left\langle (u_4')^2 \right\rangle}{2\sin(2\theta)} \text{ and } \langle v'w' \rangle = \frac{\left\langle (u_1')^2 \right\rangle - \left\langle (u_2')^2 \right\rangle}{2\sin(2\theta)} \tag{7}$$

In which the triangular brackets represent the temporal means and the primes indicate the temporal fluctuations. Quantities $\langle u'w' \rangle$ and $\langle v'w' \rangle$ represent the Reynolds stress per unit density (Eq. 7). The rate of production of TKE (P) in $\mathrm{Wm^{-3}}$ is estimated from the product of the Reynolds stress and the mean velocity shear (via Eq. 6), according to:

$$P = -\rho \left(\langle u'w' \rangle \frac{\partial u}{\partial z} + \langle v'w' \rangle \frac{\partial v}{\partial z} \right) \tag{8}$$

In Eq. 8, ρ is the water density and z is positive in the vertical direction. Using this method it is also possible to estimate the eddy viscosity, N_z:

$$\frac{\tau_x}{\rho} = N_z \frac{\partial u}{\partial z} = \langle u'w' \rangle \text{ and } \frac{\tau_y}{\rho} = N_z \frac{\partial v}{\partial z} = \langle v'w' \rangle \tag{9}$$

Quantities τ_x and τ_y are designated here to be the westward and northward components of shear stress (specifying x as westerly directed, Figure 4-2e). For a more detailed explanation of the variance method, see Stacey et al. (1999) and Lohrmann et al. (1990).

4.1.2   Estimates of Dissipation

Estimates of TKE dissipation rate (ε) can be made using the structure function method. The method was initially developed by meteorologists to estimate ε from radar measurements (Lhermitte, 1968); it was first applied to the aquatic environment by Nikora and Goring (1999) and more recently to the marine environment by Wiles et al. (2006). The method described here is based on the second-order structure function, $D(z, r)$, defined as:

$$D(z, r) = \left\langle (v'(z) - v'(z + r))^2 \right\rangle \tag{10}$$

$D(z, r)$ is the mean square of the along-beam velocity fluctuation (v') difference between two points separated by a distance r.

If the Taylor cascade theory is used to relate length scales and velocity scales to isotropic eddies, then (Wiles et al., 2006):

$$D(z, r) = C_v^2 \varepsilon^{2/3} r^{2/3} \tag{11}$$

The parameter C_v^2 is a constant with a value between 2 and 2.2 for atmospheric studies; this assumption is also used for marine studies, although it appears to underestimate the dissipation values.

The variance method has been clearly validated by Howarth and Souza (2005) and Souza and Howarth (2005) and has become a very popular technique for estimating turbulence characteristics in shelf seas and estuaries. There are now numerous examples of the use of ADCPs to estimate Reynolds stresses and turbulence production in estuaries and the coastal zone.

Generally, the highest rates of TKE production and dissipation are found near the seabed, with values decreasing exponentially with distance above the bed, although recent observations in the James River, Virginia, USA, show maximum dissipation values at the surface (Arnott et al., 2012). An example of the more classic distribution of turbulence production was observed in the Gulf of California (Souza et al., 2004). The data show a near-bed maximum value for P of the order of 10^{-2} W m^{-3} at approximately 1.5 mab (metres above bed) during both the ebb and the flood, with minimum values at slack water that are of the order of 10^{-4} W m^{-3}. The shear-generated values of P, due to bottom boundary friction, show a quarter-diurnal periodicity because they depend on the current speed; i.e., there are two peaks of current speed per semidiurnal tidal cycle. There is an apparent asymmetry between flood and ebb, with higher values of P and greater extension upward into the water column during the ebb (Souza et al., 2004).

Estimates of eddy viscosity (N_z) were calculated from hourly averages of Reynolds stress and shear for the Gulf of California study (Souza et al., 2004). These estimates of eddy viscosity showed some variability between 10^{-3} and 10^{-2} m^{2}s^{-1} in the bottom half of the water column, with maximum values around peak flow and low values around slack water. The mean eddy viscosity showed a typical profile, with values slightly increasing from the bottom (of order 3×10^{-3} m^{2}s^{-1}) to approximately 3.8×10^{-3} m^{2}s^{-1} at about 3 mab, followed by a continuous decrease to near zero at 12 mab (Souza et al., 2004).

4.2   Estimates of Turbulence Using an ADV

The techniques used to estimate turbulence parameters (Eq. 7–11) are becoming common practice with ADCPs. However, nowadays, the Acoustic Doppler Velocimeter (ADV) is the standard instrument used to carry out turbulence measurements from bottom-mounted moorings. The ADV adopts a direct method of measuring the Reynolds stresses that involves rapid sampling of the three components of velocity in a small sampling volume, so that terms of the type $\langle u'w' \rangle$ can be calculated directly from the covariance of u and w. This approach was pioneered by Bowden and Fairbairn (1956), who used electro-magnetic current meters (ECMs). With this method it is possible to derive not only the Reynolds stresses, but also the TKE per unit mass from direct estimates of the variances of u', v', and w':

$$K = \frac{1}{2}\left(\left\langle u'^2 \right\rangle + \left\langle v'^2 \right\rangle + \left\langle w'^2 \right\rangle \right) \qquad (12)$$

In recent years the use of ECMs has been superseded by ADVs, which are simpler to use and can sample smaller water volumes, depending on the frequency and the manufacturer. These instruments are highly accurate (long-term error of the order of 3×10^{-6} m^2s^{-2}) and have become the standard for boundary studies in laboratory and field experiments.

The advantage of having direct estimates of TKE and Reynolds stress is that it is then possible to investigate the relationships between them; it is generally considered that Reynolds stresses are proportional to TKE with a proportionality constant of approx. 0.19. Fieldwork in Liverpool Bay showed very close agreement with this factor, the empirically derived value being 0.21 (Howarth and Souza, 2005).

Another advantage of using ADVs is that the inertial dissipation method can be used to calculate the TKE dissipation rate. This method was first used in meteorology and was originally described by Deacon (1959) and reviewed more recently by Huntley (1988). The inertial dissipation method uses the spectra of turbulent fluctuations; it is based on the assumption that the wave numbers at which turbulence is produced are distinct from the wave numbers at which TKE is dissipated by viscosity. The range of wave numbers that lies between production and dissipation is known as the inertial subrange. In this range, the flux of energy from high to low wave numbers must be equal to the dissipation rate – if it is assumed that there are no local sources or sinks of TKE. Following Tennekes and Lumley (1972), the 3-D inertial spectrum is given by:

$$E(k) = \alpha \varepsilon^{2/3} k^{-5/3} \tag{13}$$

where k is the wave number and α the 3D Kolmogorov constant. In practice, estimates of the 3-D wave number spectrum are not available; instead, there are one-dimensional (1-D) spectra, which are functions of the along-flow wave number, k_1. These 1-D spectra, $\varphi_{ii}(k_1)$, where $i = 1, 2, 3$, are the longitudinal, transverse and vertical turbulence fluctuations, so that:

$$\varphi_{ii}(k_1) = \alpha_i \varepsilon^{2/3} k_1^{-5/3} \tag{14}$$

Eq. 14 is similar to Eq. 13, but with a different Kolmogorov constant. In practice, the vertical flow spectra tend to be used because they are likely to be less contaminated by wave motion within the inertial subrange (Grant et al., 1984; Huntley, 1988). Because turbulence data are usually collected in the form of time series, it is necessary to use Taylor's frozen turbulence concept to relate the frequency spectrum to the wave-number spectrum, as follows:

$$\varphi_{ii}(k) = \frac{\varphi_{ii}(f)}{2\pi/\langle u \rangle} \tag{15}$$

4.3   An Example of ADCP and ADV Turbulence Estimates

A comparison between estimates of TKE production and dissipation that were measured for the near-bed region (approx. 1 mab) in Liverpool Bay, UK, shows large differences when production is measured by the ADCP and dissipation by the ADV (Figure 4-3a).

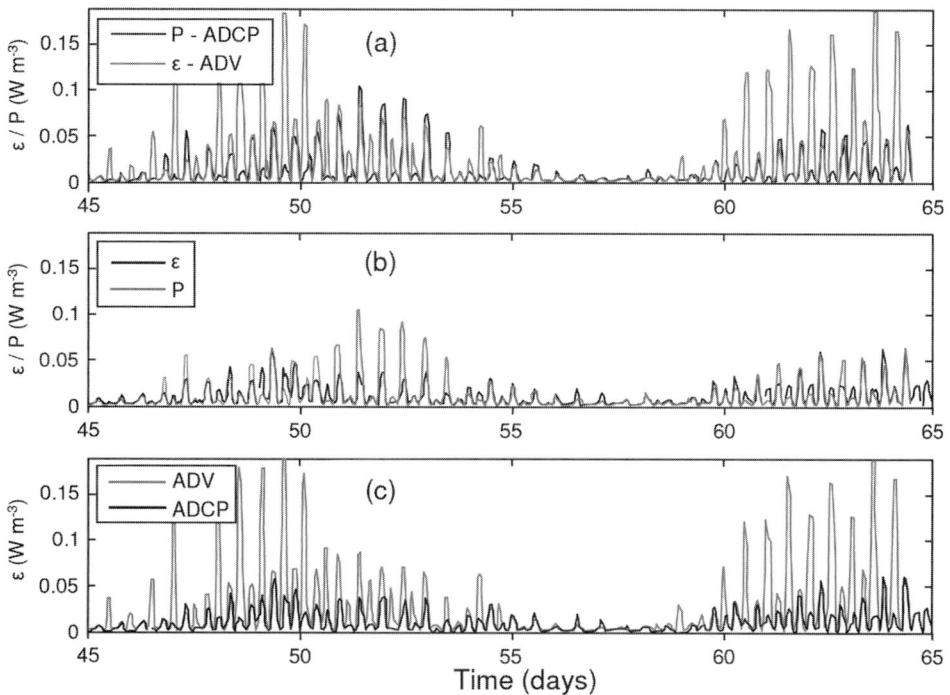

Figure 4-3 Comparison of estimates for TKE production and dissipation in Wm^{-3}: (a) ADCP production (black) and ADV dissipation (grey), (b) ADCP production (grey) and ADCP dissipation (black), and (c) dissipation from ADCP (black) and ADV (grey).

Generally, there is a close balance when both production and dissipation are estimated by the ADCP (Figure 4-3b), although an exception occurs around day 53, when the production appears to be greater than the dissipation (Figure 4-3b), probably due to the averaging used in the structure-function method (Eq. 10 and 11). When estimates of dissipation from ADVs and ADCP are compared (Figure 4-3c) it is seen that those for ADV dissipation appear to be up to an order of magnitude larger than those for the structure function method during the ebb. This could be because of the large volume and averaging of the structure function, or it could be due to the fact that the ADV 'captures' extra energy at higher frequencies, or it could even be the effect of the frontal or eddy-like structures associated with the stratification present in Liverpool Bay at the time of observations.

5   Integrating Vehicles for Quasi-Synoptic Measurements

As mentioned, to achieve a quasi-3-D picture of the water-column structure and stability in estuaries and the coastal ocean, CTDs, ADCPs and a suite of ancillary instruments can be mounted on roving instruments that are able to provide a profile of the water column as they move horizontally (see also Chapter 10). The simplest such systems are the Undulators pioneered by the Guildline SeaBat. These instruments are connected to the ship via a conductive cable that transmits instructions, data and power; the water column is profiled as the ship moves and data are transmitted in real time, thereby achieving a horizontal resolution of approximately 200 m in the coastal zone (e.g., Joordens et al., 2001). Further discussion on commercially available undulators can be found in ACT (2007b)

Although such undulator devices provide an efficient use of ship time, deployment is still expensive because the ship and full crew must be present while the measurements are made. In recent years, advances in communications, battery technology and electronics have helped the development of autonomous vehicles that can make these measurements independently. There are primarily two kinds of autonomous vehicles: Autonomous Underwater Vehicles (AUVs), which are propelled by electric motors and may be deployed for a few days at most, with independence between their horizontal and vertical movements, and gliders, which are driven by buoyancy changes. For these vehicles to move horizontally they have to dive and rise in a sawtooth manner; the advantage of using them is that they can be deployed for several weeks (Figure 4-4a), although, as yet, they cannot work efficiently in shallow estuaries and coastal waters where density gradients, which affect their buoyancy engine, are large.

6   Remote Sensing

Remote sensing techniques have matured to provide useful descriptions of ocean wind, waves, temperature, ice conditions, suspended sediments, chlorophyll, eddy and frontal locations (see also Chapter 11). Unfortunately, these techniques provide only sea-surface values, and in situ observations are necessary both for vertical profiling and to correct for atmospheric distortions during the calibration procedures. Some examples of remote sensing techniques are discussed in this section.

6.1   Satellite and Aircraft

6.1.1   Surface Currents

Currents can be identified in thermal infrared images as a result of gradients in sea surface temperature, SST (Figure 4-4b), and in ocean pigment distributions due to the differences between water masses of different origin. In Synthetic Aperture Radar (SAR) images, current characteristics are mapped given the changes in surface roughness across fronts (Figure 4-4c). In waters near the coast, land-based Doppler HF radar has proven to be a suitable system that can provide quantitative measurements of surface currents (Crombie, 1955; Lipa and Barrick, 1986; Andersen and Smith 1989; Prandle, 1991; Shay et al., 1993), and more recently X-Band radar has started to show promising results. Large-scale monitoring studies have suggested that it is possible to observe currents using over-the-horizon radar, which uses ionospheric reflections. Harlan and Georges (1997) successfully mapped surface currents during Hurricane Hortense using this technique.

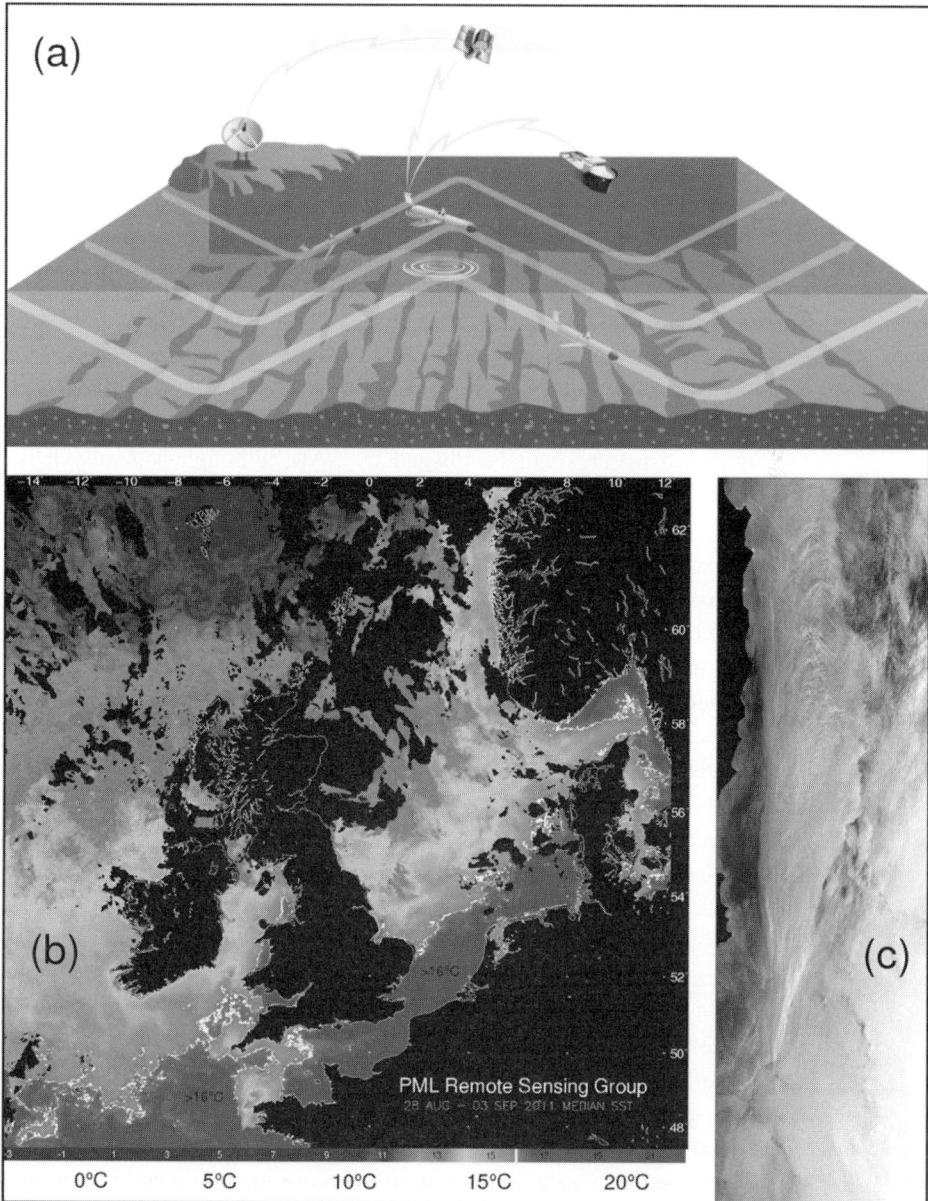

Figure 4-4 Remotely sensed data and associated instrumentation. (a) The Teledyne Web glider (reproduced with permission from Slocum Gliders and APEX Profiling Floats); (b) Sea surface temperature satellite image for the northwest European Shelf modified for greyscale presentation; the in-water white pixels delineate the 16°C isotherm, such that a darker greyscale inside that contour represents higher temperatures, oppositely to that outside of the contour (original colour image reproduced and modified, with permission, from NEODAAS, Plymouth Marine Laboratory, Plymouth, UK); (c) ERS-1 SAR image from the east coast of Australia that shows the East-Australian Current, which is the western boundary current of the South Pacific. The current is clearly visible as a broad bright band (image produced from ESA remote sensing data and processed by IFM Hamburg; with permission).

SAR (Figure 4-4c) can image the surface expressions of eddies, meanders, fronts and jets, thereby providing qualitative information on their structure and evolution (Lyzenga, 1991; Johannessen et al., 1991, 1994). Remotely sensed products that may be considered for operational use include manually interpreted images and geographical coordinates of the relevant observed features.

Although SAR is capable of cloud penetration and does not require daylight, it is unable to provide images of circulation features at very low or high wind speeds, and its output can also be degraded in the presence of heavy rain. However, given that these circulations reflect water masses of different origins, they frequently also have an expression in the surface temperature and ocean colour fields. Hence, they may be detectable using visible and infrared radiometers under cloud-free and/or daylight conditions. Thus, products combining information from these different types of sensors may be useful under varied environmental conditions and, therefore, more suited to operational use.

Circulation features can be imaged by SAR, although generally it is not possible to make a good quantitative estimate of the magnitudes of the currents involved. However, there are a number of numerical models that aim to predict the radar back-scatter variations produced in association with, e.g., various types of surface current pattern, oceanic fronts and internal waves. Interferometric SAR analysis can, in the right circumstances, provide direct quantitative measurements of surface currents (Shemer et al., 1993; Moller et al., 1998), but at present this technique is deployed only on aircraft for marine applications.

The use of satellite radar altimetry to determine geostrophic currents by measuring the sea-surface slope is possible for the determination of large-scale, time-variant current fields (although, of course, it breaks down near the equator). The variance in the elevation gradients can be used to obtain an estimate of the eddy kinetic energy of the circulation, which is a useful variable for quantifying mesoscale eddies (those with a diameter of typically less than 100 km).

6.1.2   Temperature and Salinity

The distribution of SST provides significant information that is related to a wide range of marine processes and phenomena, such as ocean currents, fronts, mesoscale eddies and upwelling phenomena. This fact allows the use of satellite-derived SST information in the mapping of ocean circulation (Johannessen et al., 1991, 1993, 1996, 1997), fisheries (Pettersson, 1990), algal blooms (Johannessen et al., 1989) and in the assimilation of SST data in physical circulation models (Stanev, 1994). SST is observed from space by thermal infrared imagery during cloud-free conditions, using the thermal infrared channels of the NOAA/AVHRR (e.g., Figure 4-4b) and from the ERS ATSR sensor systems. These instruments measure the SST distribution at a spatial resolution of 1 km and an accuracy of 0.5°C or better (Kidwell, 1995). The structure of mesoscale ocean circulation features in the North Sea tidal front and the Norwegian coastal current were documented some time ago through use of these types of Earth observation data (Johannessen et al., 1986, 1989).

Measuring salinity from space is still challenging, but techniques using passive microwave radiometry have been under development for some time (Lagerloef et al., 1995).

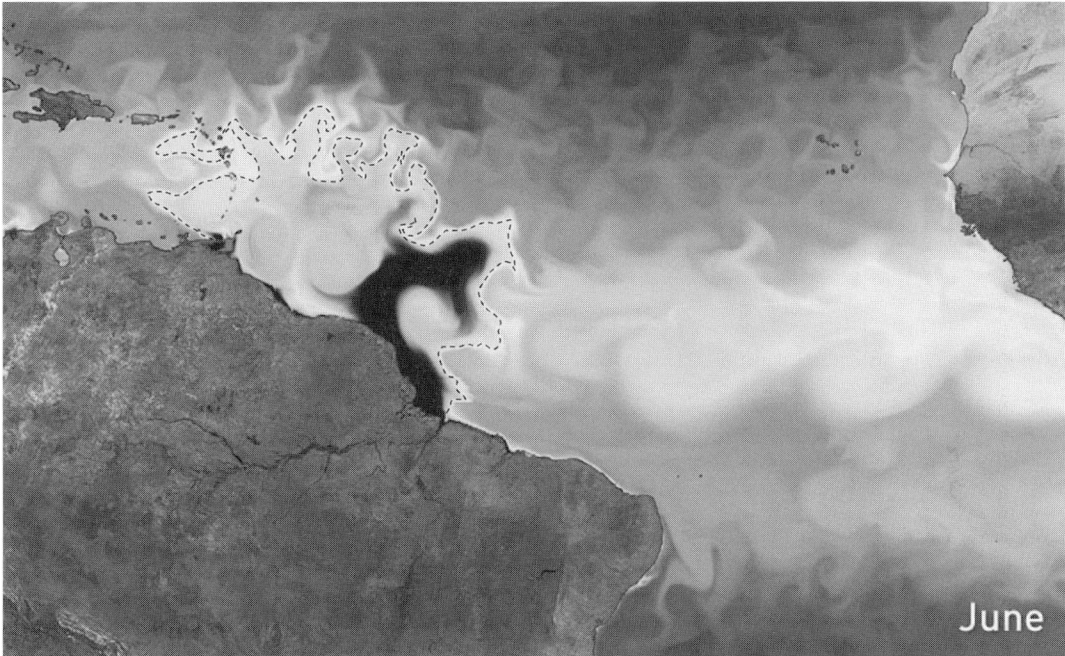

Figure 4-5 SMOS salinity image, modified for greyscale presentation, showing the Amazon freshwater plume as it discharges into the Atlantic (the original colour image courtesy of MyOcean and generated using EU Copernicus Marine Service Information; with permission). Darker areas show higher salinity, except within the dashed contours, where darker regions correspond to fresher waters (appearing black within the core of the plume and its tendril arms, and the narrow band hugging the coastline to the north).

The first measurements from space were carried out by the Skylab S-194 radiometer (Lerner and Hollinger, 1977). To date, successful results have been obtained by airborne L-band radiometry in the coastal zone (Miller, Goodberlet, et al., 1998), and satellite systems have been proposed using this method (Kerr, 1998). Currently, there are two missions attempting to measure salinity from space: the ESA Soil Moisture and Ocean Salinity (SMOS) was launched in 2009 (Figure 4-5), and the NASA Aquarius was launched in June 2011; both instruments are able to measure salinity with a resolution of 0.2. These systems use L-Band scatterometers.

Other attempts to measure salinity from space have been made using a combination of X- and C-band radiometers (Reul et al., 2009). In coastal areas, CDOM can be measured using a visible band at 440 nm (Binding and Bowers, 2003), and the relationship between salinity and optical properties at 440 nm (Figure 4-6a) shows great promise. However, some caution must be exercised because the relationship for the mid-latitudes suggests that high concentrations of CDOM are related to freshwater, while in Mediterranean climates the reverse is true, so that high levels of CDOM appear to be related to high salinities for inverse estuaries (Figure 4-6b).

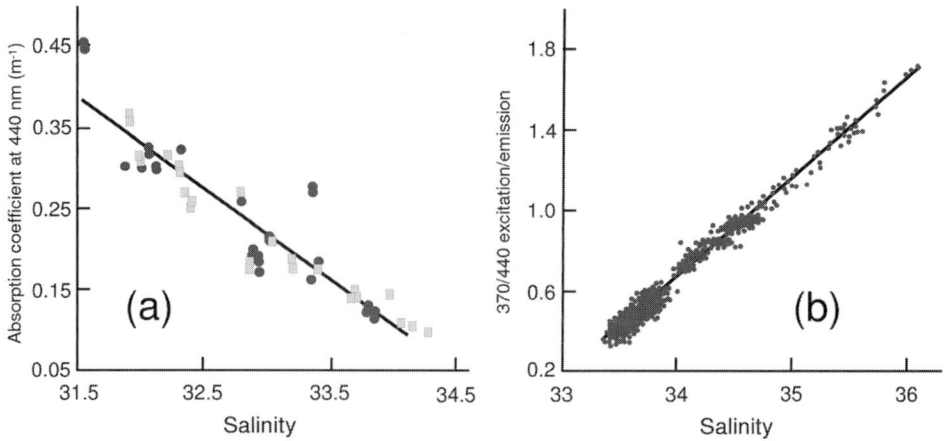

Figure 4-6 Salinity and CDOM: (a), absorption coefficient at 440 nm versus salinity in Liverpool Bay, UK (different plotting symbols correspond to different surveys), and (b) CDOM versus salinity in San Quintin, Baja California, Mexico (data courtesy of Dr Victor Camacho; with permission).

6.2   Land-Based Radar

Land-based radar systems are statically deployed in coastal areas overlooking the sea, as opposed to airborne or space-borne platforms. The most commonly used radars are HF radar and the standard marine radar, which works on the X Band.

6.2.1   HF Radar

HF radar was developed during the cold war (e.g., Crombie, 1955; Barrick, 1972) and operates in the 3–30 MHz frequency range of the electromagnetic spectrum; it is capable of obtaining information about the ocean at ranges of up to 150 km. The change in frequency between the transmitted and the backscattered signal, due to the Doppler effect, allows the determination of surface currents, wind direction and the ocean wave spectrum over a large area of the ocean surface (Barrick, 1972). Therefore, the large coverage obtained by radar is a major advantage over in situ wave measurements (e.g., by wave buoys, Chapter 5). The radar information is contained within the Doppler spectrum, which is characterised by two large peaks, known as the first-order Bragg peaks, which correspond to ocean waves of half the radar wavelength moving directly toward and away from the radar. The Bragg peaks are used to measure wind direction and radial surface currents (e.g., Figure 4-7 shows HF radar-derived tidal currents for the Rhine ROFI – region of freshwater influence). Ocean wave information is determined from the continuum that surrounds the Bragg peaks, known as the second-order effects. Barrick (1977) calculated the second-order radar cross-section for deep waters, which gives the relationship between the Doppler spectrum and the ocean wave number spectrum. A mathematical analysis of second-order ocean wave interactions is given by Lipa and Barrick (1986), Holden and Wyatt (1992) and Weber and Barrick (1977).

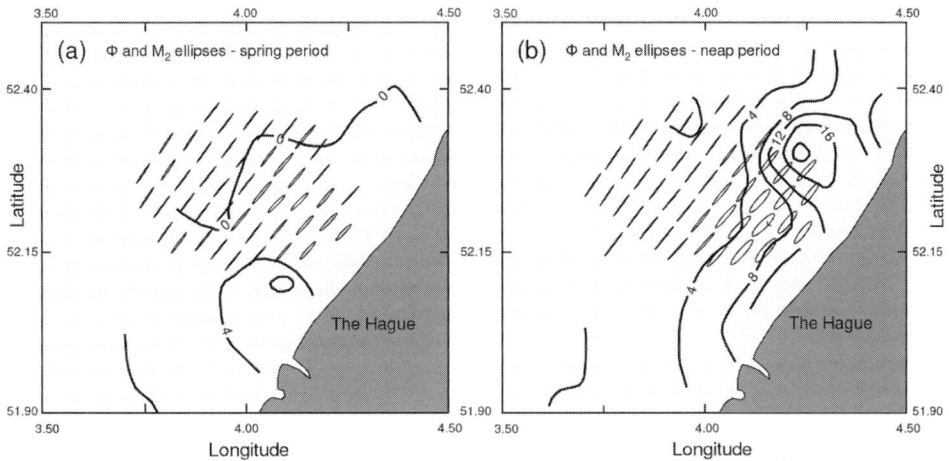

Figure 4-7 HF radar derived M_2 tidal ellipses and the influence of stratification in the Rhine ROFI: (a) spring tide conditions (7–12 October 1990), and (b) neap tide conditions (13–18 October 1990). The contours represent the distribution of potential energy anomaly derived by Simpson et al. (1993), ϕ (J m^{-3}).

6.2.2   X-Band Radar

X-band radar operates on a different principle to HF radar. It uses standard high-specification marine radar with a rotating antenna. A short pulse of 10 GHz microwave energy is transmitted in a narrow 1 degree beam, and anything that is capable of reflecting the beam, including rocks, ships and waves, bounces the energy back to the transceiver. The range of any particular target is determined from the time it takes the reflected energy to arrive back at the radar. The practical utility of this is that it produces a 360° plan view of the sea surface and anything on it every time the antenna rotates. The technique has been used for some years to estimate wave characteristics and bathymetry (e.g., Bell, 1999; Wolf and Bell, 2001). More recently, X-band radars have been used to estimate currents. The currents are resolved by assuming that the Doppler shift that occurs in the observed wave spectra is due to ocean currents (Bell et al., 2012).

In the presence of a current, the surface wind waves also experience a Doppler shift, further altering the wave propagation characteristics. Therefore, if the wavelength and speed can be measured for a range of wave periods, it is possible to fit the equations governing wave behaviour to the observations to determine the currents.

7   Final Remarks

This chapter has provided a description of the technologies that are used to measure temperature, salinity and velocity within estuaries and their coastal waters. In addition, a historical review has been given of the instrumentation that was traditionally used to measure temperature and salinity, and thence density, because of its importance to current state-of-the-art instruments and their future developments.

Remote sensing techniques for the measurement of surface temperature, salinity and currents have been described. These descriptions inevitably overlap to some extent the discussion on remote sensing techniques described in Chapter 5, although the emphasis there is on waves, and the derivation and analysis of wave properties from remotely sensed data differ substantially from those for currents. In addition, this chapter has highlighted the advances in reliability and accuracy of velocity measurements that have facilitated the estimation of turbulence and turbulent quantities. The techniques that are used to extract physical and biological in-water data from aircraft and satellites are described in Chapter 11, and those descriptions complement those that have been discussed here.

Acknowledgements

This work was funded by NERC through NOC's National Capability (NC) funding. The author wishes to thank Dr Victor Camacho from the UABC, Mexico, for the salinity and CDOM data from San Quintin, Mexico, which were collected during a joint project funded by SEP-CONACYT-México, Grant No. 40144. The author also wishes to thank ACT, ESA-IFM Hamburg, NERC-NEODAAS, NASA and MyOCEAN for the figures provided.

References

ACT, 2007a. State of technology for *in situ* measurements of salinity using conductivity-temperature sensors, Workshop proceedings, Alliance for Coastal Technologies, ACT-07-05.

ACT, 2007b. Towed vehicles: Undulating platforms as tools for mapping coastal processes and water quality assessment, Workshop proceedings, Alliance for Coastal Technologies, ACT-07-01.

ACT, 2009. Performance verification statement for the JFE ALEC CTW and CTW-FS salinity sensor, Alliance for Coastal Technologies, ACT-VS06-09. *UMCES Technical Report Series: Ref No. [UMCES] CBL* 09–033.

Andersen, C., Smith, P. C., 1989. Oceanographic observations on the Scotian Shelf during CASP. *Atmosphere-Ocean* 27, 130–156.

Arnott, K., Valle-Levinson, A., Chant, R., Li, M., 2012. *Temporal variability of dissipation from channel to channel slope in a coastal plain estuary. Physics of Estuaries and Coastal Seas Conference*. New York, NY.

Baker, D. J., 1981. Ocean instruments and experiment design. In: Warren, B. A. and Wunsch, C. (eds.), *Evolution of Physical Oceanography*. Cambridge, MA: MIT Press, 396–433.

Baker, D.J., 2007. Ocean instruments and experiment design. In: Warren, B. A. and Wunsch, C. (eds.), RES.12–000 *Evolution of Physical Oceanography*, Spring 2007. (MIT OpenCourse-Ware, Cambridge, MA: Massachusetts Institute of Technology). [online] Available at: http://ocw.mit.edu/resources/res-12-000-evolution-of-physical-oceanography-spring-2007/part-3/ (Accessed April 2016).

Balfour, C. A., Howarth, M. J., Jones, D. S., Doyle, T., 2013. The design and development of an Irish Sea passenger-ferry-based oceanographic measurement system. *Journal of Atmospheric and Oceanic Technology* 30, 1226–1239.

Barrick, D. E., 1972. First order theory and analysis of MF/HF/VHF scatter from the sea. *IEEE Transactions on Antennas and Propagation* AP-20, 2–10.

Barrick, D. E., 1977. The ocean wave height nondirectional spectrum from inversion of the HF sea-echo Doppler spectrum. *Remote Sensing of Environment* 6, 201–227.

Bell, P. S., 1999. Shallow water bathymetry derived from an analysis of X-band marine radar images of waves. *Coastal Engineering* 37, 513–527.

Bell, P. S., Lawrence, J., Norris, J. V., 2012. Determining currents from marine radar data in an extreme current environment at a tidal energy test site. Geoscience and Remote Sensing Symposium (IGARSS), *2012 IEEE International*, 7647–7650.

Binding, C. E., Bowers, D. G., 2003. Measuring the salinity of the Clyde Sea from remotely sensed ocean colour. *Estuarine, Coastal and Shelf Science* 57, 605–611.

Bolaños, R., Brown, J. M., Amoudry, L. O., Souza, A. J., 2013. Tidal, riverine and wind influences on the circulation of a macrotidal estuary. *Journal of Physical Oceanography* 43, 29–50.

Bowden, K. F., Fairbairn, L. A., 1956. Measurements of turbulent fluctuations and Reynolds stresses in a tidal current. *Proceedings of the Royal Society of London* A237, 422–438.

Brown, N. L., Hamon, B. V., 1961. An inductive salinometer. *Deep-Sea Research* 8, 65–75.

Brown, N. L., 1974. A precision CTD microprofiler. In Ocean 74 Record, 1974 IEEE Conference on Engineering in the Ocean Environment, IEEE Publication 74 CHO 873–0 OEC, *Institute of Electrical and Electronics Engineers, New York*, 2, 270–278. doi:10.1109/OCEANS.1974 .1161443

Burchard, H., Hetland, R. D., 2010. Quantifying the contributions of tidal straining and gravitational circulation to residual circulation in periodically stratified tidal estuaries. *Journal of Physical Oceanography* 40, 1243–1262.

Buijsman, M. C., Ridderinkhof, H., 2008. Variability of secondary currents in a weakly stratified tidal inlet with low curvature. *Continental Shelf Research* 28, 1711–1723.

Candela, J., Beardsley, R. C., Limeburner, R., 1992. Separation of tidal and subtidal currents in ship-mounted acoustic Doppler current profiler observations. *Journal of Geophysical Research* 97, 769–788.

Carrillo, L., Souza, A. J., Hill, A. E., Brown, J., Ferdinand, L., Candela, J., 2005. Detiding ADCP data in a highly variable shelf sea. *Journal of Atmospheric and Oceanic Technology* 22, 84–97.

Crombie, D. D., 1955. Doppler spectrum of sea echo at 13.56 Mc./s. *Nature* 175, 681–682.

Deacon, E. L., 1959. The measurements of turbulent transfer in the lower atmosphere. *Advances in Geophysics* 6, New York: Academic Press, 211–228.

Grant, W. D., Williams, A. J., Glenn, S. M., 1984. Bottom stress estimates and their prediction on the northern continental shelf during CODE-1: The importance of wave current interaction. *Journal of Physical Oceanography* 14, 506–527.

Hamon, B. V., 1955. A temperature-salinity-depth recorder. *Journal du Conseil* 21, 72–73.

Hamon, B. V., 1956. A portable temperature-chlorinity bridge for estuarine investigations and seawater analysis. *Journal of Scientific Instruments* 33, 329–333.

Hamon, B. V., Brown, N. L., 1958. A temperature-chlorinity-depth recorder for use at sea. *Journal of Scientific Instruments* 35, 452–458.

Hansen, D. V., Rattray, M., 1965. Gravitational circulation in straits and estuaries. *Journal of Marine Research* 23, 104–122.

Harlan, J. A., Georges, T. M., 1997. Observations of Hurricane Hortense with two over-the-horizon radars. *Geophysical Research Letters* 24, 3241–3244.

Hill, A. E., Durazo, R., Smeed, D., 1994. Observations of a cyclonic gyre in the Irish Sea. *Continental Shelf Research* 14, 479–490.

Horiuchi, T., Wolk, F., Macoun, P., 2010. Long-term stability of a new conductivity-temperature sensor tested on the VENUS cabled observatory. *Proc. OCEANS '10 Conference*, Sydney, NSW, Australia: IEEE, 1–4.

Holden, G. J., Wyatt, L. R., 1992. Extraction of sea state in shallow water using HF radar. *IEEE Proceedings Radar Signal Processing* 139, 175–181.

Howarth, M. J., Souza, A. J., 2005. Reynolds stress observations in continental shelf seas. *Deep Sea Research II* 52, 1075–1086.

Huntley, D. A., 1988. A modified inertial dissipation method for estimating seabed stresses at low Reynolds numbers, with application to wave/current boundary layer measurements. *Journal of Physical Oceanography* 18, 339–346.

Jacobsen, A. W., 1948. An instrument for recording continuously the salinity, temperature, and depth of sea water. *Transactions of the American Institute of Electrical Engineers* 67, 714–722.

Johannessen, O. M., 1986. Brief overview of the physical oceanography. In: *The Nordic Seas* (chapter 4), New York: Springer Verlag, 103–127.

Johannessen, J. A., Svendsen, E., Sandven, S., Johanessen, O. M., Lygre, K., 1989. Three dimensional structure of mesoscale eddies in the Norwegian Coastal Current. *Journal of Physical Oceanography* 19, 3–19.

Johannessen, J. A., Shuchmann, R., Johannessen, O. M., Davidson, K. L., Lyzenga, D. R., 1991. Synthetic aperture radar imaging of upper ocean circulation features and wind fronts. *Journal of Geophysical Research* 96, 10411–10422.

Johannessen, J. A., Røed, L. P., Johannessen, O. M., Evensen, G., Hackett, B., Petterson, L. H., Haugan, P. M., Sandven, S., Shuchman, R., 1993. Monitoring and modeling of the marine coastal environment. *Photogrammetric Engineering and Remote Sensing* 59, 351–361.

Johannessen, J. A., Digranes, G., Espedal, H., Johannessen, O. M., Samuel, P., Browne, D., Vachon, P., 1994. *SAR Ocean Feature Catalogue*. Noordwijk, The Netherlands: Publications Division, ESTEC, ESA-SP-1174, ISBN 92-9092-133-1.

Johannessen, J. A., Shuchman, R. A., Digranes, G., Wackerman, C., Johannessen, O. M., Lyzenga, D., 1996. Coastal ocean fronts and eddies imaged with ERS-1 SAR. *Journal of Geophysical Research* 101, C3, 6651–6667.

Johannessen, O. M., Pettersson, L. H., Bjørgo, E., Espedal, H., Evensen, G., Hamre, T., Jenkins, A., Korsbakken, E., Samuel, P., Sandven, S., 1997. A review of the possible applications of earth observation data within EuroGOOS. In: Stel, J., Behrens, H. W. A., Borst, J. C., Droppert, L. J., van der Meulen, J. P. (eds.), *Operational Oceanography – The Challenge for European Cooperation, Proceedings of the 1st International Conference on EuroGOOS*. Amsterdam: Elsevier, 192–205. ISBN 0-444-82892-3.

Joordens, J. C. A., Souza, A. J., Visser, A. W., 2001. The influence of tidal straining and wind on suspended matter and phytoplankton dynamics in the Rhine outflow region. *Continental Shelf Research* 21, 301–325.

Kerr, Y. H., 1998. *The SMOS (Soil Moisture and Ocean Salinity) Mission: MIRAS on RAMSES*. A proposal to the call for Earth Explorer Opportunity Mission. Proposal, CESBIO, Toulouse (France).

Kidwell, K. B., comp. and ed., 1995, *NOAA Polar Orbiter Data (TIROS-N, NOAA-6, NOAA-7, NOAA-8, NOAA-9, NOAA-10, NOAA-11, NOAA-12, and NOAA-14) Users Guide*. Washington, DC: NOAA/NESDIS.

Lagerloef, G. S. E., Swift, C. T., LeVine, D. M., 1995. Sea surface salinity: the next remote sensing challenge. *Oceanography* 8, 44–50.

Lerner, R. M., Hollinger, J. P., 1977. Analysis of 1.4GHz radiometric measurements from Skylab. *Remote Sensing of Environment* 6, 251–269.

Lhermitte, R. M., 1968. Turbulent air motion as observed by Doppler radar. *Proceedings 13th Conference Radar Meteorology,* American. Meteorological Society, 498–503.

Lipa, B. J., Barrick, D. E., 1986. Extraction of sea state from HF radar sea echo: Mathematical theory and modelling. *Radio Science* 21, 81–100.

Lohrmann, A., Hackett, B., Røed, L. P., 1990. High resolution measurements of turbulence, velocity and stress using pulse-to-pulse coherent sonar. *Journal of Atmospheric and Oceanic Technology* 7, 19–37.

Lohrmann, A., Cabrera, R., Gelfenbaum, G., Haines, J., 1995. Direct measurements of Reynolds stress with an acoustic Doppler Velocimeter. *Proceedings of the IEEE 5th Working Conference on Current Measurements*, 205–210.

Lu, Y., Lueck, R. G., 1999. Using a broadband ADCP in a tidal channel. Part II: Turbulence. *Journal of Atmospheric and Oceanic Technology* 16, 1568–1579.

Lwiza, K. M. M., Bowers, D. G., Simpson, J. H., 1991. Residual and tidal flow at a tidal mixing front in the North Sea. *Continental Shelf Research* 11, 1379–1395.

Lyzenga, D. R., 1991. Synthetic aperture radar imaging of ocean circulation features and wind fronts. *Journal of Geophysical Research* 96, 10411–10422.

Miller, J. L, Goodberlet, M. A., Zaitzeff, J. B., 1998. Airborne salinity mapper makes debut in coastal zone. *Earth and Space Science News* 79, 173–177.

Moller, D., Frasier, S. J., Porter, D. L., McIntosh, R. E., 1998. Radar-derived interferometric surface currents and their relationship to subsurface current structure. *Journal of Geophysical Research* 103, C6, 12839–12852.

Nikora, V., Goring, D., 1999. On the relationship between Kolmogorov's and generalized structure functions in the inertial subrange of developed turbulence. *Journal of Physics A: Mathematical and General* 32, 4963–4969.

O'Donnell, J., Dam, H. G., Bohlen, W. F., Fitzgerald, W., Gay, P. S., Houk, A. E., Cohen, D. C., Howard-Strobel, M. M., 2008. Intermittent ventilation in the hypoxic zone of western Long Island Sound during the summer of 2004. *Journal of Geophysical Research* 113, C9, C09025. doi:10.1029/2007JC004716

Pettersson, L. H., 1990. *Application of Remote Sensing to Fisheries*, Vol. 1. Technical Report EUR 12867 EN, Commission of the European Communities. Joint Research Centre, Ispra, Italy.

Prandle, D., 1991. A new view of near-shore dynamics based on H.F. radar. *Progress in Oceanography* 27, 403–438.

RDI, 1996. *Acoustic Doppler Current Profiler Principles of Operation – A Practical Primer*, San Diego: RDI.

Rippeth, T. P., Williams, E., Simpson, J. H., 2002. Reynolds stress and turbulent energy production in a tidal channel. *Journal of Physical Oceanography* 32, 1242–1251.

Reul, N., Saux-Picart, S., Chapron, B., Vandemark, D., Tournadre, J., Salisbury, J., 2009. Demonstration of ocean surface salinity microwave measurements from space using AMSR-E data over the Amazon plume. *Geophysical Research Letters* 36, L13607. doi:10.1029/2009GL038860.

Sea-Bird Scientific, 2017. 13431 NE 20th Street, Bellevue, Washington 98005, USA (www.seabird.com).

Schmidt, R. W., 2008. Salinity in Argo. *Oceanography* 21, 56–67.

Shay, L. K., Graber, H. C., Ross, D. B., Chemi, L., Peters, N., Hargrove, J., Vakkayil, R., Chamberlain, L., 1993. Measurement of ocean surface currents using an H.F. Radar during HIRES-2. Technical Report 93–007, Rosentiel School of Marine and Atmospheric Sciences, University of Miami.

Shemer, L., Marom, M., Markman, D., 1993. Estimates of currents in the nearshore ocean region using interferometric synthetic aperture radar. *Journal of Geophysical Research* 98, C4, 7001–7010.

Simpson, J. H., Bos, W. G., Schirmer, F., Souza, A. J., Rippeth, T. P., Jones, S. E., Hydes, D., 1993. Periodic stratification in the Rhine ROFI Freshwater influence Tidal mixing Periodic stratification Coastal region in the North Sea. *Oceanologica Acta* 16, 23–32.

Souza, A. J., Simpson, J. H., Schirmer, F., 1997. Circulation in the Rhine ROFI. *Journal of Marine Research* 55, 277–292.

Souza, A.J., Alvarez, L.G., Dickey, T., 2004. Tidally induced turbulence and suspended sediment. *Geophysical Research Letters* 31, L20309. doi:10.1029/2004GL021186

Souza, A. J., Howarth, M. J., 2005. Estimates of Reynolds stress in a highly energetic shelf sea. *Ocean Dynamics* 55, 490–498. doi:10.1007/s10236-005-0012-7

Souza, A. J., Bolaños, R., Wolf, J., Prandle, D., 2011. Measurement technologies: Measure what, where, why, and how? *Reference Module in Earth Systems and Environmental Sciences: Treatise on Estuarine and Coastal Science* 2, 361–394. doi:10.1016/B978-0–12-374711-2.00215–1.

Stacey, M. T., Monismith, S. G., Burau, J. R., 1999. Measurement of Reynolds stress profiles in unstratified tidal flow. *Journal of Geophysical Research* 104, C5, 10933–10949.

Stanev, E. V., 1994. Assimilation of sea surface temperature data in a numerical ocean circulation model. A study of the water mass formation. In: eds. Brasseur, P. P., Nihoul, J. C. J. *Data Assimilation: Tools for Modelling the Ocean in a Global Change Perspective*, Vol. I19, Berlin: Springer, NATO ASI, 33–58.

Sverdrup, H. U., Johnson, M. W., Fleming, R. H., 1942. *The Oceans: Their Physics, Chemistry, and General Biology*. Englewood Cliffs, NJ: Prentice Hall.

Teledyne RD Instruments, 2016. *Teledyne RD Instruments, San Diego Facility, 14020 Stowe Drive*, Poway, CA, USA. http://rdinstruments.com [accessed August 2016].

Tennekes, H., Lumley, J. L., 1972. *A First Course in Turbulence*. Cambridge, MA: MIT Press.

Weber, B. L., Barrick, D. E., 1977. On the non-linear theory for gravity waves on the ocean's surface. Part I: Derivations. *Journal of Physical Oceanography* 7, 3–10.

Wiles, P. J, Rippeth, T. P., Simpson, J. H., Hendicks, P. J., 2006. A novel technique for measuring the rate of turbulence dissipation in the marine environment. *Geophysical Research Letters* 33, L21608, doi:10.1029/2006GL027050.

Wilson, T. C., Lwiza, K. M. M., Allen, G. L., 1997. Performance comparison of RDI ADCPS: broadband versus narrowband. In: *Oceans '97 MTS/IEEE Conference Proceedings*, Washington, DC: IEEE, Vol. 1, 120–125.

Wolf, J., Bell, P. S., 2001. Waves at Holderness from X-band radar. *Coastal Engineering* 43, 247–263.

5 Measurement and Analysis of Waves in Estuarine and Coastal Waters

J. Wolf

1 Introduction

Nothing illustrates the power of the sea better than an image of waves crashing against the shore or a rolling breaker ridden by surfers. The power of these waves can lead to coastal erosion and damage to coastal infrastructure (Figure 5-1). The regular depressions passing over the NW European continental shelf in winter from the Atlantic can produce storm force winds that generate surges and waves. Wind-stress is particularly effective in piling up water against the coast in the shallow waters of the continental shelf because the effect is inversely proportional to water depth: depths in the southern North Sea and the eastern Irish Sea are only about 40 m on average. The UK, in common with many other countries, has assets worth billions of pounds at risk from coastal floods, river floods and coastal erosion (Wolf, 2008). The effects of coastal flooding may be exacerbated by wave overtopping of natural or man-made coastal defences (e.g., Seelig, 1979; Seelig and Ahrens, 1981; d'Angremond et al., 1996; Pullen et al., 2007). Most sea defences are weaker on the landward side, thus overtopping can lead to failure in some cases.

As sea level rises due to global warming, and the rate of rise accelerates, low-lying coastal regions may be inundated, allowing waves to penetrate further inland, causing

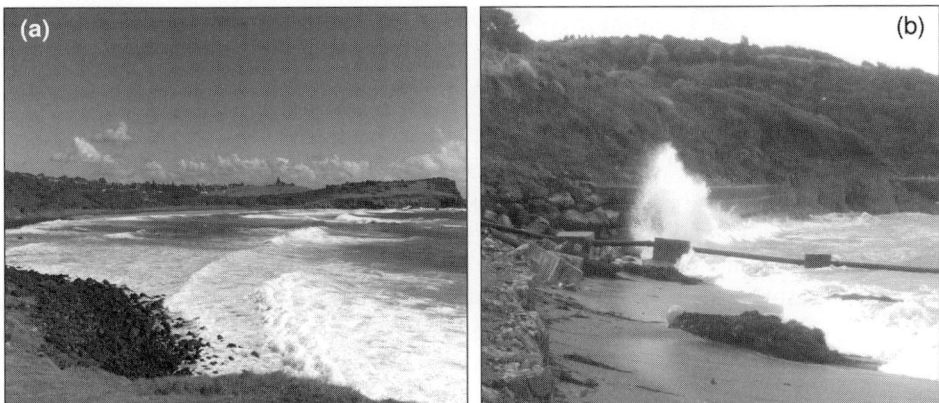

Figure 5-1 Coastal waves: (a) shoaling waves near Byron Bay, east coast of Australia; (b) storm waves impacting with sea defences and a disused waste pipe, Batten Beach, Plymouth, southwest UK. Photographs: R. J. Uncles (with permission).

further damage. Although estuaries are usually sheltered from the larger sea waves by ebb and flood shoals at the mouth, where waves break, they can still be affected by local wave generation, and the effect on mudflats and saltmarshes can be locally important. When the wind is against the tide, the shorter, steeper waves can be important for resuspension of sediment (e.g., Wolanski and Spagnol, 2003). Very small waves in shallow waters, such as over mud flats and reef surfaces, can readily suspend the fine sediment (e.g., Lambrechts et al., 2010; Uncles and Stephens, 2010).

Another reason to be interested in the power of waves is the harnessing of wave power for electricity generation. To the west of Scotland there is an annual average of 50 kW of wave power per metre of coastline (at the 50-m contour; Tucker and Pitt, 2001). The peak power in severe storms is approximately 300 kW m^{-1}. Harnessing 50 per cent of this power would provide 2.5 MW per kilometre of coastline. On the island of Islay, in the Sea of the Hebrides, there is a coastal power station that uses a wave power device called LIMPET (Land Installed Marine Powered Energy Transformer) that employs oscillating water-column technology.

A long time series of wave observations (at least 10 years, ideally 30 years) is required to reliably describe the wave climate (Wolf et al., 2011). Such knowledge is required to characterise coastal vulnerability and plan coastal management strategies. We also know that there is a large amount of interannual and interdecadal variability. There are only a few long-term wave data sets in UK waters; for example, the UK Meteorological Office (Met Office) Marine Automatic Weather Station (MAWS) system, which consists of various met-ocean recording systems, some of which have been maintained for several decades (Hawkes et al., 2001). Data collected from the Seven Stones Light Vessel since 1962 led to the earliest observation of an increase in wave height in the North Atlantic in recent years (Bacon and Carter, 1991). This observation has since been validated and extended using altimeter wave data and models and attributed largely to observed changes in the North Atlantic atmospheric circulation patterns, principally the North Atlantic Oscillation (NAO, Woolf et al., 2002; 2003). Due to the lack of long-term data sets, numerical models are often used to extend the time series. Many global and regional wave hind-casts and reanalyses are now available. The benefit of a reanalysis is that it incorporates state-of-the-art modelling with the available data. A coastal wave atlas has been developed for European waters (Pontes, 1998) using the Simulating WAves Nearshore (SWAN) wave model (described in Section 6) for wave transformation in the coastal zone. More recently, global ocean wave atlases have been derived and validated against satellite and buoy data (e.g., the WorldWaves database, Barstow et al., 2003).

In this chapter we will cover the properties of wind-generated surface waves in estuaries and open coastal waters, appropriate wave measuring instrumentation and shallow water wave modelling. The measuring of waves may be done by in situ or remote-sensing methods (satellite, airborne and ground-based). Waves in the nearshore zone are treated by a combination of mathematical wave theory and other more empirical coastal engineering methods. Linear wave theory is used to explain basic wave processes, with some reference to nonlinear interactions, higher order wave theories and their area of applicability. The main wave parameters of interest are significant wave height, mean and peak period and direction, which usually are derived from wave spectra that describe the full frequency and directional distribution of wave energy in a given sea state. Details of waves in the surf zone and the effects of harbours and coastal structures, being within the province of coastal engineering, are not given here, although a brief description of some of the processes is included. These are treated in various specialist texts, and some references are given for further reading.

2   Definitions and Linear Wave Theory

Coastal and estuarine waters generally are regarded as shallow (< 100 m) compared with the deep ocean (> 1000 m). However, water depth for wind waves is defined in relation to the wavelength; thus, waves can be in deep water close to the coast if they are very short. For deep-water waves the water may be regarded as infinitely deep because these waves are fundamentally a phenomenon of the air–sea interface. In finite water depth, where the water depth is less than half the wavelength, the wave orbital velocity at the seabed is nonzero, so that shoaling waves then interact with the seabed.

Some basic definitions are provided in this section, which allow a discussion of the transformation of waves as they progress from deep waters to the shore in shallow waters. The processes involved include shoaling, refraction, diffraction and energy dissipation due to breaking and bottom friction. Local generation of waves may also occur, and this 'wind sea' can interact with swell waves propagating into the area. The interaction of waves, currents and turbulence in shallow waters with complex bathymetry of channels and shoals can make coastal processes difficult to interpret, and this topic is discussed in Section 3.

Waves are often divided into wind-sea or swell, depending on whether they are being actively generated by the local wind or have moved out of the active generation area. Wind-driven waves grow in height and period with increasing surface wind-speed (usually specified at 10 m above the surface), fetch (distance over which the wind has blown) and the duration of the wind, up to a limit known as a 'fully developed sea'. The dynamic fetch effect can increase waves further than a static fetch, if the wind-generation area moves with similar speed to the waves propagating away from the generation area; e.g., in a moving storm such as a tropical cyclone or mid-latitude depression, because in this case energy from the wind continues to be fed into the wave field. As waves grow, their steepness (ratio of wave height to wavelength) may exceed a critical value, causing wave breaking and energy dissipation. In deep water this is called 'white-capping'. In shallow water, depth-limited breaking and bottom friction also cause wave energy dissipation.

Wind waves have wavelengths from a few centimetres to several hundred metres and periods from 1–100 s. The common range of dominant wave periods in coastal waters is 1–10 s, with swell waves in areas exposed to the open ocean having periods of up to about 20 s. In most of this range the waves are termed 'gravity waves' because gravity is the restoring force acting on the surface displacement, but very short waves (wavelengths < 0.1 m, periods < 0.25 s) also are affected by surface tension and are called

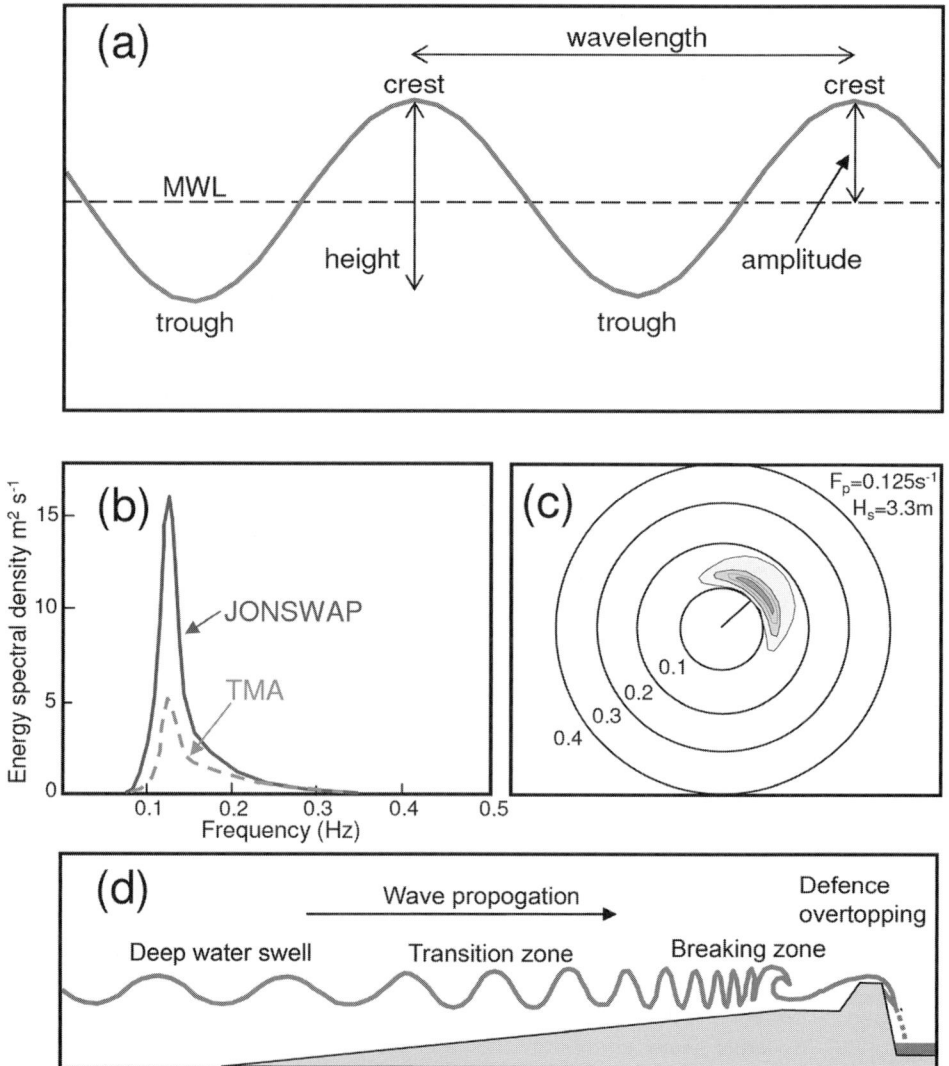

Figure 5-2 Wave definitions and examples that illustrate wave properties; (a) wave parameter definitions are illustrated for variables and terminology used in the text; (b) a typical 1-D wave frequency spectra, showing the JONSWAP and TMA spectral shapes; (c) a 2-D (frequency-direction) wave energy spectrum generated by a 10 ms^{-1} wind from the northeast, with a significant wave height of 3.3 m and a peak frequency of 0.125 Hz; (d) the transformation of waves as they approach shallow water from deep water. Illustrations (b)–(d) are reproduced with minor modifications from Wolf et al. (2011), with permission from Elsevier.

'capillary waves'. Most of the energy is in the gravity wave part of the spectrum, but capillary waves can be important in terms of the surface roughness and the remote sensing characteristics of waves.

Water waves can be characterised by a few basic parameters (Figure 5-2a): wave height (H, crest to trough, which is twice the wave amplitude, a), wavelength (λ, crest to

where the wave height is $H=2a$. Nonlinear waves may be approximated, e.g., by higher-order Stokes theory (Stokes, 1847), where higher harmonics are added to the first-order linear solution:

$$\zeta_0 = \varepsilon\zeta_1 + \varepsilon^2\zeta_2 + \varepsilon^3\zeta_3 + \varepsilon^4\zeta_4 + \varepsilon^5\zeta_5 + \ldots \tag{11}$$

i.e., an expansion in terms of the wave steepness, where the first-order solution $\varepsilon\zeta_1$ is the linear wave, obtained by replacing the sine wave by a cosine wave in Eq. 2c, which makes the notation for the higher harmonics simpler. The second-order correction $\varepsilon^2\zeta_2$ is then given by:

$$\varepsilon^2\zeta_2(\mathbf{x}, t) = ka^2 \frac{\cosh(kh)}{4\sinh^3(kh)}(2 + \cosh(2kh))\cos(2(\mathbf{k}.\mathbf{x} - \sigma t - \phi)) \tag{12}$$

Further higher order corrections become rapidly more complex. Other higher order wave theories are Dean's stream-function theory (Dean, 1965) and cnoidal theory, also known as KdV theory, after Korteweg and de Vries (1895), which is an expansion in terms of a different parameter, $\beta = a/h$:

$$\zeta_0 = \beta\zeta_1 + \beta^2\zeta_2 + \beta^3\zeta_3 + \beta^4\zeta_4 + \beta^5\zeta_5 + \ldots \tag{13}$$

Cnoidal waves are expressed with standard mathematical functions in terms of Jacobian elliptic functions. See Appendix B of Holthuijsen (2007) for more details of the expansions.

For long waves ($\lambda \gg h$) with small Ursell number, $U_r \ll 32\pi^2/3 \approx 100$, linear wave theory is applicable. Otherwise a nonlinear theory for fairly long waves ($\lambda > 7h$), e.g., the KdV equation or Boussinesq equations, has to be used. Hedges (1995) shows the physical significance of the value $32\pi^2/3$ and provides evidence to demonstrate that a suitable boundary between linear wave theory and nonlinear theories for long waves is $U_r = 40$. Stokes's higher-order theories are more appropriate than linear wave theory if $U_r < 40$ but $H/\lambda > 0.04$. Only the cnoidal theory is applicable in very shallow water ($h < H$). The addition of bound higher harmonics (which travel at the same speed as the fundamental wave) makes the wave surface more asymmetrical, with sharper peaks and flatter troughs.

3.1 The Surf Zone

Waves break more intensively in the surf zone due to depth-limited breaking, and their energy is dissipated at the coast. This leads to net mass transport by waves, contributing to the mean circulation, wave setup and longshore drift (e.g., Brown et al., 2011; 2013). Waves here can move considerable amounts of sediment, especially noncohesive sediment, building a mobile beach where the coast is of erodible material (Bolaños et al., 2012). The rate at which the wave energy is dissipated depends on the slope of the beach, which in turn is determined by the coarseness of the sediment. Gravel beaches may be much steeper than sand beaches.

Details of the types of breaker and the mechanism of breaking waves and surf zone dynamics can be found in Holthuijsen (2007) and Elfrink and Baldock (2002). The total dissipation due to depth-induced wave-breaking in the surf zone can be modelled by the dissipation of a bore, applied to the breaking waves in a random wave field in shallow water (Battjes and Janssen, 1978).

There are many interesting phenomena at the shore induced by waves, such as wave groups, wave run-up and the swash zone, rip currents and undertow, solitary waves and edge waves (water waves that are trapped at the shoreline by refraction). Surf beat is the long period oscillation of the water line (typically of several minutes period, known as infra-gravity waves), which may be induced by wave groups or edge waves. Most of these phenomena can be explained by the combination of incident waves with shoreline features and nonlinear feedback effects. There is not room to explore all these processes here, and the reader is referred to texts on surf zone hydrodynamics and beach processes, e.g., Battjes (1988), Arcilla and Lemos (1990), Svendsen (2006) and Nielsen (2009).

3.2   Harbours and Coastal Structures

Harbours are structures generally designed to provide shelter from large waves at the coast; however, sometimes a seiche can occur in which a resonant frequency is triggered that may cause problems with large water level disturbances. The vertical walls of the harbour can lead to the reflection of wave energy, although details of the seiche-generation mechanism are unclear. Another process that may become locally important is diffraction (e.g., Dalrymple et al., 1984). Diffraction is the turning of waves toward areas with lower amplitudes due to amplitude changes along the wave crest and may be induced by obstacles such as islands, headlands and breakwaters, which can shelter an area of water from waves. For long-crested harmonic waves propagating over a horizontal bottom, Huygens' principle can be used to compute the diffraction pattern (Holthuijsen, 2007). Essentially, this states that every point on a wave front is a source of new wave fronts; i.e., new wave fronts are generated at each point on the crest of a travelling wave, which propagate out spherically in every direction, like ripples generated by throwing a pebble into a pond.

Overtopping of coastal structures is an important issue for coastal management. Various methods exist for modelling overtopping, e.g., Hedges and Reis (1998), whose method has recently been implemented by the UK Environment Agency in a new combined wave and water level flood warning system, initially for the NE of England from Berwick-upon-Tweed to the Humber Estuary (Lane et al., 2008). In the LEACOAST2 project the interaction of waves, water levels and current with shore-detached breakwaters was studied using various observational and modelling techniques (Pan et al., 2010). The joint probability of waves and water levels must be taken into account because waves and surges are driven by the same storm forcing, but an increase in water level can result in a reduction in wave dissipation and lead to higher waves (Hawkes et al., 2002).

The effect of projected climate change on offshore waves has been investigated, e.g., by Leake et al. (2009), and impacts of these future waves and water levels on the East Anglian coast are discussed in Chini et al. (2010).

4   Observing Waves in Shallow Water

Wave observations are required as a fundamental part of understanding wave growth, transformation, dissipation, and interactions with currents, offshore and coastal structures, and the coast. Wave observations are also critically important for validation of wave models, which can then provide a larger spatial and temporal coverage of wave information in terms of hind-casts and forecasts.

The most basic wave measurements are observations of the vertical displacement of the water surface, such as those from wave staffs, which take no account of the wave direction. Directional wave measurements usually provide basic wave variables, such as the wave height, period and direction for a range of wave frequencies. The peak period provides the dominant frequency of the wave energy. The mean or peak wave direction indicates which way the most energetic waves are propagating (different conventions may be used: wave 'to' or wave 'from' direction). Observations of waves can be raw time series or 1-D or 2-D spectra derived from high-frequency observations.

There are various instruments for collecting wave data, including deployment of in situ instruments such as directional wave buoys, arrays of current meters and pressure sensors, each of which has advantages and disadvantages. Waves can also be measured by remote sensing of wave characteristics, such as the surface roughness, using microwave radar or optical sensing systems deployed on aircraft, satellites or ground-based platforms. Some instruments give more direct measurements, while others measure wave-related phenomena or, as in remote sensing, a signature of the radio waves reflecting from waves using an empirical algorithm. Different instruments can have different responses to different parts of the wave spectrum, and it can be useful to compare their different characteristics; e.g., Wolf (1996b) and Krogstad et al. (1999).

In very shallow waters some observing methods are no longer applicable; e.g., wave buoys cannot be deployed in much less than about 20-m depth, satellite altimeter systems are affected by the presence of land in the footprint, while other methods come into use, such as bottom pressure measurements, which cannot be used in deep waters (greater than about 30 m). In this section, various measuring systems are briefly described. For further details about instrumentation and analysis, the reader is referred to Tucker and Pitt (2001), COST (2005), Marshall and Bishop (1984), and Gower (1981).

4.1   In Situ Wave Measurements

Visual wave observations can be very useful, although they are rather labour intensive. These require the observing and counting of wave crests over a set time interval to determine the mean wave period and the estimation of the crest to trough height. In fact, voluntary observing ships have provided much useful data going back many years that have contributed to wave climate studies (e.g., Gulev and Grigorieva, 2004), and it is relatively easy to train an observer to estimate wave height at the shore or from a vessel.

Instrumental methods include wave staffs, wave buoys, bottom pressure sensors, combined pressure and velocity measurements (PUV), the wave-enabled acoustic

Doppler current profiler (ADCP, Chapter 4), and other high-frequency acoustic measurements. Most instruments have some limitations in the measurement of waves.

The most commonly accepted wave measurements are from surface wave-following buoys, although these cannot perfectly follow the sea surface, especially in extreme wave conditions, and mooring interference may occur in strong currents. Another limitation is that in relatively shallow waters (< 20 m) they are not always easy to deploy because the mooring requires a minimum depth of water. Ideally, it is recommended to include two elastic strops to avoid interference of the mooring with the surface-following properties of the buoy. In shallow waters it can also be difficult to separate the effects of waves, currents and turbulence using single-point velocity observations (Trowbridge, 1998; Wolf, 1999; Souza and Howarth, 2005).

Note that a wave staff, at a fixed horizontal position, and a wave buoy, free to move on the sea surface (except as constrained by a mooring), measure different things, although both are intended to record the sea surface displacement as a function of time (James, 1986). The differences may be expected to become greater as the waves become steeper and more nonlinear. In particular, the skewness of the sea surface (associated with the sharp crests and flat troughs of steep waves) is not reproduced in the wave buoy record. This phenomenon is discussed by Tucker and Pitt (2001); the physical interpretation is that the buoy moves forward in the crests and backwards in the trough, which cancels out the asymmetry of a real sea wave. The effect of the mooring may distort this simple picture somewhat.

4.1.1   Wave Staffs and Fixed Water Level Measurement

The most direct means of measuring waves is to observe the high frequency (> 1 Hz) changes in water level. Various methods are described by Tucker and Pitt (2001), including resistance-wire and capacitance-wire gauges. These can be deployed on fixed platforms or poles in shallow water. The staff, mounted on a fixed structure, needs to be designed so that there is little disturbance of the wave field by the structure itself (at 10 diameters away from any substantial structural member). The direct interaction of the water surface with the resistance or capacitance wires acts to change their electrical properties, which is then used to detect the water level. Resistance wave gauges are usually formed from two parallel wires, separated by a fixed distance, and the water surface provides a short circuit across the two wires. Due to fouling, these instruments are not very successful in the sea and are usually used in wave tanks (Tucker and Pitt, 2001). In capacitance wire gauges, an insulated wire is mounted vertically through the water surface, and the capacitance between the wire and the sea is measured. A group of three wave poles at the corners of a small triangle can be used to estimate the slope of the water surface and hence the direction of the waves, similarly to the directional wave buoy. The ASIS buoy acts as an array of wave capacitance gauges (Graber et al., 2000).

4.1.2   Wave Buoys

A wave buoy is a surface-following instrument, normally anchored to the seabed by a mooring, which can measure the vertical acceleration or the water surface elevation. An accelerometer mounted within the buoy registers the rate at which the buoy is rising or falling with the waves. The most commonly used instrument for measuring waves, in intermediate to deep water, is the 'heave, pitch, and roll buoy' that measures the surface height and slope in two orthogonal directions, for example the Datawell Directional Waverider buoy (Tucker and Pitt, 2001). This measures the three orthogonal component accelerations of the buoy, which are integrated to yield its horizontal and vertical displacements. The hull and mooring of this buoy have been designed to give it good wave-following characteristics, thereby allowing the buoy displacements to approximate the displacements of an actual water particle at the sea surface.

The correct mooring of a wave buoy is essential to measuring wave parameters accurately. The design of an appropriate mooring requires knowledge of the current speed and profile, the depth, tides, wave height and sometimes the seabed structure. Datawell has developed a standard mooring layout that applies to a wide range of situations (e.g., Figure 5-3c for shallower waters). The buoy is attached to a swivel by means of an elastic cord that provides the flexibility to ride waves up to 40 m in height. Buoys can also contain an electronic compass and additional accelerometers to measure the directional components of the wave field.

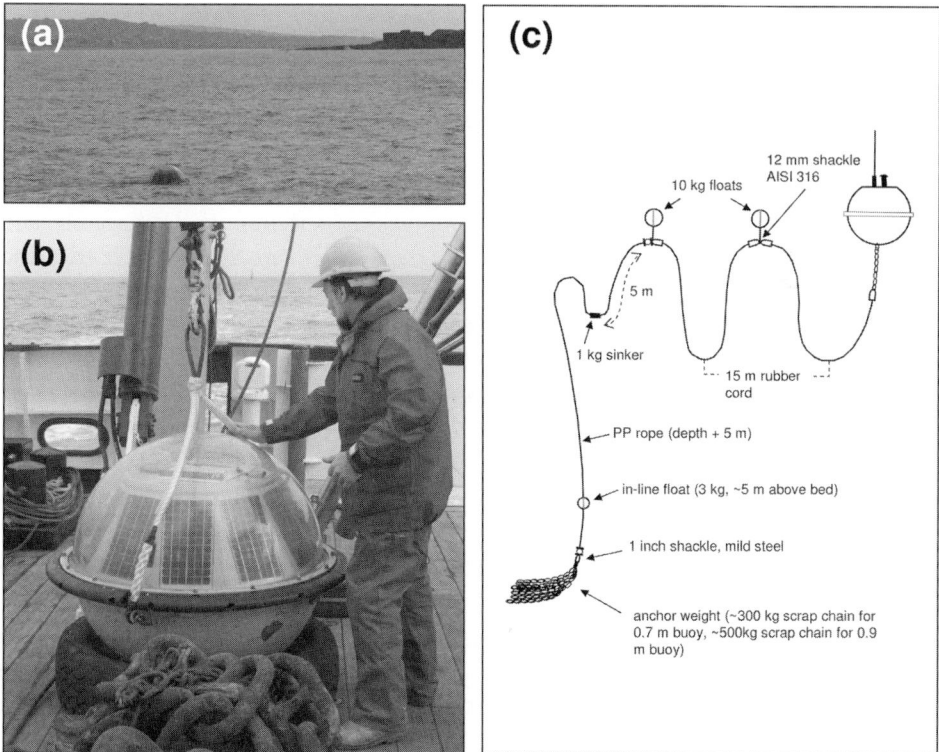

Figure 5-3 Examples of a wave-buoy mooring and deployments; (a, b) Triaxys wave buoy deployment in the Dee Estuary, northeast English-Welsh border region, UK. Photographs: R. D. Cooke, NOC (with permission); (c) shallow water mooring layout for a Datawell Waverider buoy for water depths between 8 and 17 m (redrawn with modifications from the Datawell Manual; Datawell, 2009; with permission).

Other wave buoys are also currently available with a variety of sizes and instrumentations (such as anemometers and current meters). A similar buoy to the Directional Waverider is the Triaxys buoy, developed by Axys Environmental (Figure 5-3a and b). This includes three accelerometers, three rate gyros and a fluxgate compass and is solar powered. The Wave Sentry Buoy is a small-diameter discus buoy, which is lightweight, easily transported, deployed, and retrieved by hand from a small boat with a one- or two-person crew. The electronics and battery housings can be removed from the flotation in a matter of seconds, which allows the user to replace battery packs at sea without disturbing the mooring. New versions of wave buoys are being developed that can also measure surface currents.

An alternative type of buoy is the spar buoy, which combines the surface-following properties described earlier with wave staffs, as in the previous section; e.g., the air–sea interaction spar (ASIS) buoy (Graber et al., 2000). The ASIS buoy is especially designed to produce little surface disturbance and to be a partial wave follower at low frequencies. The relative location of the interface is measured by an array of capacitance wave staffs, while the motion of the buoy is monitored using linear accelerometers and rate gyros. This combination permits high-resolution measurement of wave directional properties from 0.1 m to 600 m.

4.1.3   Bottom Pressure Sensors and the PUV Method

The pressure induced by a wave is in phase with the water surface elevation (Eq. 3 and 4); therefore, a measurement of subsurface pressure can be used to provide information on the surface waves. Linear wave theory is usually used to compute the depth attenuation and convert the bottom pressure spectrum to a surface elevation spectrum. However, the pressure decays with depth more rapidly for higher frequency waves, so that bottom pressure sensors can be used to measure the full wave spectrum only in relatively shallow waters (less than about 20 m).

Pressure measurements combined with current measurements can provide extra directional information by using the PUV (pressure and orthogonal current components) method; e.g., Krogstad (1991) and Wolf (1997). This method uses the velocity and pressure spectra recorded at the depth of the instrument to calculate the surface elevation spectra. The wave direction is calculated from the cross-spectra of pressure and velocity at each spectral frequency. One such instrument that simultaneously observes high-frequency bottom pressure and velocity is the InterOcean S4DW (Figure 5-4a) and an example of the analysis of such data is given in Section 4.3 (Wolf, 1997). Other problems associated with measuring waves using a bottom pressure recorder can be caused by the Bernoulli Effect, which is the generation of pressures due to wave currents. This must be minimised by correct instrument design. It has been suggested that the linear theory is inadequate for the conversion of bottom-to-surface pressures, but Bishop and Donelan (1987) show that it should be accurate to within 5 per cent, provided the analysis is carried out on the spectra rather than on a wave-by-wave basis. Lee and Wang (1984) demonstrate the effects of currents and nonlinearities; nonlinear effects are likely to be small in depths greater than 10 m.

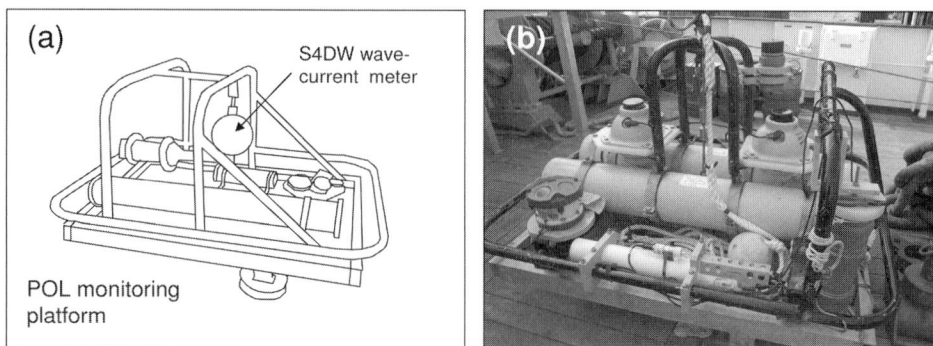

Figure 5-4 A schematic and photograph of a POL (NOC, Liverpool) monitoring platform that illustrate the deployment of: (a) a bottom pressure recorder, S4DW (redrawn with minor modifications from Prandle et al., 1996; with permission); (b) an RDI ADCP. Photograph: R. D. Cooke, NOC (with permission).

4.1.4 Acoustic Doppler Current Profilers

The cross-spectra of a current meter that measures the orthogonal current components can provide the same wave information as that obtained from directional buoy measurements. The ADCP (Figure 5-4b) measures the velocity of water using the Doppler shift in frequency due to the ambient current (Chapter 4) and may be mounted to measure currents in the horizontal or vertical. It has been used for measurement of the current profile for over 20 years, and ADCPs are now routinely deployed around the world (Chapter 4). The observations are provided in a series of bins along three or more (near-vertical) beams and then combined to infer the velocity profile encompassed by these beams. ADCPs are generally applied over long ranges of 10–100 m for mean velocity measurements, where large bin widths are used. The resulting trade-off between resolution and velocity accuracy requires, statistically, velocity measurements to be averaged over several minutes. More recently, ADCPs have also been adapted to capture wave data. For wave measurements the vertical profile configuration is used, and high-frequency data must be collected (at several Hz) to allow resolution of the wave periods. An example of wave data collection in the Irish Sea and its comparison with other instruments is given by Wolf et al. (2011).

Three methods are available for obtaining the wave spectra from the ADCP: a PUV method using bottom pressure together with currents (as earlier), a surface tracking method, and a wave array method using the near-surface orbital velocities obtained from the separate beams. The latter method provides the most reliable data. The ADCP can profile the water volume all the way to the surface; thus, it can be mounted in much deeper water than a traditional pressure instrument. The power spectra are calculated from the time series of velocities. To derive a surface wave height spectrum, the velocity spectrum is converted to surface displacement using linear wave theory. To calculate directional spectra, phase information must be preserved. Each bin beam is considered an independent sensor in an array that makes a measurement of one component of the wave field velocity. The cross-spectrum is then calculated between each sensor and every other sensor in the array. The result is a cross-spectral matrix that contains phase information along the path between each sensor and every other sensor at each frequency band, which is linearly related to the directional spectrum at each frequency; the directional spectrum is obtained by inverting this matrix.

4.2   Remote Sensing

Remote sensing is usually taken to mean observations made without direct contact between the sensor and the medium being observed (Chapter 11). Thus, electromagnetic radiation (light and radio waves) that is reflected or emitted from an object is the usual source of remote-sensing data. Sound waves are also used in the sea (acoustic methods). Waves may be imaged by direct sensing of a wave property, such as surface elevation or wave orbital velocity, or more indirectly; e.g., by sensing a change in surface roughness due to waves. Capillary wave 'ripples' tend to converge at gravity wave crests, and this can lead to an imaging of long wave crests. Many imaging mechanisms depend on the Bragg wave resonance.

Remote sensing may include space-borne or airborne sensors, such as the synthetic aperture radar (SAR) and microwave altimeter, as well as ground-based radars such as X-band radar and HF Doppler radar (Chapter 4). Space-borne sensors have a lower resolution (of the order of a few tens of metres to a few kilometres), although new SAR systems can achieve higher resolution. On the other hand, ground-based radar is suitable for monitoring the waves in near offshore or shallow zones, with the wave field width of a few tens of centimetres to a few hundred metres.

4.2.1   Satellite and Airborne Sensors

SAR can be used to obtain wave spectra, although this is mostly successful for long waves in deep water and tends to be ineffective in shallow coastal areas. Satellite (and airborne) radars can be used to obtain wave height (and period) as well as water level. The shape of the reflected altimeter wave form can be related to the significant wave height. While altimetry over the open ocean is a mature discipline, in the coastal zone (the strip within a few tens of kilometres from the coast) data are often discarded because land effects interfere with the altimetric waveforms and/or lack adequate corrections for various effects, such as path delays, coastal tides and high-frequency atmospheric signals. The European Space Agency Coastal Altimetry (COASTALT) project addressed the reprocessing of these data to provide much more useful information in the nearshore zone (Gommenginger et al., 2011).

The water's surface roughness due to short waves has also been shown to provide an imaging mechanism, due to the modulation of the roughness by ambient depth and currents, thereby allowing the assessment of water depth and current. Coupled wave and current models of the Liverpool Bay/Dee Estuary (e.g., Li et al., 2011), and Helwick and Nash Banks in the Bristol Channel, UK, were used to explore the current and wave modulation by bathymetry that leads to the capability of SAR images to show signatures of shoals, e.g., Macklin et al. (2006).

4.2.2   Ground-Based Radar Systems

4.2.2.1   *Radar Ranging (Mounted on a Structure)*

A radar water-level sensor, often mounted on a seawall or jetty as a tide gauge, may be able to provide high-frequency water level data that can be used to obtain wave information. Such a high-frequency radar gauge is available from MIROS and has been used in the Spanish REDMAR network (Pérez, 2016). Although this technology is relatively new, radar tide gauges are being installed by a number of agencies as replacements for older instruments or for completely new networks. They are as easy to operate and maintain as acoustic sensors, without the disadvantage of a high dependence on the air temperature. Also, the need for accurate knowledge of the water density is not required because these are direct measurements of the water surface. However, many gauges only sample at intervals greater than about 20 s, which is not sufficiently discriminating for wave analysis.

A radar gauge is mounted several metres above the surface of the sea. Some radars measure changes in sea level by means of the time of travel of the radar pulse to the surface and back to the sensor. Others use a Frequency Modulated Continuous Wave (FMCW) system, in which transmitted radar waves are mixed with signals reflected from the surface to determine the phase shift between the two waves and hence the distance. The instruments are supplied with the necessary hardware and software to convert the radar measurements into a sea-level height. Usually, the installation will be on a fixed platform; e.g., a harbour wall, pier, jetty or offshore structure.

The MIROS FMCW radar sensor was selected for the REDMAR network (Martín et al., 2005). One advantage is that the 2 Hz original raw data sampling allows wind-wave or agitation (short waves inside the harbour) parameters to be estimated, which are needed in some harbours. The availability of wave data also may allow for an assessment of wave bias, which can be a problem when measuring the tidal level (Woodworth and Smith, 2003).

Between 2007 and 2011, practically all the REDMAR tide gauge stations were upgraded, incorporating a MIROS radar sensor. At the same time, many new harbours that utilise this new equipment have been incorporated into the network. The network now comprises 35 multipurpose radar stations, which allow the measurement of the whole range of frequencies of sea level variation, and two AANDERAA pressure sensors (Pérez et al., 2008). The spatial coverage has been significantly improved in the Canary and the Balearic Islands (one or two tide gauges available now for each island – see later), as well as on the Mediterranean mainland Spanish coast, the Gibraltar Strait and Melilla (North Africa).

4.2.2.2   *Marine Navigation (X-Band) Radars*

X-band radar (0.03 m wavelength) is generally used for marine navigation on ships. The measurement of waves is based on the analysis of the backscatter of radar energy from the ocean surface (termed sea clutter when it interferes with the main objects of interest; i.e., coastal features and ships). This backscatter, which can be made visible on the marine radar, shows the wave patterns, particularly for longer and higher waves, because the reflections are mostly from the wave crests. In recent years, deployments at various coastal locations have been made, including Holderness, Hilbre Island and Sea Palling around the UK, which can observe the wavelength, direction and frequency of waves (Heathershaw et al., 1980; Young et al., 1985; Borge et al., 1999; Wolf and Bell, 2001).

The Wave and Surface Current Monitoring System WaMoS® II has been developed for real-time measurement of directional ocean wave spectra. A significant advantage of WaMoS® II is the continuous availability of wave data in rough seas during harsh weather conditions with limited visibility, and at night. The system uses the unfiltered output from marine X-Band radar to determine wave and surface current parameters in near real-time. WaMoS® II measures and displays all the essential wave field parameters, such as significant wave height (H_s), peak wave period (T_p) and peak wave direction (D_p), as well as surface current speed (U) and current direction (θ). It can operate automatically and unattended from moored platforms, moving vessels and coastal sites.

4.2.2.3   HF Radar

HF radar can simultaneously observe waves and currents over a spatial grid of many kilometres, which is ideal for comparison with nearshore models. Estimation of the directional wave spectrum is more difficult than acquiring current data, requiring good definition of the second-order spectrum and hence a better signal-to-noise ratio than for currents. The inversion method is outlined in Wyatt (2000). There have been UK deployments at Holderness, Liverpool Bay (Howarth et al., 2007; Wolf et al., 2007; 2011; Figure 5-5) and the Celtic Sea (Wyatt, 2005). Siddons et al. (2009) explored the possibility of assimilating HF radar wave data into the SWAN model.

4.3   Multisensor Arrays

Since 2002, the WaveNet system of nearshore wave buoys has been deployed and maintained around the coast of England and Wales for the UK Environment Agency

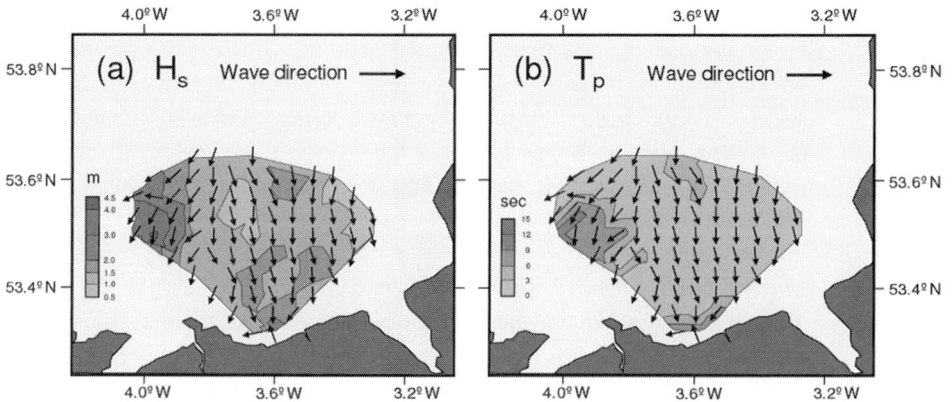

Figure 5-5 HF radar wave: (a) significant wave height and direction, (b) peak wave period and direction. Illustrations (a) and (b) are reproduced with minor modifications, including greyscale presentation, from Souza et al. (2011), with permission from Elsevier.

(EA) and the Department for Environment Food and Rural Affairs (DEFRA) by the Centre for Environment, Fisheries and Aquaculture Science (CEFAS). One of these wave buoys is in Liverpool Bay, adjacent to the National Oceanography Centre (NOC) Irish Sea Observatory (ISO) Site 'A' on a mooring near the Liverpool Bar Light in a water depth of 22 m (Figure 5-6 shows the layout of the ISO). The NOC Irish Sea Observatory ceased operations in 2012, and 10 years of its data have been archived. Access to wave data from this project is available online (accessed March 2016) from the British Oceanographic Data Centre (BODC, 2016).

As part of the ISO (Howarth et al., 2006), a coastal high-frequency radar system provided a grid of current and wave data over Liverpool Bay, and a wave buoy was deployed in the mouth of the Dee in the Hilbre Channel (Figure 5-6). Other sources of wave observations were high-frequency current and pressure data from in situ acoustic instruments, such as the Sontek ADV or wave-enabled ADCP measurements at Site 'A', between 2002 and 2012, and Site 'B', between 2005 and 2012 (Figure 5-6), and other remote sensing techniques, such as X-band radar and satellite altimetry. An example from ISO of wave measurements using high-frequency radar illustrates the extent of the spatial coverage (Figure 5-5). For a more extensive description and intercomparison of these wave data, see Wolf et al. (2011).

Various other examples of Coastal Observatories are available worldwide, e.g., FRF Duck in North Carolina (USACE), LEO-15 (Rutgers University, USA), Channel Coastal Observatory (UK), COSYNA in the German Bight and the 'Acqua Alta' tower off Venice (Cavaleri, 2000). Simultaneous measurements of waves, together with surface wind and other oceanographic measurements, is greatly desirable for interpretation of physical processes and coastal management.

4.4   Other Methods

There are many less well-used methods of observing waves. Camera-based systems include stereo photography (e.g., Holthuijsen, 1983) and the ARGUS video system, which images breakers in the surf zone (e.g., Fairley et al., 2007). The Shipborne Wave Recorder (SBWR) is nowadays a rather crude accelerometer-based system, installed in ships, which has been used for many years on Ocean Weather Ships, providing some of the first long-term wave climate observations, and is still operational on some unmanned light vessels.

Ocean waves incident on coasts generate seismic surface waves in three frequency bands via three pathways (Traer et al., 2012; Ardhuin, Balanche, et al., 2012): direct pressure on the seafloor (primary microseisms, PM), standing waves from interaction of incident and reflected waves (double-frequency microseisms, DF), and swell-transformed infragravity wave interactions (the Earth's seismic hum).

4.4.1   Ultrasound Tide/Wave Gauge and Harbour Wave-Gauge Installations

'Ultrasound' applies to frequencies that exceed the upper level of human hearing, typically 20 kHz. As an illustration, ultrasound is used for measuring tides and waves in the Port of Hamburg (Figure 5-7a). Another example of the deployment of

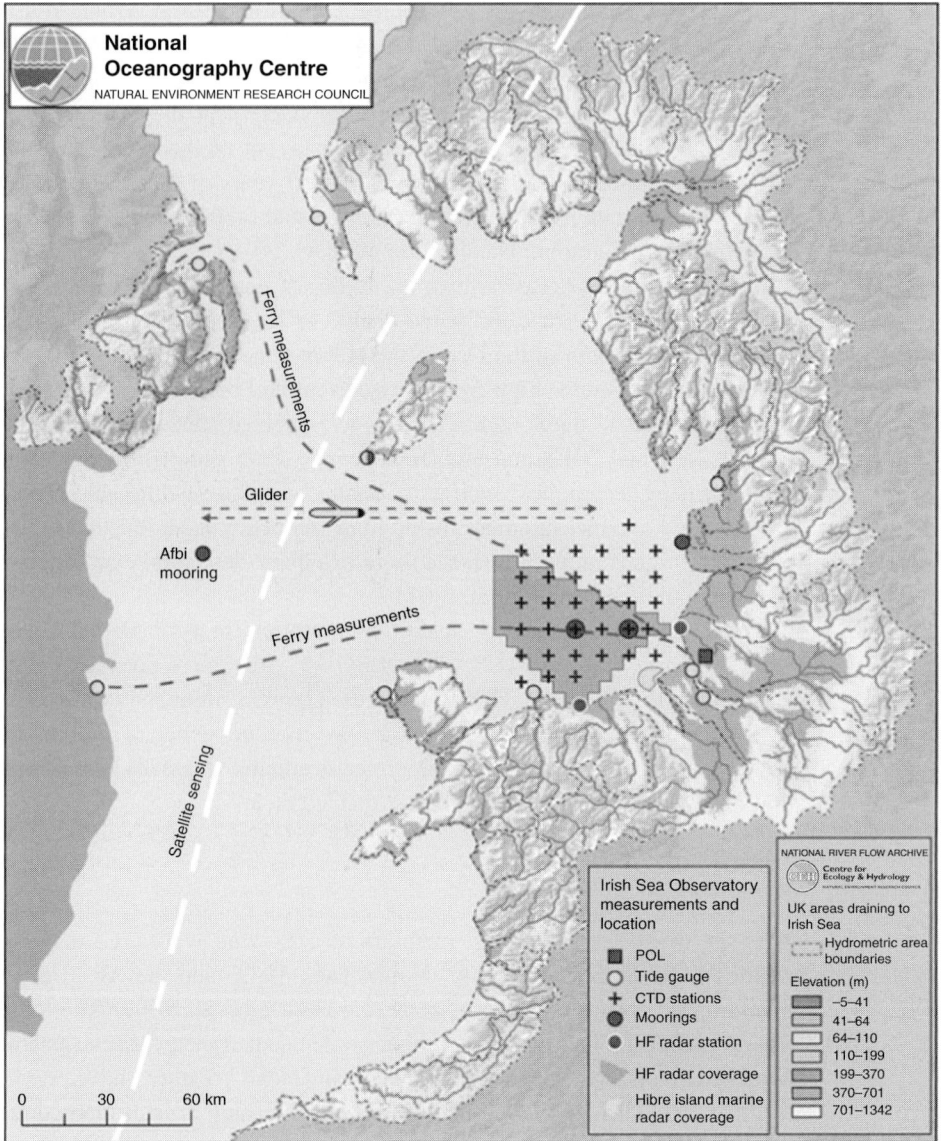

Figure 5-6 Liverpool Bay Coastal Observatory. Site 'A', referred to in the text, is the eastern-most filled circle (mooring station) on the ISO station grid, and Site 'B', referred to in the text, is the western-most filled circle (mooring station) on the ISO station grid (not the afbi mooring station or the northwest English coast coastal mooring station much farther north). Reproduced and modified for greyscale presentation from Souza et al. (2011), with permission from Elsevier.

instrumentation to measure water levels in a harbour is the MIROS radar station (Section 4.2.2.1) at Las Palmas, Gran Canaria, Canary Islands (Spain). Temporary installations of portable radar instruments are being used in smaller harbours to measure water levels, for example, near the entrance to Polperro Harbour, southwest England (Figure 5-7b), a Valeport Radar Level Sensor (VRS-20, Valeport, 2016) was

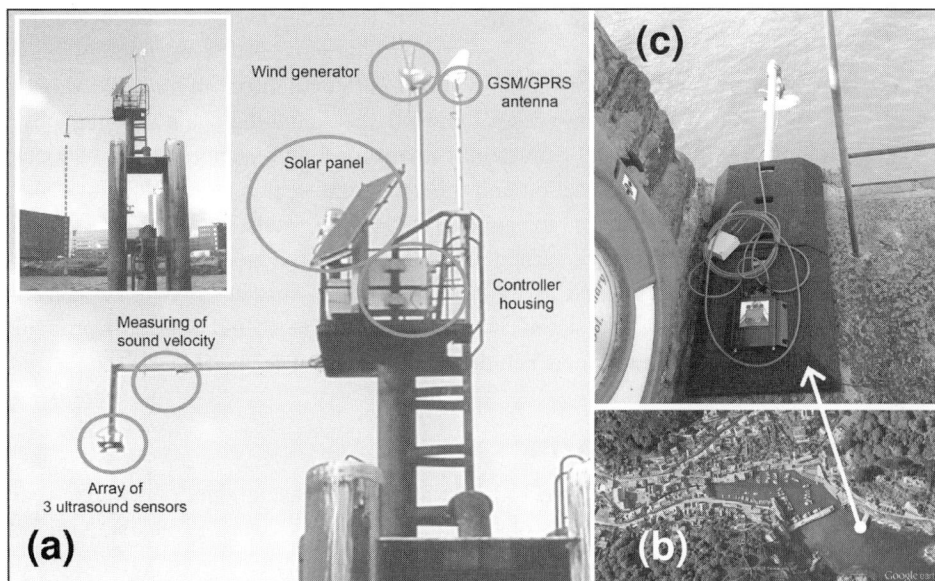

Figure 5-7 Water level and wave height measurements: (a) use of ultrasound in the Port of Hamburg, Germany, for measuring water levels and waves; (b, c) Valeport VS-20 Radar Level Sensor deployed near the entrance to Polperro Harbour, southwest England, to measure water levels (tides and long-period infragravity waves). Photographs: (a) © General Acoustics; reproduced with permission; (b) Google Earth image; (c) V. J. Abbott; with permission.

temporarily installed (Figure 5-7c) to record long-period infragravity wave variations (e.g., Uncles et al., 2014), with periodicities of minutes and longer in this deployment, and tidal variations in water levels.

4.4.2   Laser Altimeter

A laser altimeter (also known as Lidar, Chapter 3) works by emitting short flashes of laser light that travel to the surface being measured, where they are reflected. Part of the reflected laser radiation returns to the laser altimeter, where it is detected and the travel time measured. A laser altimeter can be operated from a plane, a helicopter, or a satellite. It has also been demonstrated from a nearshore platform (Irish et al., 2006). In December 1999, a nonintrusive directional Lidar wave gauge (LWG) was field tested at the U.S. Army Corps of Engineers (USACE) Field Research Facility (FRF) at Duck in North Carolina. The LWG measures water surface elevation by ranging obliquely to the surface (when the nadir angle is greater than 0°). The prototype system worked reasonably well for large waves.

5   Wave Analysis Methods

As has been shown in Sections 2 and 3, linear wave theory can be used in most cases for wave data analysis. Often the requirement for wave data is to estimate the wave climate (e.g., Woolf and Wolf, 2013). This may frequently be presented as a bivariate wave height-period histogram. In general, the wave height and period increase together up to a limiting steepness for wind-sea (where the wind is actively forcing the waves). There may be a bimodal distribution with a separate relationship for swell waves. In this case, the wave phase speed exceeds the wind-speed and there is no longer active forcing and little dissipation; swell waves may travel for long distances across the ocean but will be affected by bottom friction and other dissipative processes in shallow water.

There is often a need to estimate the 50-year return period wave, e.g., for the purpose of design of offshore structures and coastal defences. This must be obtained from extreme value theory, extrapolating from a shorter record. Further discussion on extreme waves, the statistics of individual waves and the joint probability of wave height and period is not included here (see Tucker and Pitt, 2001).

5.1   Time Series

One of the problems in recording waves is the vast amount of data that is acquired if the whole water-level record, observed at > 1 Hz, is saved. Typically, a sea state is defined from a record between 15 and 20-min long, and the wave height and period parameters are assumed constant over the record. Successive records may be at hourly or 3-hour intervals. Traditional methods of wave observations involved direct measurement of the water level and processing this time series to derive the significant wave height and period. In the early days, before fast computers, the time series were analysed by hand, identifying the mean zero-upcrossing period as the total record length divided by the number of zero-upcrossings. The significant wave height was determined by taking the first and second highest crests and the first and second lowest troughs in the record and applying the method of Cartwright (1958). Digital records and spectral analysis have made these methods obsolete.

5.2   Wave Spectra

The analysis procedure to convert the raw measurements into wave parameters presently employed by many buoys is based on Fourier analysis, in which the sea surface is described as a sum of sine and cosine waves with different amplitudes, frequencies and directions. Tucker and Pitt (2001) discuss the use of the Fast Fourier Transform (FFT), which revolutionised the processing of wave spectra. In the following, $E(f, \theta)$ is the wave energy spectrum at frequency f and direction θ.

The following wave parameters, derived from the wave spectrum, are most commonly used to describe the sea state:

Significant wave height, H_s:

$$H_s = 4\sqrt{m_0} \tag{14}$$

where the jth moment of the spectrum is defined as:

$$m_j = \iint_{f, \theta} f^j E(f, \theta) d\theta df \tag{15}$$

Mean wave period, T_{m02} (approximately equal to, and often referred to as, the zero-crossing wave period, T_z):

$$T_{m02} = \sqrt{\frac{m_0}{m_2}} \tag{16}$$

An alternative mean wave period, T_{m01}, is of interest because it is less dominated by the high-frequency part of the spectrum and is useful in overtopping studies, where the longer waves are more important:

$$T_{m01} = \frac{m_0}{m_1} \tag{17}$$

Mean wave direction:

$$D_m = \arctan \left[\frac{\iint \sin (\theta) E(f, \theta) d\theta df}{\iint \cos (\theta) E(f, \theta) d\theta df} \right] \tag{18}$$

The frequency bandwidth of the spectrum is

$$v = \left(\frac{m_0 m_2}{m_1^2} - 1 \right)^{1/2} \tag{19}$$

The theory of wave height statistics in a narrow-band sea (in the limit this is a sea state consisting of a single period wave) was derived by Longuet-Higgins (1952). It was shown that the distribution of wave heights in this case was a Rayleigh distribution. In shallow waters the spectral bandwidth tends to increase, but the Rayleigh distribution has proved resilient.

5.3 Example Analysis Procedure for S4DW Data

Here we show an example of the wave analysis carried out on S4DW data from the Holderness Experiment (1994–1996), as given in Wolf (1997). During this experiment, data from Waverider buoys were also available for validation. The analysis of bottom pressure data to obtain surface waves is critically dependent on the high frequency cut-off (see Section 5.3.1). The analysis procedure was as follows:

(i)   read the binary data file,
(ii)   calculate the east and north components of current and the pressure for each burst,
(iii)   calculate the mean depth and current,
(iv)   de-mean, de-trend and de-spike the time series,
(v)   apply a cosine taper to the first and last 10 per cent of the time series,
(vi)   perform an FFT on currents and pressures,
(vii)   correct for depth attenuation,
(viii) fit a high-frequency f^{-5} tail above the cutoff frequency, and
(ix)   calculate spectral moments and variances plus significant wave height, mean and peak period, peak direction and spread

Small-amplitude wave theory for finite depth (the Airy wave solution) is assumed as in Section 2. The neglect of nonlinear terms may not be justifiable when the waves become higher and steeper; however, in the present instance nonlinearities are likely to be small because the minimum depth considered is about 10 m. Integration of the spectra may be carried out up to infinity or some maximum frequency, such as 0.5 Hz. For the S4DW wave-current meter (where subscript 1 denotes heave, 2 denotes north current and 3 denotes east current, corrected for depth attenuation) the following cross-spectra are defined: $C_{11}(f)$, $C_{22}(f)$, $C_{33}(f)$, $C_{12}(f)$, and $C_{13}(f)$. Full definitions of these are given in Wolf (1996a), which are used to obtain the related angular harmonics of the directional wave spectrum, $E(f, \theta)$:

$$a_n = \frac{1}{\pi} \int_0^{2\pi} \cos(n\theta) \cdot E(f, \theta) d\theta \tag{20}$$

$$b_n = \frac{1}{\pi} \int_0^{2\pi} \sin(n\theta) \cdot E(f, \theta) d\theta \tag{21}$$

Thus:

$$a_1 = \frac{C_{12}}{\sqrt{C_{11}(C_{22} + C_{33})}} \tag{22}$$

$$b_1 = \frac{C_{13}}{\sqrt{C_{11}(C_{22} + C_{33})}} \tag{23}$$

from which the standard wave direction $D(f)$ and spread $s(f)$ at each frequency can be calculated:

$$D(f) = \arctan(b_1/a_1) \tag{24}$$

$$s(f) = \sqrt{(2 - 2r)} \tag{25}$$

$$\text{where } r = \sqrt{a_1^2 + b_1^2} \tag{26}$$

5.3.1   High-Frequency Tail-Fitting

The preceding analysis is critically dependent on the high frequency cutoff. Because the attenuation of pressure with depth increases rapidly at high frequencies, inverting this can enhance the noise level. One solution to this is to cut off the frequency spectrum above a critical level; another solution is to remove an estimated white noise level from the whole spectrum before inversion. There would seem to be two options for choosing this critical frequency: (i), select a fixed frequency cutoff (e.g., 0.3 Hz); or (ii), allow the cutoff to be determined by a threshold level being reached by the attenuation factor, K_p. The latter allows more of the observed spectra to be included in shallower waters where the high-frequency waves are less attenuated. Values of $K_p = 0.01$ and $K_p = 0.001$

were tested, and the latter was chosen to optimise the wave height and period values relative to the Waverider data. The f^{-5} tail fitted the observed data better than f^{-4} when compared with the Waverider spectra.

5.3.2 Effect of Currents

The data were analysed with and without a correction for the Doppler shift of apparent (observed) frequency. There was a noticeable tidal modulation of observed wave parameters. This affected both Waverider and bottom pressure recorders, but the latter were also affected by the attenuation correction. Inclusion of currents allowed more accurate calculation of wave-number and hence the attenuation. Clayson and Ewing (1988) discuss the correction of Waveriders for Doppler shift effects including frequency-varying response functions. The comparison was therefore between the spectra at apparent rather than intrinsic frequencies. Surface currents were estimated from the observed near-bed currents by applying a constant factor of 1.4. This was estimated using data from Prandle (1982) and was found to be close to the optimum correction. The test of method effectiveness was the amount of variance at tidal periods that was removed from the wave period time series, and also the reduction in the correlation between wave period and the tidal current component in the direction of the high-frequency (0.25 Hz) waves. The effect on wave height was quite small, but for wave period the current correction removed about 44 per cent of the variance between 11.9 h and 13 h periodicities (the main semidiurnal tide, centred on 0.08 cycles h^{-1}).

5.3.3 Nonlinear Analysis

The effect of correction for nonlinear terms, using the iterative method given in Lee and Wang (1984) and based on Sharma and Dean (1981), was implemented but found to give a negligible correction in this water depth. It should be noted that some errors were found in formulae taken from Sharma and Dean (1981); in depths of 5 m or less the corrections could become significant.

5.3.4 Quality Control

Tucker (1993) suggests the use of low-frequency and high-frequency spectral checks that can be carried out. Excess low-frequency energy can be due to mooring interference, while high-frequency energy occurs at both low and high wave height, indicating increased noise. The high-frequency threshold for flagging the data seemed rather low, with about 25 per cent of Waverider data triggering this value, and therefore was reset at twice the original value, at which the number of flagged Waverider values became less than 2 per cent.

The occasions for which data were flagged were examined in detail for the period of the Holderness Experiment. For the S4DW, 3.8 per cent of data points were flagged for the low-frequency check. These were mostly for H_s less than 1 m. A number of points were flagged at the high-frequency check (1.4 per cent), mostly for H_s greater than 1 m. It seemed justifiable to ignore the data points that were flagged by the low-frequency check. These were usually related to rather low H_s or recognisably bad data. The high-frequency flags were somewhat more ambiguous because higher energy could be due to real growth-related conditions, and the data points were not removed.

A parameter called the check ratio was computed for the directional spectra. The check ratio is a measure of the ratio between horizontal and vertical displacements. Full definitions for the Waverider and S4DW are given in Wolf (1996b). This ratio would be unity for a perfect wave measuring device in the absence of currents. In practice, the check ratio deviates from unity. In the case of the Waverider, the deviation is known to include the effects of finite water depth and currents, as well as the response of the buoy, which falls off at high frequencies, and possible mooring effects (most likely at low frequencies). The interpretation of this parameter can be difficult because of the combination of physical and instrumental effects causing deviation. The peak check ratio is usually close to unity, however, and can be a useful guide to data quality. For example, the peak check ratio became large for the S4DW data for the latter part of the second deployment (at which time unrealistic values for other parameters were also noted) and at the beginning and end of the third deployment, where there was not so obviously erroneous data. A value of about 1.5 for the S4DW data seemed a useful threshold value. The Waverider data was more variable because there was the known effect of $\tanh(kh)$ to be corrected for.

The S4DW and Waverider were in good agreement for wave height. The peak wave period was in good agreement for all instruments most of the time. The mean wave period, calculated by an integration of the spectrum weighted by frequency squared, was more sensitive to differences in the high-frequency end of the spectrum. The S4DW shares the limitations of any bottom-mounted pressure recorder in that a high-frequency peak will not always be detected. The mean period parameter T_{m01} is perhaps more useful because it is not so biased toward the high-frequency energy, but most wave data analysis still quotes T_z as standard. The comparison with T_z was, as expected, a general overestimate. For the peak period, T_p (which is not an integrated parameter) there was a cluster of points showing an overestimate of T_p when the Waverider detected a high-frequency peak, mostly when the recorded Waverider peak periodicity was less than about 6 s, which usually corresponds to low wave energy conditions. The agreement in mean wave direction of the spectral peak between both instruments generally was very good.

6   Modelling Waves in the Nearshore Zone

Wave modelling and forecasting have been carried out since the predictions for the D-Day landings in World War II and have been carried out operationally as part of the MetOcean system for almost as long as meteorological forecasts. The accuracy of the wave models is closely tied to the wind forecasts. Initially, the application was for the deep ocean, but increasingly regional and coastal seas have been included. In these areas the wind is fetch-limited, and the details of the coastal boundary and near-shore bathymetry become important. The interactions of the waves with the coastline, in terms of the swash zone, sediment transport and beach processes, although important, are not discussed here. Resolving individual waves (as in the Boussinesq and Mild Slope equation models) is only done for specific cases, such as wave-structure interaction or harbour oscillation problems, but is seldom used for extended coastal areas. Other wave models are based on the spectral (phase-averaged) approach. Because it is the action density that is conserved in the presence of currents, these spectral wave models are based on the wave action-density balance equation:

$$\frac{\partial}{\partial t}N + \frac{\partial}{\partial x}c_x N + \frac{\partial}{\partial y}c_y N + \frac{\partial}{\partial \sigma}c_\sigma N + \frac{\partial}{\partial \theta}c_\theta N = \frac{S}{\sigma} \tag{27}$$

where $N(\sigma, \theta) = E(\sigma, \theta)/\sigma$ is the spectral action density of each wave spectral component, x and y are the independent spatial (Cartesian) coordinates (note the equations can also be expressed in spherical polar, i.e., latitude and longitude, coordinates); c_x, c_y, c_σ and c_θ are the wave propagation velocities in x, y, σ and θ space, respectively.

The first term on the left-hand side of Eq. 27 represents the local rate of change of action density with time; the second and third terms represent propagation of action in geographical space. The fourth term represents shifting of the relative frequency due to variations in depths and currents, and the fifth term represents depth-induced and current-induced refraction, which change the wave propagation direction. The expressions for these propagation speeds are taken from linear wave theory (Section 2; Dingemans, 1997). On the right-hand side, $S(\sigma, \theta)$ includes all the sources and sinks of energy. The assumption is that the properties of the medium (water depth and current), as well as the wave field itself, vary on time and space scales that are much larger than those of an individual wave. Specification of the source terms (wave growth due to the wind, wave–wave interactions and wave dissipation by deep-water breaking/white-capping, depth-limited breaking and bottom friction) is an area of active research.

The present third-generation (3G) spectral wave models became possible due to the development of the discrete interaction approximation (DIA, Hasselmann and Hasselmann, 1985), which allows a full evolution of the spectrum, including wave–wave interactions, rather than using a prescribed spectral shape, as was done in previous generations of wave models. The global, deep water, WAM model was implemented operationally at the European Centre for Medium-Range Weather Forecasts (ECMWF) in the early 1990s (Komen et al., 1994).

Accurate wave modelling in coastal areas is more complex than in the open ocean due to the number of interacting processes. Coastal wave modelling demands high spatial resolution due to the complexity of the bathymetry and coastline and thus often also requires a high temporal resolution. Land/sea effects in the atmospheric boundary layer and nearshore zone can change rapidly. The effect of wave growth in limited fetch introduces further complexity (e.g., Alves and Banner, 2003) because the effective fetch can be dominated by the alongshore wind component when winds are blowing offshore at an angle to the coast. In certain areas, wind seas develop in opposing swell conditions.

The introduction of the SWAN model (Booij et al., 1999) started a new era for the application of spectral wave modelling in the coastal zone. Many SWAN model applications still make use of the concept of stationarity (no variation in time, reaching a steady state), which is a concept that is valid for restricted coastal areas only, but the model can also be run in a nonstationary mode. SWAN is an open source code that can easily be implemented (SWAN, 2016a) and is, without doubt, the spectral wave model code that is most widely used by both the scientific and the engineering communities. SWAN is a 3G wave model using the same formulations for the source terms as WAM, plus some additional formulations primarily for shallow water; however, the numerical solution techniques are very different.

The SWAN code can model many of the processes affecting wind-generated waves that are known today: wave generation by wind; wave propagation; current and depth refraction; frequency shifting due to currents and nonstationary water depth; wave dissipation by white-capping; bottom friction; vegetation; mud; turbulence and depth-induced breaking; nonlinear interactions (both four- and three-wave interactions); wave-induced setup; and transmission and reflection by obstacles and diffraction. For many of the processes one can choose between different formulations (e.g., the different combinations for wind input and white-capping/deep water dissipation terms). Table 5-1 shows the source term options available in SWAN version 41.10 (SWAN, 2016b). These source terms are widely used in other models, although, as previously mentioned, new ones have been added recently.

Table 5-1 An overview is shown of the physical processes and generation mode in SWAN (from SWAN User Manual Cycle III version 41.10, courtesy of Delft University of Technology; reproduced with permission).

Process	Authors	Generation mode		
		1st	2nd	3rd
Linear wind growth	Cavaleri and Malanotte-Rizzoli (1981) (modified)	X	X	
	Cavaleri and Malanotte-Rizzoli (1981)			X
Exponential wind growth	Snyder et al. (1981) (modified)	X	X	
	Snyder et al. (1981)			X
	Janssen (1989, 1991)			X
	Yan (1987) (modified)			X
Whitecapping	Holthuijsen and De Boer (1988)	X	X	
	Komen et al. (1984)			X
	Janssen (1991)			X
	Alves and Banner (2003)			X
Quadruplets	Hasselmann et al. (1985)			X
Triads	Eldeberky (1996)	X	X	X
Depth-induced breaking	Battjes and Janssen (1978)	X	X	X
Bottom friction	JONSWAP Hasselmann et al. (1973)	X	X	X
	Collins (1972)	X	X	X
	Madsen et al. (1988)	X	X	X
Obstacle transmission	Seelig (1979), d'Angremond et al. (1996)	X	X	X
Wave-induced setup		X	X	X
Vegetation dissipation	Dalrymple et al. (1984)	X	X	X
Mud dissipation	Ng (2000)	X	X	X
Turbulence dissipation		X	X	X

The WISE group (Cavaleri et al., 2007) summarised the (then) state of the art in wave modelling and covered developments since the inception of the 3G wave models. The latter have been very successful, especially on oceanic scales, but at that time had various limitations, especially in shallow waters. Various efforts since 2007 have addressed these areas, notably the work of Babanin and others on measuring and modelling the white-capping dissipation term (van der Westhuysen et al., 2007; Babanin et al., 2010a; 2010b; 2011) and development of new semi-empirical source terms (Ardhuin et al., 2010). Investigations of wave growth in fetch- and depth-limited conditions have been carried out by Bottema and van Vledder (2008, 2009) and van der Westhuysen (2010). A further development of the WAM model allowed more efficient application in shallow water coastal regions with smaller spatial and time scales (Monbaliu et al, 2000).

Coastal modelling generally requires high spatial resolution, and this may utilise one-way or two-way nesting. The use of unstructured grids, with either finite element or finite volume numerical methods, is particularly useful for coastal applications and is now available in SWAN (Zijlema, 2010) and WAVEWATCH III™ (Tolman, 2009). These allow a seamless grid refinement rather than nesting of models. As well as the open scientific community codes, there are also commercial and semicommercial developments. Well-known ones are the TOMAWAC model (Benoit et al., 1996; Benoit, 2005; part of the TELEMAC suite – note that the TELEMAC software was recently released as public software) and the MIKE 21 spectral wave model (Sørensen et al., 2004) as part of the MIKE suite of models, respectively, and the STWAVE module as part of the CEDAS-system (e.g., VE, 2016). It is noted that for some purposes parametric wave models may be adequate and very fast (e.g., Dearing et al., 2006).

To support high-resolution coastal modelling, there is a need for good nearshore bathymetry; this is becoming more readily available, with coastal lidar surveys filling in the gaps that previously resulted from intertidal areas. An example of output from the SWAN wave model, applied in stationary mode to Liverpool Bay at 180-m resolution, shows that wave heights reach about 5-m offshore with a steady NW wind of 15 ms^{-1} (Figure 5-8). Further developments are continuing, such as the full inclusion of the effects of wave setup and longshore drift due to radiation stress, which are mainly confined to the very nearshore zone (Brown et al., 2011; 2013), as well as the effects of nonuniform, unsteady 3-D currents and vertical shear (Bolaños et al., 2011). The effect of currents on high-frequency waves may cause wave blocking, producing wave reflection or breaking (e.g., Chawla and Kirby 2002), which may be significant in tidal inlets and areas such as the Bristol Channel, UK, where tidal currents can exceed 2 ms^{-1} (e.g., Uncles, 2010).

7   Recent Developments and Final Remarks

Since this chapter was first written there have been further advances in the observation and modelling of coastal and estuarine environments, a few of which are

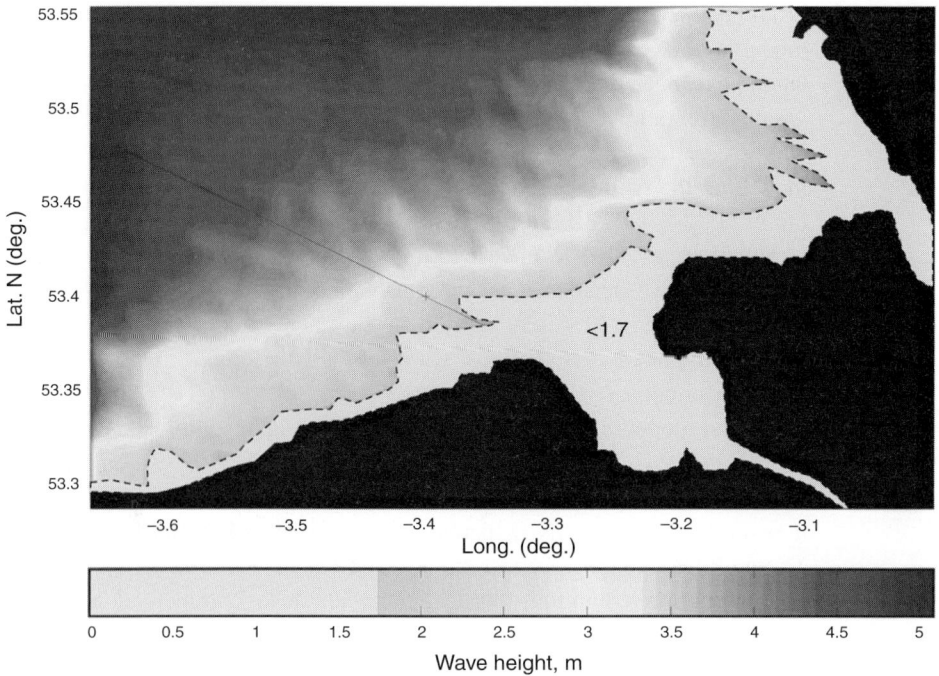

Figure 5-8 Wave height in Liverpool Bay derived from a 180-m resolution SWAN model forced by a steady 15 ms⁻¹ wind from the NW.

mentioned here. Coastal altimetry has made further advances and will become increasingly useful; e.g., Passaro et al. (2015) describe successful wave validation for the German Bight.

With respect to coastal and estuarine wave modelling, further work has been done on waves in the Dee Estuary as part of a coupled model system (Brown et al., 2014). Waves within the estuary's channels are able to reach 2.2 m. A study of the devastating storm events of 2013–2014 focussed on two coastal regions: the Sefton coast in Liverpool Bay (the Bay's west-facing coastline, Figure 5-6) and Sizewell on the Suffolk coast of the southern North Sea (Wadey et al., 2015) and identifies the relative importance of waves in causing coastal damage. A paper focussing on Sizewell (Prime et al., 2016) shows how different combinations of waves and water levels can lead to the same joint probability event but with different outcomes for flooding and coastal erosion. The greatest flood hazard to a location is due to low swell waves with the longest periods during extreme water levels, rather than large short-period wind waves occurring at lower water levels.

Finally, the subject of wind-waves is a large one, and one that is important both in coastal engineering and coastal management. This chapter has aimed to provide an introduction to its key concepts and methods. However, many topics have not been covered here that may be of interest, and further reading and various textbooks can be recommended, such as Phillips (1977), Dingemans (1997), Tucker and Pitt (2001) and Holthuijsen (2007).

References

Alves, J. H. G. M., Banner, M. L., 2003. Performance of a saturated based dissipation-rate source term in modeling the fetch-limited evolution of wind waves. *Journal of Physical Oceanography* 33, 1274–1298.

d'Angremond, K., van der Meer, J. W., de Jong, R. J., 1996. Wave transmission at low-crested structures, Proc. *25th Int. Conf. Coastal Engineering*, Orlando: ASCE, 2418–2427.

Arcilla, A. S., Lemos, C. M., 1990. *Surf-Zone Hydrodynamics*. Barcelona: Centro Internacional de Métodos Numéricos en Ingeneria. 310pp.

Ardhuin, F., Rogers, E., Babanin, A., Filipot, J.-F., Magne, R., Roland, A., van der Westhuysen, A., Queffeulou, P., Lefevre, J.-M., Aouf, L., Collard, F., 2010. Semi-empirical dissipation source functions for ocean waves. Part I: Definitions, calibration and validations. *Journal of Physical Oceanography* 40, 1917–1941.

Ardhuin, F., Roland, A., Dumas, F., Bennis, A.-C., Sentchev, A., Forget, P., Wolf, J., Girard, F., Osuna, P., Benoit, M., 2012. Numerical wave modeling in conditions with strong currents: Dissipation, refraction and relative wind. *Journal of Physical Oceanography* 42, 2101–2120.

Ardhuin, F., Balanche, A., Stutzmann, E., Obrebski, M., 2012. From seismic noise to ocean wave parameters: General methods and validation, *Journal of Geophysical Research* 117, C05002, doi:10.1029/2011JC007449.

Babanin, A. V., Chalikov, D., Young, I. R., Savelyev, I., 2010a. Numerical and laboratory investigation of breaking of steep two-dimensional waves in deep water. *Journal of Fluid Mechanics* 644, 433–463.

Babanin, A. V., Tsagareli, K. N., Young, I. R., Walker, D. J., 2010b. Numerical investigation of spectral evolution of wind waves. Part 2. Dissipation function and evolution tests. *Journal of Physical Oceanography* 40, 667–683.

Babanin, A. V., Waseda, T., Kinoshita, T., Toffoli, A., 2011. Wave breaking in directional fields. *Journal of Physical Oceanography* 41, 145–156.

Bacon, S., Carter, D. J. T., 1991. Wave climate changes in the North Atlantic and North Sea. *International Journal of Climatology* 11, 545–558.

Barstow, S., et al., 2003. WORLDWAVES – High quality coastal and offshore wave data within minutes for any global site, Proc. 2003 Int. Conference on Offshore Mechanics and Arctic Engineering (OMAE 2003), paper 37297, Cancun, Mexico, 2003.

Battjes, J. A., Janssen, J. P. F. M., 1978. Energy loss and set-up due to breaking of random waves. *Proc. 16th Int. Conf. Coastal Engineering*, ASCE, 569–587.

Battjes, J. A., 1988. Surf-zone Dynamics. *Annual Review of Fluid Mechanics* 20, 257–293.

Benoit, M., Marcos, F., Becq, F., 1996. Development of a third generation shallow-water wave model with unstructured spatial meshing. *Proceeding 25th International Conference on Coastal Engineering*, ASCE, Orlando-USA, 465–478.

Benoit, M., 2005. TOMAWAC software for finite element sea state modelling, release 5.5-User Manual, EDF-LNHE, Chatou, France (distributed by HR Wallingford, UK).

Bishop, C. T., Donelan, M. A., 1987. Measuring waves with pressure transducers. *Coastal Engineering* 11, 309–328.

BODC, 2016. Wave data from the ISO project is available online [accessed March 2016] at: www.bodc.ac.uk/data/information_and_inventories/edmed/report/224/.

Bolaños, R., Osuna, P., Wolf, J., Monbaliu, J., Sanchez-Arcilla, A., 2011. Development of POLCOMS-WAM model. *Ocean Modelling* 36, 102–115.

Bolaños, R., Thorne, P. D., Wolf, J., 2012. Comparison of measurements and models of bed stress, bedforms and suspended sediments under combined currents and waves. *Coastal Engineering* 62, 19–30.

Booij, N., Ris, R. C., Holthuijsen, L. H., 1999. A third-generation wave model for coastal regions, Part I. Model description and validation. *Journal of Geophysical Research* 104(C4), 7649–7666. doi:10.1029/98JC02622

Borge, J.-C. N., Reichert, K., Dittmer, J., 1999. Use of marine radar as a wave monitoring instrument. *Coastal Engineering* 37, 331–342.

Bottema, M., van Vledder, G. P., 2008. Effective fetch and nonlinear four-wave interactions during wave growth in slanting fetch conditions. *Coastal Engineering* 55, 261–275.

Bottema, M., van Vledder, G. P., 2009. A ten-year data set for fetch- and depth-limited wave growth. *Coastal Engineering* 56, 703–725.

Bouws, E., Gunther, H., Rosenthal, W., Vincent, C. L., 1985. Similarity of the wind wave spectrum in finite depth water 1. Spectral form. *Journal of Geophysical Research* 90, C1, 975–986.

Brown, J., Wolf, J., 2009. Coupled wave and surge modelling for the eastern Irish Sea and implications for model wind-stress. *Continental Shelf Research* 29, 1329–1342.

Brown, J. M., Bolaños, R., Wolf, J., 2011. Impact assessment of advanced coupling features in a tide-surge-wave model, POLCOMS-WAM, in a shallow water application. *Journal of Marine Systems* 87, 13–24. doi:10.1016/j.jmarsys.2011.02.006.

Brown, J. M., Bolaños, R., Wolf, J., 2013. The depth-varying response of coastal circulation and water levels to 2D radiation stress when applied in a coupled wave–tide–surge modelling system during an extreme storm. *Coastal Engineering* 82, 102–113.

Brown, J. M., Bolaños, R., Souza, A. J., 2014. Process contribution to the time-varying residual circulation in tidally dominated estuarine environments. *Estuaries and Coasts* 37, 1041–1057.

Cartwright, D. E., 1958. On estimating the mean energy of sea waves from the highest wave in the record. *Proceedings of the Royal Society A* 247, 22–48.

Cavaleri, L., Malanotte-Rizzoli, P., 1981. Wind wave prediction in shallow water: Theory and applications. *Journal of Geophysical Research* 86, No. C11, 10961–10973.

Cavaleri, L., 2000. The oceanographic tower Acqua Alta – activity and prediction of sea states at Venice. *Coastal Engineering* 39, 29–70.

Cavaleri, L., Alves, J.-H., Ardhuin, F., Babanin, A., Banner, M., Belibassakis, K., Benoit, M., Donelan, M., Groeneweg, J., Herbers, T. H. C., Hwang, P., Janssen, P. A. E. M., Janssen, T., Lavrenov, I. V., Magne, R., Monbaliu, J., Onorato, M., Polnikov, V., Resio, D., Rogers, W. E., Sheremet, A., McKee Smith, J., Tolman, H. L., van Vledder, G., Wolf, J., Young, I. (i.e. the WISE group), 2007. Wave modelling – the state of the art. *Progress in Oceanography* 75, 603–674.

Chawla, A., Kirby, J. T., 2002. Monochromatic and random wave breaking at blocking points. *Journal of Geophysical Research* 107, C7. doi:10.1029/2001JC001042.

Chini, N., Stansby, P., Leake, J., Wolf, J., Roberts-Jones, J., Lowe, J., 2010. The impact of sea level rise and climate change on extreme inshore wave climate: A case study for East Anglia (UK). *Coastal Engineering* 57, 973–984.

Clayson, C. H., Ewing, J. A., 1988. Directional wave data recorded in the southern North Sea. IOSDL Report no. 258.

Collins, J. I., 1972. Prediction of shallow water spectra, *Journal of Geophysical Research* 77 (15), 2693–2707.

COST, 2005. *Measuring and Analysing the Directional Spectra of Ocean Waves*, (eds.) D. Hauser, K. Kahma, H. E. Krogstad, S. Lehner, J. A. J. Monbaliu and L. R. Wyatt, Luxembourg: Office for Official Publications of the European Communities.

Dalrymple, R. A., Kirby, J. T., Hwang, P. A., 1984. Wave diffraction due to areas of energy dissipation. *Journal of Waterways, Ports, Harbours and Coastal Engineering* 110, 67–79.

Datawell, 2009. Datawell Waverider Reference Manual: WR-SG, DWR-MkIII, DWR-G. [online]. Available at: https://cdip.ucsd.edu/documents/index/gauge_docs/mk3.pdf [Accessed March 2016].

Dean, R. G., 1965. Stream function representation of nonlinear ocean waves. *Journal of Geophysical Research* 70, C18, 4561–4572.

Dearing, J. A., Richmond, N., Plater, A. J., Wolf, J., Prandle, D., Coulthard, T. J., 2006. Modelling approaches for coastal simulation based on cellular automata: The need and potential. *Philosophical Transactions of the Royal Society A* 364, 1051–1071. doi:10.1098/rsta .2006.1753.

Dingemans, M. W., 1997. Water wave propagation over uneven bottoms, Part I – Linear wave propagation. *Singapore, World Scientific* 13, 1–471.

Eldeberky, Y., 1996. Nonlinear transformation of wave spectra in the nearshore zone, PhD thesis, Delft University of Technology, Department of Civil Engineering, The Netherlands.

Elfrink, B., Baldock, T., 2002. Hydrodynamics and sediment transport in the swash zone: A review and perspectives. *Coastal Engineering* 45, 149–167.

Fairley, I., Davidson, M., Kingston, K., 2007. Video monitoring of overtopping of detached breakwaters in a mesotidal environment. *Coastal Structures '07*, Vol. 2. Hackensack, NJ: World Scientific.

Gommenginger, C., Thibaut, P., Fenoglio-Marc, L., Quartly, G., Deng, X., Gómez-Enri, J., Challenor, P., Gao, Y., 2011. Retracking altimeter waveforms near the coasts – A review of retracking methods and some applications to coastal waveforms. In: *Coastal Altimetry* S. Vignudelli, A. Kostianoy. P. Cipollini, J. Benveniste (eds.), New York: Springer.

Gower, J. F. K., 1981. Oceanography from space. *Marine Science* 13, New York: Plenum Press.

Graber, H. C., Terray, E. A., Donelan, M. A., Drennan, W. M., Van Leer, J. C., Peters, D. B., 2000. ASIS—A new air–sea interaction spar buoy: Design and performance at sea. *Journal of Atmospheric and Oceanic Technology* 17, 708–720.

Grant, W. D., Madsen, O. S., 1979. Combined wave and current interaction with a rough bottom. *Journal of Geophysical Research* 84(C4), 1797–1808. doi:10.1029/JC084iC04p01797.

Gulev, S. K., Grigorieva, V., 2004. Last century changes in ocean wind wave height from global visual wave data. *Geophysical Research Letters* 31, L24302.

Hasselmann, K., Barnett, T. P., Bouws, E., Carlson, H., Cartwright, D. E., Enke, K., Ewing, J. A., Gienapp, H., Hasselmann, D. E., Kruseman, P., Meerburg, A., Müller, P., Olbers, D. J., Richter, K., Sell, W., Walden, H., 1973. Measurements of wind-wave growth and swell decay during the Joint North Sea Wave Project (JONSWAP), *Dtsch. Hydrogr. Z. Suppl.* 12, A8.

Hasselmann, S., Hasselmann, K., 1985. Computations and parametrizations of the nonlinear energy transfer in a gravity-wave spectrum. part 1: A new method for efficient computations of the exact nonlinear transfer integral. *Journal of Physical Oceanography* 15, 1369–1377.

Hawkes, P. J., Atkins, R., Brampton, A. H., Fortune, D., Garbett, R., Gouldby, B. P., 2001. WAVENET: Nearshore Wave Recording Network for England and Wales: Feasibility Study, *HR Wallingford Report TR* 122.

Hawkes, P. J., Gouldby, B. P., Tawn, J. A., Owen, M. W., 2002. The joint probability of waves and water levels in coastal defence design. *Journal of Hydraulic Research* 40, 241–251.

Heathershaw, A. D., Blackley, M. W. L., Hardcastle, P. J., 1980. Wave direction estimates in coastal waters using radar. *Coastal Engineering* 3, 249–267.

Hedges, T. S., 1995. Regions of validity of analytical wave theories. *Proceedings of the Institution of Civil Engineers – Water Maritime and Energy Journal* 112, 111–114.

Hedges, T. S., Reis, M. T., 1998. Random wave overtopping of simple seawalls: A new regression model. *Proceedings of the Institution of Civil Engineers – Water, Maritime and Energy Journal* 130, 1–10.

Holthuijsen, L. H., 1983. Stereophotography of ocean waves. *Applied Ocean Research* 5, 204–209.

Holthuijsen, L. H., 2007. *Waves in Oceanic and Coastal Waters*. Cambridge: Cambridge University Press.

Holthuijsen, L. H., De Boer, S., 1988. Wave forecasting for moving and stationary targets. In: Schrefler, B. Y., Zienkiewicz, O. C. (eds.), *Computer Modelling in Ocean Engineering*. Rotterdam, The Netherlands: Balkema, 231–234.

Howarth, M. J., Proctor, R., Knight, P. J., Smithson, M. J., Mills, D. K., 2006. The Liverpool Bay Coastal Observatory – towards the goals. *Proceedings of Oceans 2006, 18–21 September 2006*, Boston: IEEE.

Howarth, M. J., Player, R. J., Wolf, J., Siddons, L. A., 2007. HF radar measurements in Liverpool Bay, Irish Sea. 6pp. In: Oceans '07 IEEE Aberdeen, conference proceedings. *Marine Challenges: Coastline to Deep Sea*. Aberdeen, Scotland: IEEE.

Irish, J. L., Wozencraft, J. M., Cunningham, A. G., Giroud, C., 2006. Nonintrusive measurement of ocean waves: Lidar wave gauge. *Journal of Atmospheric and Oceanic Technology* 23, 1559–1572.

James, I. D., 1986. A note on the theoretical comparison between wave staffs and Waverider buoys in steep gravity waves. *Ocean Engineering* 13, 209–214.

Janssen, P. A. E. M., 1989. Wave-induced stress and the drag of air flow over sea waves. *Journal of Physical Oceanography* 19, 745–754.

Janssen, P. A. E. M., 1991. Quasi-linear theory of wind-wave generation applied to wave forecasting. *Journal of Physical Oceanography* 21, 1631–1642.

Komen, G. J., Cavaleri, L., Donelan, M., Hasselmann, K., Hasselmann, S., Janssen, P. A. E. M., 1994. *Dynamics and Modelling of Ocean Waves*. Cambridge: Cambridge University Press.

Korteweg, D. J., de Vries, G., 1895. On the change of form of long waves advancing in a rectangular canal, and a new type of long stationary waves. *Philosophical Magazine, Series 5* 39, 422–443.

Krogstad, H. E., 1991. Reliability and resolution of directional wave spectra from heave, pitch and roll data buoys. In: *Directional Wave Spectra* R. C. Beal (ed.), Baltimore, MD: Johns Hopkins University Press, 66–71.

Krogstad, H. E., Wolf, J., Thompson, S. P., Wyatt, L., 1999. Methods for the intercomparison of wave measurements. *Coastal Engineering* 37, 235–257.

Lambrechts, J., Humphrey, C., McKinna, L., Gourge, O., Fabricius, K. E., Mehta, A. J., Lewis, S., Wolanski, E., 2010. The importance of wave-induced bed fluidisation in the fine sediment budget of Cleveland Bay, Great Barrier Reef. *Estuarine, Coastal and Shelf Science* 89, 154–162.

Lane, A., Hu, K., Hedges, T. S., Reis, M. T., 2008. New north east of England tidal flood forecasting system. *Proceedings of FLOODrisk 2008*, London: Taylor and Francis.

Leake, J., Wolf, J., Lowe, J., Hall, J., Nicholls, R., 2009. Response of marine climate to future climate change: Application to coastal regions. *Proceedings of ICCE 2008*. Hamburg, August 31–September 5, 2008.

Lee, D.-Y., Wang, H., 1984. Measurement of waves from subsurface gage. In: *Proceedings of the 19th Coastal Engineering Conference*, Sept. 3–7 1984, Houston, Texas (Ed. B. L. Edge), 271–286.

Li, M., Raymond, I., Wolf, J., Chen, X., Burrows, R., 2011. Numerical investigation of wave propagation in the Liverpool Bay, NW England. *Acta Oceanologica Sinica* 30, 1–13.

Longuet-Higgins, M. S., 1952. On the statistical distribution of the heights of sea waves. *Journal of Marine Research* 11, 245–266.

Longuet-Higgins, M. S., Stewart, R. W., 1962. Radiation stress and mass transport in gravity waves, with applications to "surf beats". *Journal of Fluid Mechanics* 13, 481–504. doi:10.1017/S0022112062000877.

Macklin, T., Wolf, J., Wakelin, S., Gommenginger, C., Ferrier, G., Elliott, A., Neill, S., 2006. The benefits of combining coupled wave-current models with SAR observations for the interpretation of ocean-surface currents. In: *Proceedings of SEASAR 2006: Advances in SAR oceanography from Envisat and ERS Missions SeaSAR* (H. Lacoste, ed.). ESA SP-613, April 2006, Frascati, Italy.

Madsen, O. S., Poon, Y.-K., Graber, H. C., 1988. Spectral wave attenuation by bottom friction: theory, In: *Proc. 21st Int. Conf. Coastal Engineering (Malaga)*. New York: ASCE, 492–504.

Marshall, D. E., Bishop, J. M., 1984. *Practical Guide to Ocean Wave Measurement and Analysis*. Marion, MA: Endeco.

Martín Míguez, B., Pérez Gómez, B., Alvarez Fanjul, E., 2005. The ESEAS-RI sea level test station: Reliability and accuracy of different tide gauges. *International Hydrographic Review* 6, 44–53.

Monbaliu, J., Padilla-Hernandez, R., Hargreaves, J. C., Carretero Albiach, J. C., Luo, W., Sclavo, M., Günther, H., 2000. The spectral wave model WAM adapted for applications with high spatial resolution. *Coastal Engineering* 41, 41–62. doi:10.1016/S0378-839(00)00026-0.

Ng, C.-O., 2000. Water waves over a muddy bed: A two-layer Stokes' boundary layer model. *Coastal Engineering* 40, 221–242.

Nielsen, P., 2009. *Coastal and Estuarine Processes*. Hackensack, NJ: World Scientific Publishing.

Osuna, P., Wolf, J., 2005. A numerical study on the effect of wave-current interaction processes in the hydrodynamics of the Irish Sea. In: *Proceedings of the 5th International Conference on Ocean Wave Measurement and Analysis: WAVES2005, 3–7 July, 2005*, Madrid, Spain.

Ozer, J., Padilla Hernandez, R., Monbaliu, J., Alvarez Fanjul, E., Carretero Albiach, J. C., Osuna, P., Yu, J. C. S., Wolf, J., 2000. A coupling module for tides, surges and waves. *Coastal Engineering* 41, 95–124.

Pan, S., O'Connor, B., Vincent, C., Reeve, D., Wolf, J., Davidson, M., Dolphin, A., Thorne, P., Bell, P., Souza, A., Chesher, T., Johnson, H., Leadbetter, A., 2010. Larger-scale morphodynamic impacts of segmented shore-parallel breakwaters on coasts and beaches: An overview of the LEACOAST2 Project. *Shore and Beach (Journal of ASBPA)* 78/79, 35–43.

Passaro, M., Fenoglio-Marc, L., Cipollini, P., 2015. Validation of significant wave height from improved satellite altimetry in the German Bight. *IEEE Transactions on Geoscience and Remote Sensing* 53, 2146–2156.

Peregrine, D. H., Jonsson, I. G., 1983. *Interaction of Waves and Currents*. US Army Corps of Engineers Miscellaneous Reports, MR83-6.

Pérez, B., 2016. Personal Communication. Puertos del Estado, Avda. del Partenón 10, Campo de las Naciones, 28042 Madrid, Spain. Tel: +34 91 5245500; Fax:+34 91 5245504; Website: www.puertos.es [accessed April 2016].

Pérez, B., Vela, J., Alvarez-Fanjul, E., 2008. A new concept of multi-purpose sea level station: example of implementation in the REDMAR network. In: *Proceedings of the Fifth International Conference on EuroGOOS,* May 2008*: Coastal to Global Operational Oceanography: Achievements and Challenges.* Exeter, UK.

Phillips, O. M., 1977. *The Dynamics of the Upper Ocean.* Cambridge: Cambridge University Press.

Pontes, M. T., 1998. Assessing the European Wave Energy Resource. *Journal of Offshore Mechanics and Arctic Engineering* 120, 226–231.

Prandle, D., 1982. The vertical structure of tidal currents. *Geophysical and Astrophysical Fluid Dynamics* 22, 29–49.

Prandle, D., Ballard, G., Banaszek, A., Bell, P. S., Flatt, D., Hardcastle, P., Harrison, A., Humphery, J. D., Holdaway, G., Lane, A., Player, R. J., Williams, J. J., Wolf, J., 1996. *The Holderness Coastal Experiment '93–'96.* Birkenhead, UK: Proudman Oceanographic Laboratory, Report No. 44. NERC Open Research Archive (NORA). Available: http://nora.nerc.ac.uk [accessed November 2016].

Prime, T., Brown, J. M., Plater, A. J., 2016. Flood inundation uncertainty: The case of a 0.5% annual probability flood event. *Environmental Science and Policy* 59, 1–9.

Pullen, T., Allsop, N. W. H., Bruce, T., Kortenhaus, A., Schüttrumpf, H., van der Meer, J. W., 2007. *EurOtop – Wave Overtopping of Sea Defences and Related Structures: Assessment Manual.* www.overtopping-manual.com

Seelig, W. N., 1979. Effects of breakwaters on waves: Laboratory tests of wave transmission by overtopping. *Proceedings of the Conference on Coastal Structures* 79, 941–961.

Seelig, W. N. and Ahrens, J. P., 1981. *Estimation of Wave Reflection and Energy Dissipation Coefficients for Beaches, Revetments, and Breakwaters,* Technical Paper No. 81–1, Ft. Belvoir, VA: U.S. Army Corps of Engineers, Coastal Engineering Research Center.

Sharma, J. N., Dean, R. G., 1981. Second-order directional seas and associated wave forces. *Society of Petroleum Engineers Journal* 21, 129–140.

Siddons, L. A., Wyatt, L. R., Wolf, J., 2009. Assimilation of HF radar data into the SWAN wave model. *Journal of Marine Systems* 77, 312–324.

Snyder, R. L., Dobson, F. W., Elliott, J. A., Long, R. B., 1981. Array measurement of atmospheric pressure fluctuations above surface gravity waves. *Journal of Fluid Mechanics* 102, 1–59.

Sørensen, O. R., Kofoed-Hansen, H., Rugbjerg, M., Sørensen, L. S., 2004. A third-generation spectral wave model using an unstructured finite volume technique. *Proceedings of the 29th International Conference of Coastal Engineering (Lisbon), World Scientific,* Singapore, 894–906.

Souza, A. J., Howarth, M. J., 2005. Estimates of Reynolds stress in a highly energetic shelf sea. *Ocean Dynamics* 55, 490–498.

Souza, A. J., Bolaños, R., Wolf, J., Prandle, D., 2011. Measurement technologies: Measure what, where, why, and how? *Reference Module in Earth Systems and Environmental Sciences: Treatise on Estuarine and Coastal Science* 2, 361–394. doi:10.1016/B978-0–12-374711-2.00215–1.

Stokes, G. G., 1847. On the theory of oscillatory waves. *Transactions of the Cambridge Philosophical Society* 8, 441–455. *Reprinted in Mathematical and Physical Papers, London* 1, 314–326.

Svendsen, J. A., 2006. *Introduction to Nearshore Hydrodynamics.* Advanced Series on Ocean Engineering. Singapore: World Scientific, 24.

SWAN, 2016a. SWAN – Simulating WAves Nearshore (homepage). http://swanmodel.source forge.net/ [accessed August 2016].

SWAN, 2016b. *SWAN User Manual*; SWAN Cycle III version 41.10. Delft, The Netherlands: The SWAN team, Delft University of Technology, Faculty of Civil Engineering and Geosciences, Environmental Fluid Mechanics Section. http://swanmodel.sourceforge.net/online_doc/swanuse/swanuse.html [accessed August 2016].

Tolman, H. L., 2009. *User Manual and System Documentation of WAVEWATCH III™ version 3.14*. NOAA / NWS / NCEP / MMAB Technical Note 276.

Traer, J., Gerstoft, P., Bromirski, P. D., Shearer, P. M., 2012. Microseisms and hum from ocean surface gravity waves. *Journal of Geophysical Research* 117, B11307. doi:10.1029/2012JB009550.

Trowbridge, J. H., 1998. On a technique for measurement of turbulent shear stress in the presence of surface waves. *Journal of Atmospheric and Oceanic Technology* 15, 290–298.

Tucker, M. J., 1993. Recommended standard for wave data sampling and near real-time processing. *Ocean Engineering* 20, 459–474.

Tucker, M. J., Pitt, E. G., 2001. *Waves in Ocean Engineering*. Amsterdam: Elsevier Ocean Engineering Book Series. Elsevier.

Uncles, R. J., 2010. Physical properties and processes in the Bristol Channel and Severn Estuary. *Marine Pollution Bulletin* 61, 5–20. doi:10.1016/j.marpolbul.2009.12.010.

Uncles, R. J., Stephens, J. A., 2010. Turbidity and sediment transport in a muddy sub-estuary. *Estuarine, Coastal and Shelf Science* 87, 213–224. doi:10.1016/j.ecss.2009.03.041.

Uncles, R. J., Stephens, J. A., Harris, C., 2014. Infragravity currents in a small ría: Estuary-amplified coastal edge waves? *Estuarine, Coastal and Shelf Science* 150, 242–251. http://dx.doi.org/10.1016/j.ecss.2014.04.019.

van der Westhuysen, A. J., Zijlema, M., Battjes, J. A., 2007. Nonlinear saturation-based whitecapping dissipation in SWAN for deep and shallow water. *Coastal Engineering* 54, 151–170.

van der Westhuysen, A. J., 2010. Modeling of depth-induced wave breaking under finite depth wave growth conditions. *Journal of Geophysical Research-Oceans* 115, C01008, doi:10.1029/2009JC005433.

VE, 2016. Veritech Enterprises. www.veritechinc.com/products/cedas [accessed March 2016].

Wadey, M. P., Brown, J. M., Haigh, I. D., Dolphin, T., Wisse, P., 2015. Assessment and comparison of extreme sea levels and waves during the 2013/14 storm season in two UK coastal regions. *Natural Hazards and Earth System Sciences* 15, 2209–2225.

Wiegel, R. L., 1964. *Oceanographical Engineering*. Englewood Cliffs, NJ: Prentice Hall,.

Wolanski, E., Spagnol, S., 2003. Dynamics of the turbidity maximum in King Sound, tropical Western Australia. *Estuarine, Coastal and Shelf Science* 56, 877–890.

Wolf, J., Hubbert, K. P., Flather, R. A., 1988. A feasibility study for the development of a joint surge and wave model. *Proudman Oceanographic Laboratory Report no. 1*.

Wolf, J., 1996a. The Holderness Project wave data. *Proudman Oceanographic Laboratory Internal Document no. 89*.

Wolf, J., 1996b. *The Intercomparison of Wave Data from Moored Instruments*, POL Internal Document no. 103, 17pp.

Wolf, J., 1997. The analysis of bottom pressure and current data for waves. In: *Proceedings of the 7th International Conference on Electronic Engineering in Oceanography*, Southampton, June 1997, Conference Publication 439, London: IEE, 1997, 165–169.

Wolf, J., 1999. The estimation of shear stresses from near-bed turbulent velocities for combined wave-current flows. *Coastal Engineering* 37, 529–543.

Wolf, J., Prandle, D., 1999. Some observations of wave-current interaction. *Coastal Engineering* 37, 471–485.

derived from the catchment, local cliffs or offshore deposits, secondary minerals derived in situ (e.g., iron oxyhydroxides or sulphides) and organic material derived either externally from fluvial and/or marine sources or internally from the death and decay of estuarine organisms. Deposited sediments are primarily characterised in terms of their size, shape, composition and structure. The size of bed particles is largely controlled by local hydrodynamics and can vary both spatially and temporally across the estuarine environment, dependent on flow conditions, whilst composition is dependent on local geology and physicochemical gradients within the water and sediment environment. Particle size (and to a lesser extent shape and composition) influences the nature of the substrate in terms of its erodibility (e.g., Grabowski et al., 2012), transport characteristics, and suitability as a habitat for benthic organisms (e.g., Ysebaert et al., 2003 and references therein). In addition, because many contaminants preferentially bind to fine-grained sediments (Kersten and Smedes, 2002) particle size can also influence sediment quality. Particle size also influences matrix porosity, with impacts on sediment bulk density and the transport of water and solutes (including gases) through the deposited material, thus affecting both biological and chemical activity.

The purpose of this chapter is to provide an outline of the main methods that are fit-for-purpose for collecting deposited sediment samples and an overview of the laboratory techniques that are currently used for measuring the physical characteristics of deposited sediment. Suspended sediment properties, including flocculation, are covered in Chapters 7 and 8, and bed-load and suspended sediment transport are discussed in Chapter 9.

2   Sampling Methods

Estuarine studies frequently involve the collection of surface samples comprising recently settled material or sediment cores that represent historically or even geologically accumulated material. Selecting which type of sample is required depends on the purpose of the study. The nature and composition of surface sediments are characteristic of present conditions within the estuary, and consequently surface sediments are often collected during monitoring or baseline environmental assessments. Analysis of sediment cores can provide information on sediment at depth, and depending on the degree of postdepositional sediment disturbance, either a time-integrated or chronological record of past environmental conditions such as sediment supply or contaminant input (e.g., Birch et al., 2013). Generally, approximately 1 kg of material is considered sufficient to allow for a range of physical and chemical measurements, although frequently much smaller sample sizes suffice.

The physical nature of estuaries, including their morphology, bathymetry and hydrodynamics, and the deposited sediment within them, are highly variable, and hence a wide range of both sampling equipment and techniques have been developed. The main practical criteria for determining which equipment to use are water depth, flow conditions, and sediment composition. For example, intertidal mudflats, saltmarshes and mangroves, or low-relief tidal flats that may only be covered by shallow water for part of the tidal cycle, can be sampled directly by hand or from the side of a small boat. Subtidal sediments can be collected using grab samplers or corers relatively easily in slow currents, but this can become very difficult, if not impossible, in fast currents. In deeper waters, such as fjords, although currents may be slow, substantial water depths mean that mechanical winches and more substantial survey vessels may be required. The nature of the sediment to be collected is also an important consideration. Sampling equipment for consolidated materials is fairly standardised, and finer-grained material (silts and sands) can be sampled with relatively lightweight equipment. However, shelly, coarser material needs more robust sampling equipment, and it may be very difficult to sample very fine-grained muds or unconsolidated material without disturbing sediment structure, particularly at the sediment–water interface.

2.1   Surface Sediment Samples

It is relatively simple to collect surface sediments where there is no need to know the exact depth from which sediment was collected, or to preserve sediment structure. Where sampling locations can be accessed easily and safely by foot or small boat, samples can be collected by hand using a trowel or scoop, allowing very accurate location of sampling sites with GPS. Where sediments are sampled from very shallow water depths (of order 1 m) it is also possible to use lightweight grab samplers, such as an Ekman grab, that have been modified by the addition of short extension rods. For deeper waters there is a range of grab samplers available for use in estuarine and marine environments, and details of their advantages and disadvantages, and selection of sampler type according to water depth, current strength, sediment type and sediment volume required are reported widely (e.g., Environment Canada, 1994; U.S. EPA, 2001; IAEA, 2003). Information on individual grab sampler types can be found via manufacturers' websites, and an overview of recommended samplers is given here (shown in Figure 6-1), although this is not an exhaustive list.

Grab samplers all work on the same basic principle, in which a specialised bucket or scoop is lowered through the water column (usually on a wire or rope) until it impacts the bed and a release mechanism is activated enabling sample collection and withdrawal. They are simple to use, but their improper use can destroy the integrity of the sample and sediment structure. The grab should be lowered and raised slowly and vertically through the water column, avoiding free fall (approximately 0.3 m s^{-1}) to minimise disturbance at the sediment surface on impact and to minimise the loss of fine-grained material as the grab is raised through the water column (Environment Canada, 1994; Simpson et al., 2005). Fine sediments are easily lost or disturbed when using Ponar, Van Veen and mini-Shipek grab samplers. Collecting samples in deep waters or using heavy equipment for harder substrates may require the use of a mechanical winch (e.g., Smith-McIntyre and Petersen grab samplers). Upon collection, the grab sampler should be carefully washed to remove any additional material that has adhered to the equipment. Samples can then be transferred to labelled plastic bags or bottles.

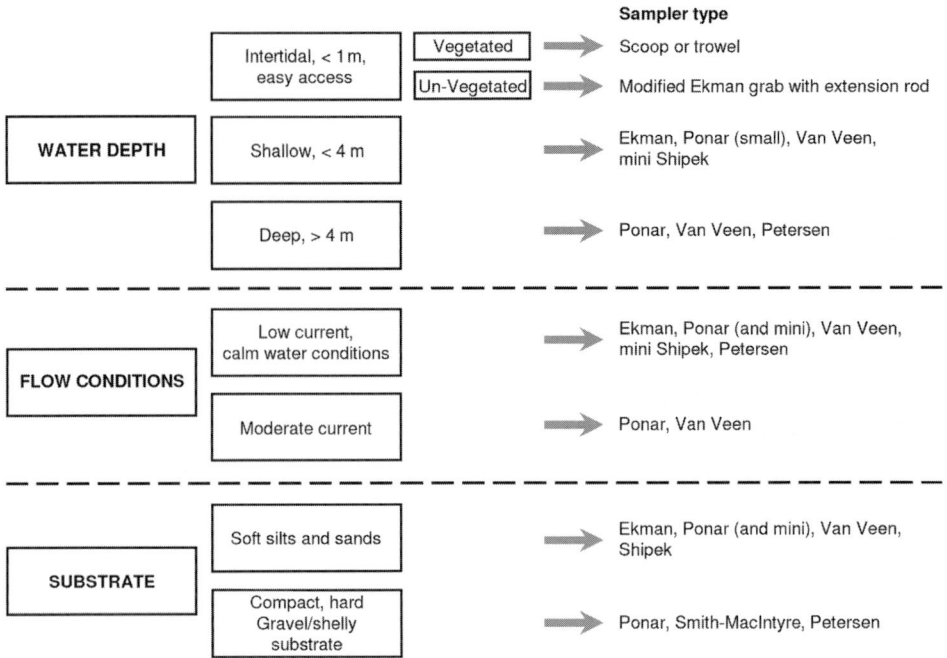

Figure 6-1 Selection of appropriate grab samplers based on water depth, flow conditions and sediment substrate.

2.2   Core Samples

Where data on vertical changes in composition beneath the sediment surface are required, or where information on sediment structure is needed, the use of coring devices becomes necessary. Many studies have focused on the examination of environmental data in intertidal sediments, such as saltmarshes, because low rates of sediment deposition and minimal postdepositional disturbance mean that these are excellent recorders of environmental change (e.g., Teasdale et al., 2011). Such sites can be easily cored by hand using PVC or polycarbonate tubes, which are pushed into the sediment and/or using manually extracted sediment monoliths. Handheld tubes may be bevelled to ease entry into the sediment and can include a core catcher for retaining finer-grained or less consolidated material. Russian peat corers can also be used to collect short cores in soft, unconsolidated sediment (e.g., saltmarshes and mudflats), although dense root material may make this impossible. It is usually only possible to collect short cores (< 1 m) by hand.

There is a wide range of coring devices that can be used below the water surface and for collecting deeper cores, and again these have been thoroughly reviewed in the literature; the choice of corer depends on physical estuarine conditions, water depth and sediment type (Weaver and Schultheiss, 1990; Blomqvist, 1991; Mudroch and MacKnight, 1994). An overview of corer types and selection criteria is given here (shown in Figure 6-2); however, the reader is encouraged to examine manufacturers'

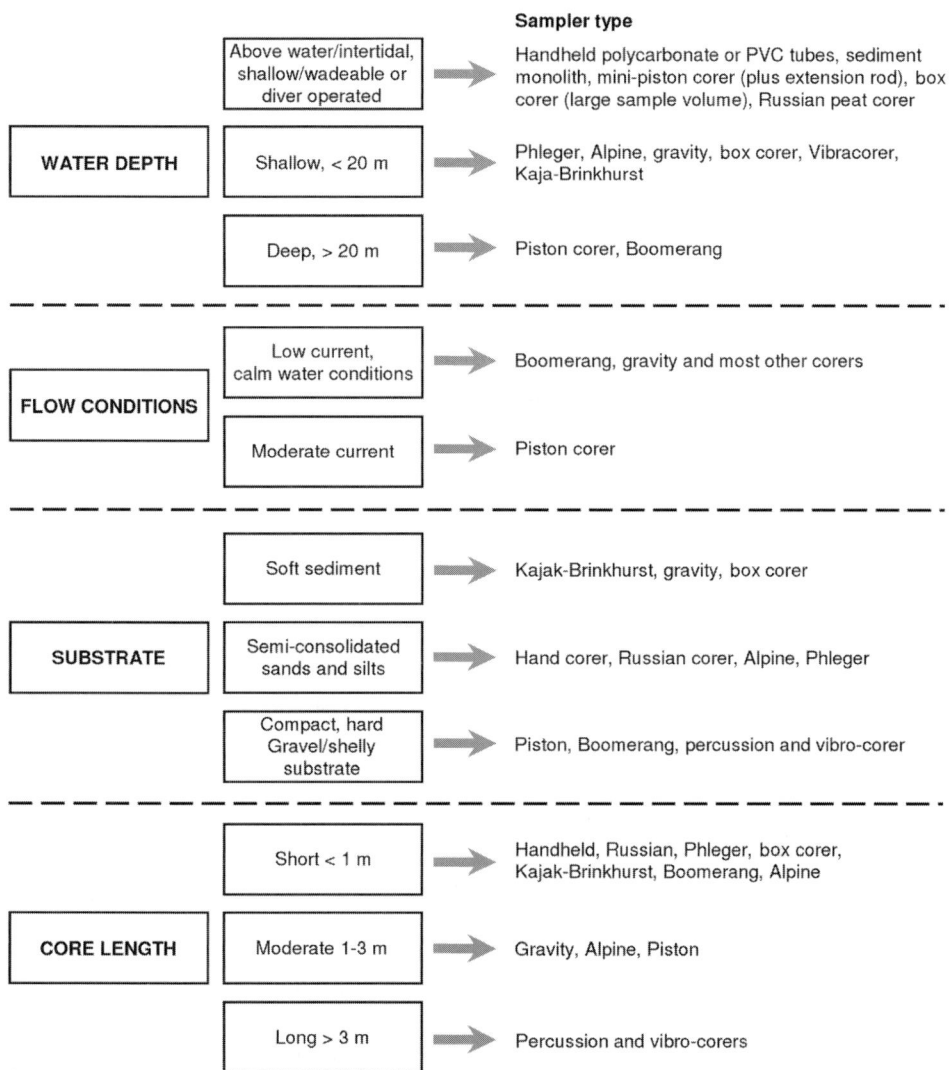

Figure 6-2 Selection of appropriate corers based on water depth, flow conditions, sediment substrate and length of core required.

websites for samplers because modifications frequently enable more flexible use. Essentially, corers fall into four main groups: gravity corers, piston corers, vibro- and percussion corers, and box corers.

The simplest gravity corer consists of a weighted metal tube with stabilising fins at the upper end. It is allowed to free-fall through the water column and penetrates the sediment surface on impact, sometimes with the aid of a sharpened core cutter. The core cutter has a slightly smaller diameter than the core chamber, reducing compression and internal friction as the corer enters the sediment. Most gravity corers have messenger-operated chamber closure, and this reduces the water depth in which they are effective.

Gravity corers (e.g., the Phleger corer, Kajak-Brinkhurst or Alpine corer) can collect cores up to 2 m in length, and depth of water, mass of corer and length of core tube all dictate how deep a core can be extracted. Vertical free-fall is required to ensure the gravity corer penetrates the sediment surface and that the core provides a vertical cross-section through the sediment; therefore, gravity corers are inappropriate for use in moderate to strong currents.

Piston corers (e.g., the Kullenberg corer) are widely used in deep sea surveying and provide long, continuous sediment cores and large volumes of sediment. The corer is lowered, and as it impacts the sediment surface, a piston is released that creates a vacuum inside the core chamber, aiding both penetration of the corer into the sediment and retention of sediment as the core is extracted. Piston corers can easily be damaged if they encounter bedrock or gravelly deposits and hence are best used for soft, unconsolidated sediment, where long cores can be easily extracted (IAEA, 2003). The piston corer has also been modified for use in shallow water conditions, and mini-piston corers have been designed for collecting lake sediments in low flow conditions (Glew, 1991; Glew et al., 2002) and can also be used effectively in saltmarsh environments (e.g., Macreadie et al., 2013), or can be diver-operated in shallow marine conditions, reducing disturbance at the sediment water interface (e.g., Sansone et al., 1994).

Percussion and vibro-corers use mechanical means to enable penetration and retrieval of continuous sediment cores of up to 10 m in length from consolidated and coarse material (Jones et al., 1992; Adachi et al., 2010). Penetration is achieved through either percussive force from a hand- or air-operated hammer tool or petrol/diesel powered vibration techniques. The weight of the equipment is such that usually both a substantial survey vessel and a support team are required in the field, although sample collection can be very fast.

The box corer is used for collecting large amounts of surface material (0.5 m x 0.5 m x 0.5 m) where there is a need to preserve internal sediment structure. A large, steel box is driven into the sediment, and a blade then cuts through the sediment and covers the open end of the box. A monolith of sediment can then be lifted clear and examined.

Coring can result in loss of sediment structure and compaction, and this issue has been discussed at length in the literature (e.g., Wright, 1991, 1993; Blomqvist, 1991; Glew et al., 2002; Lane and Taffs, 2002). As the corer enters the substrate, and as the core is extruded, friction can result in disturbance to sediment structure, which is exacerbated in vegetated intertidal sediments, where root material is present. In addition, smearing along the edges of the sediment core will result in contamination across vertical sediment subsamples. In unconsolidated sediment, and where pore space and organic matter content is high (resulting in low bulk densities), sediment can also be significantly compacted, resulting in core shortening, changes to sediment structure, and underestimates of sediment accumulation rates (Turner et al., 2006). Increasing the core diameter can reduce friction, but this makes core retrieval more difficult. Problems of core compaction have been addressed for shallow, vegetated sediment cores using various sediment monolith techniques (e.g., Stoodley, 1998; Inglett et al., 2004), although artefacts associated with sediment coring remain a significant problem; disturbance can be further compounded by subsampling, transport and storage of sediment samples.

2.3   Sampling at the Sediment–Water Interface

Increasingly there is a desire to collect fine-grained (silts and clays) and poorly consolidated material, including flocculated material, at the sediment–water interface because
understanding the physical and chemical behaviour of these particles can be important
for understanding cohesive sediment transport, bed stability and erodibility, and contaminant behaviour (Milligan and Law, 2013). There have been many claims that mini
piston corers and gravity corers can retrieve undisturbed surface sediments (e.g., Barnett
et al., 1984; Pedersen et al., 1985; Jahnke and Knight, 1997; Xu et al., 2011), and whilst
gross sediment structure may be largely preserved, unconsolidated material at the
sediment surface is very fragile and will be lost (Muschenheim and Milligan, 1996).
This unconsolidated material is very difficult to sample and can be easily disturbed or
lost due to the formation of a bow wave as the sampler reaches the sediment surface, as
the core is retrieved through the water column, or as sediments drain following collection (Blomqvist, 1991; Lane and Taffs, 2002). Recent developments that attempt to both
preserve the sediment water interface and retain unconsolidated material include the
USS (Undisturbed Sediment Sampler, U.S. EPA, 2005) and the Slo-corer (Milligan and
Law, 2013), as well as devices for carefully subsectioning sediment cores following
collection (Kornijów, 2013).

3   Sample Design and Frequency

Choice of sampling strategy, including selection of sample site locations, a decision on
spatial and temporal sampling frequency, and the need for replication depends on the
requirements of the project. For example, sediments may be collected for monitoring
purposes to establish the current state of the estuarine environment, to demonstrate
statutory compliance or to inform better management decisions, e.g., dredging activity.
Estuaries are highly heterogeneous compared with other aquatic environments (Caeiro
et al., 2003). Hydrodynamics can result in variability in sediment grain size over spatial
distances from metres to hundreds of metres, whilst sediments disturbed by roots and
burrows (e.g., saltmarsh and mangrove sediments) display heterogeneity in sediment
structure and physical composition over much smaller spatial scales (cm and less).

Sample designs can be random, systematic or stratified, and sampling designs
associated with the collection of spatial data are explained in many general earth and
environmental statistic texts (e.g., Webster, 1999; Schuenemeyer and Drew, 2011).
Random sampling may provide unbiased results, but in estuarine systems where sample
areas are frequently large and inaccessible, it may be difficult to capture variability
unless very large numbers of samples are collected (Caeiro et al., 2003). Regular,
systematic grids or transects (for example, along a longitudinal salinity gradient) are
frequently used in estuarine studies, but these can become biased if the grid becomes
offset by the distribution of a variable of interest (Caeiro et al., 2003), or due to practical
constraints such as site access. A stratified approach can be used where sample locations
are subdivided into more homogenous subsets (e.g., subdividing the intertidal zone into

biogeomorphological units such as bare mud, vegetated saltmarsh, creek, saltpan) that are less variable than the original population. In reality, it is frequently desirable to combine characteristics from these three approaches. For example, random stratified sampling approaches are commonly used for estuarine monitoring programmes (e.g., U.S. EPA, 2001).

4   Sample Storage, Preparation and Pretreatment

Although some simple observations and measurements may be made in the field, it is likely that sediment samples will be transported to the laboratory and stored prior to analysis. Samples should be carefully labelled with key information such as date, location, nature of sample, and worker, and each sample should be allocated a unique identifier. Core samples should be kept vertical to avoid compaction and disturbance, and samples should be sealed so that they do not dry out or spill in transit. Samples that also require biological or chemical analysis need to be stored at 4°C or frozen (usually at approx. −18°C). Once returned to the laboratory, samples should be further audited to make sure that packaging and labels are intact.

The nature and extent of sample preparation will depend on the type of sediment samples that have been collected, the analyses required, and the purpose of the study. For the examination of density, porosity and structure, the sediments need to be disturbed as little as possible; even freezing cores intact may disrupt sediment fabric. The most common physical characterisation of sediment is the determination of its grain-size distribution, providing information on texture and enabling the prediction of hydraulic properties, such as transport and deposition rates. In this instance, and particularly for the fine-grained cohesive fraction (i.e., particles that interact with each other and their transporting medium), it may be more valuable to look at the distribution of particle aggregates; however, care must be taken because any sample pretreatment is likely to alter the size and shape of these aggregates (Chapter 8) and may not be desirable. The grain size of the coarse fraction is most frequently determined by sieving, whilst sieving, gravimetric and automated techniques (e.g., laser granulometry) are most commonly used for the fine fraction. These will be discussed later. A range of pretreatments exists prior to grain-size analysis that enables the removal of salts, biogenic calcite and organic matter, and aggregate dispersion.

4.1   Washing

It may be necessary to remove salts from the sediment because precipitated salts can overestimate the percentage of fine-grained sediment in the coarse fraction (> 63 μm; Poppe, Eliason, Fredericks, et al., 2003 and Chapter 7) as well as influencing the bulk chemistry of a sample. However, in general, it is rarely necessary to remove salts from sand or gravelly material that is required for mechanical grain-size analysis. Removing salts from the coarse fraction is relatively easy to achieve by stirring the sediment repeatedly in large volumes of water, allowing the sediment to settle and decanting the supernatant water. A final rinse with deionised water will remove low concentrations of soluble salts. As a general rule, three repeat washes in one litre of water will remove salts from 200 g of sediment. Care should be taken if the fine-grained fraction is also to be retained for chemical analysis by using deionised water for sieving and a chemically inert sieve (Simpson et al., 2005) and by ensuring that fine sediment is not lost during decantation. For finer sands it may be preferential (and quicker) to remove dissolved salts and supernatant water via centrifugation.

Fine-grained sediments that include clays and, to a certain extent, silts are cohesive, and flocs (fragile aggregates, Chapter 8) are held together, in part, by electrostatic forces between the particles in saline waters (Winterwerp and van Kesteren, 2004). Washing the sample will remove these dissolved salts and cause the flocs to disaggregate, so that any subsequent grain-size analysis will provide a record of the distribution of primary particles, rather than flocs, with consequences for the interpretation and prediction of the dynamic properties of the sediment.

4.2   Removal of Biogenic Calcite and Organic Matter

Sediment grain size is frequently measured to make predictions regarding the hydro-dynamic properties of sediment, such as its erosion and deposition characteristics. Such calculations are based on measurements of the diameter and density of individual grains and assume that particles are near-spherical with densities approaching that of quartz. Whilst this may in part hold true for the mineral component of sediment, it is less relevant for the organic matter fraction, which has low density and for which fragments of plant/root material may be fibrous in nature. Shelly material (calcium carbonate) may also be a significant component of estuarine sediment, which is derived either in situ or transported over long distances and accumulated due to prevailing local hydrodynamics, e.g., the formation of shelly chenier plains and ridges attributed to storm surges (Otvos, 2000). Therefore, the grain-size characteristics of shell material may not be hydrodynamically representative of the depositional environment, and it may be sensible to remove both organic material and biogenic calcite prior to grain-size analysis.

Live shells and larger material can be removed manually, or by passing the sample through a coarse sieve (e.g., > 2 mm). Any remaining calcium carbonate can then be removed by acidifying the samples in dilute hydrochloric acid or sodium acetate (which may be preferred because hydrochloric acid can attack the clay mineral lattice); this is usually carried out cold, but can be accelerated by heating (Percival and Lindsay, 1997). The dissolved carbonate can then be removed by multiple decantation or centrifugation with deionised water. Organic matter is removed through oxidation, usually using 30 per cent hydrogen peroxide at room temperature and for up to one week to ensure the removal of all recalcitrant material (Kunze and Dixon, 1982; Lewis, 1984; Allen and Thornley, 2004; Gray et al., 2010).

4.3   Dispersion

If full dispersion of the sample is required, this can be achieved by a combination of chemical and physical means. The aim of chemical dispersion is to create and maintain repulsive forces between the sediment grains, and this can be achieved through the addition of 0.5 per cent sodium hexametaphosphate (Na-HMP), although the volume and solution strength required will depend on the nature of the sediment and whether pretreatments such as those described earlier have also been used (Poppe, Eliason, Fredericks, et al., 2003). Sediments are physically dispersed by either shear action or turbulent mixing, using various mechanical shakers for up to 30 min. Ultrasonic baths or probes can also be used to break apart aggregates by physical means, which is highly efficient but can break up quartz particles, so that short ultrasonification times are recommended, e.g., 3 min (Di Stefano et al., 2010).

4.4   Drying

Once samples have been pretreated, they can be dried prior to analysis by placing them in open evaporating dishes in a drying oven. The temperature and duration of drying is largely dependent on the size of the particles involved. Drying can remove structural water from the mineral lattice of some clay minerals and result in 'caking' of the clay minerals, which may be impossible to redisperse. As a result, drying of fine sediments (< 63 μm) is not recommended and, if necessary, freeze-drying is preferred (Mudroch and MacKnight, 1994; Percival and Lindsay, 1997). Once dried, sediments can be stored for an indeterminate period of time without any change to their physical and chemical characteristics.

5   Grain-Size Analysis

Grain size is a fundamental property of sediment that affects its dynamic behaviour; it can provide information on sediment provenance, transport history and pathways, and depositional conditions and is a key input variable for the prediction and modelling of sediment transport. The fine-grained sediment fraction (< 63 μm) provides an important sorption site for contaminants, nutrients and radionuclides, due to its large surface area to volume ratio and high cation exchange capacity. The relationship between grain size and contaminant distribution has been thoroughly explored for marine and estuarine sediments (e.g., Loring, 1991).

There are multiple methods for the analysis of grain size, with the availability of mechanical methods of sieving for coarser materials (gravels and sands, and to a lesser extent silts) and a range of methods including gravimetry, centrifugation, heavy-liquid flotation and magnetic separation for the finer materials (sits and clays; Clifton et al., 1999). In addition, there is increasing use of automated techniques such as electro-resistance particle size analysers (e.g., the Coulter Counter), X-ray granulometry (e.g., the SediGraph) or the use of laser diffraction granulometry, as instruments have become cheaper and more readily available. Many of these techniques and their applications in estuarine and marine environments have been thoroughly reviewed by Syvitski (2007).

These various techniques provide overlap in terms of the size fraction they can measure, and as a consequence it is not uncommon for more than one technique to be used in a single study (Figure 6-3). There is also no optimum technique, and the choice

Millimetres (mm)	Micrometres (μm)	Phi	Wentworth size class
4096		-12.0	Boulder
256		-8.0	Cobble
64		-6.0	Pebble
4.00		-2.0	Granule
2.00		-1.0	Very coarse sand
1.00		0.0	Coarse sand
1/2 0.50	500	1.0	Medium sand
1/4 0.25	250	2.0	Fine sand
1/8 0.125	125	3.0	Very fine sand
1/16 0.0625	63	4.0	Coarse silt
1/32 0.031	31	5.0	Medium silt
1/64 0.0156	15.6	6.0	Fine silt
1/128 0.0078	7.8	7.0	Very fine silt
1/256 0.0039	3.9	8.0	Clay
0.00006	0.06	14.0	

(Right-hand axis labels: Gravel, Sand, Silt, Mud; analytical ranges: Sieve analysis, Gravimetric analysis, Electro-resistance analysis, Laser granulometry.)

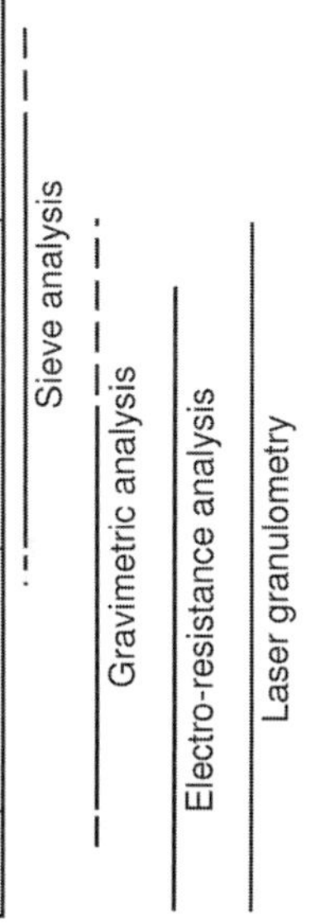

Figure 6-3 This table shows the Wentworth (1922) grain-size classification and the analytical range of various grain-size analysis techniques.

of method depends on the requirements of the study, the nature of the sediment and the amount of material available. In addition, practicalities such as the time required for sample preparation and the complexity of the protocol and data processing may be more or less important to different workers.

It is also important to understand what variable has been measured. The size reported for any given sediment is essentially a single dimension that is an idealised descriptor of the particle, which is method dependent and may, for example, be a measure of volume or diameter. At this point it is also worth noting the difference between the terms *grain* and *particle*. A grain usually refers to a single-mineral particle, although in many cases sediment comprises not only single-mineral grains, but also aggregates or 'flocs' of loosely bound organic and inorganic material (Chapter 8) that can be more adequately described as particles. As a result, although the term *grain size* is widely used and accepted in sedimentology, some workers prefer the use of *particle size* (Blott and Pye, 2012). This also clarifies the use of *effective* and *absolute* grain-size measurements, where absolute grain size refers to sediment that has been fully dispersed and is a measure of the individual sediment grains.

5.1   Dry Sieving

Dry sieving has been widely used and is generally suitable in the size range 0.063–250 mm diameter, although in practical terms once the mesh size is > 32 mm the sample and sieve size must be very large to gain a representative sample. Methods have changed little since those described in older texts (e.g., Allen, 1990) and therefore will not be discussed here in detail. If necessary, the silt and clay fraction may be removed and retained prior to dry sieving, and samples are then dried and gently disaggregated with a mortar and pestle. Dry sieving is unsuitable for the silt and clay fraction because the clays will cake with drying, and fine particles will tend to adhere to the mesh rather than pass through the sieve aperture.

Sieves usually comprise fine wires arranged in a square mesh, and sieves are stacked with progressively smaller apertures at, e.g., ¼ phi intervals. Sieves are available with smaller mesh sizes, particularly with the availability of electro-formed mesh (see various sieve manufacturers' websites for details), although these are fragile, expensive and easily damaged, and there are far superior techniques available for the analysis of silts and clays. The sieve stack is shaken, allowing the separation of progressively smaller grain-size fractions. Particles pass through a square mesh; therefore, sieving provides a physical description of the particles as a population of spheres, with the particle diameter being equivalent to the internal dimension of the square mesh (Blott and Pye, 2006). The length of time required for shaking depends on the shape of the particle; e.g., elongate particles may only be able to pass through the aperture in one orientation. Shaking is usually automated, and shaking times vary in the literature, but are usually around 10–15 min, although longer times are sometimes reported (Bartholdy et al., 2010).

The amount of sample required for dry sieve analysis depends on the average grain size of the sediment; gravels require up to 10 kg, whilst coarse sands require 100–250 g, and fine sands 10–100 g. Once agitation of the sample is complete, the fraction in each sieve can be weighed and the distribution between size classes calculated. Sieves should be kept clean with a wire brush, although care should be taken not to damage the mesh during the cleaning process.

5.2   Wet Sieving

Wet sieving is essentially used to separate the fines (silt and clay) from the coarse fraction, and although sieves are manufactured with much finer mesh sizes (e.g., 20 μm) other techniques such as gravimetry and laser granulometry are more effective and much quicker for sizing the silt fraction. The sediment is first dried and dispersed with Na-HMP, as described earlier, and washed through a 2-mm sieve nested inside a 63-μm sieve. The sample is added gradually with additions of water, and sieving continues until the water runs clear. The coarse fraction can then be dried and the total amount of fines calculated from the weight difference before and after sieving. The coarse fraction can then be analysed as described earlier. However, this results in the loss of the fine fraction, and if the salt content is not calculated it can result in the overestimation of fine content. If further investigation of the fine fraction is required it is possible to retain the material that passes through the 63-μm sieve and dry it to remove excess water, but this can result in the aggregation of clay minerals.

5.3   Pipette Method

There are a number of indirect methods for measuring grain size based on gravimetric settling including the pipette and hydrometer methods. Results from these tend to be in agreement as long as the same sample pretreatment techniques are used (Eshel et al., 2004). These methods are based on Stokes' Law and the assumption that as particles settle through a column of water, those with a larger diameter will settle more quickly. The pipette method first requires separation of the < 63-μm fraction from the coarse fraction by wet sieving. The sediment is usually dispersed through the addition of Na-HMP to provide a measure of absolute grain size, but in theory it may also be possible to look at effective grain size provided that the particle concentration and ionic strength of the solution are suitable, although the flocculation dynamics of the sediment in suspension are unlikely to be representative of the in situ estuarine environment, and this has rarely been explored in the literature.

Sediments are added to a graduated cylinder, stirred and allowed to settle. If sediments are dispersed, all solutions should contain Na-HMP to ensure samples do not begin to flocculate. A set volume of the suspension is withdrawn with a pipette from the same position in the settling column; according to Stokes' Law, the withdrawn sediment should have a particle size below a specific diameter. Careful selection of sampling times allows the removal of sediment fractions with a predetermined specific size interval (Clifton et al., 1999). The withdrawn fluid is evaporated and the sample weighed to determine the mass of suspended sediment in each size fraction interval. A particle distribution curve can then be calculated.

Stokes' Law assumes that: (1) particles are smooth, solid spheres; have a known, uniform density (usually estimated at 2650 kg m^{-3}); (2) do not interact with each other or the cylinder walls; and (3) they are falling through a nonturbulent fluid (Konert and Vandenberghe, 1997). In reality, natural estuarine particles deviate from this ideal in terms of both shape and density. For example, clay particles tend to be plate-like in nature, and particle density can range from 1660 to 2990 kg m^{-3}, depending on their composition (Konert and Vandenberghe, 1997; Clifton et al., 1999). As such, the settling times for clays and low-density organic matter are increased, and gravimetric methods are widely reported to overestimate the proportion of the finest sediment fraction (e.g., Lu et al., 2000; Di Stefano et al., 2010). Even at low particle concentrations the displacement of water leads to upward-moving columns of the suspension, leading to settling convection (Kuenen, 1968), whilst fluctuations in temperature result in thermal convection currents. All gravimetric methods tend to be unreliable for particles < 1 μm due to the influence of Brownian motion on settling rates (Eshel et al., 2004). Suspended particle concentrations used vary in the literature between 4 and 50 kg m^{-3}. If particle concentrations are too high there will be an increased incidence of interparticle collisions, which could result in flocculation, whilst at lower concentrations weighing errors will be high and results may not be reproducible. The time required for the finest particles to settle out of suspension can run into days; consequently, this technique is time consuming and can require relatively large sample volumes.

Table 6-1 Particle settling times in pure water at a temperature of 20°C and for a particle specific gravity of 2.65 using the Stokes' relationship derived by Krumbein and Pettijohn (1938): $w = 3.57 \times 10^4 (d/2)^2$, with settling velocity, w, in units of cm s^{-1} and particle diameter, d, in units of cm. Note that clay-sized diameters are defined to be $\leq 3.9 \mu m$ on the Wentworth scale and $\leq 2 \mu m$ on the international (ISO 14688–1:2002) scale.

Particle Settling Times (particle specific gravity of 2.65 in water at 20°C)

Diameter (Phi)	Diameter (μm)	Sediment type	Settling depth (m)	Settling time (h)
4.0	62.5	silt	0.2	0.02
5.0	31.2		0.1	0.03
6.0	15.6		0.1	0.13
7.0	7.8		0.1	0.51
8.0	3.9	clay	0.1	2.05
9.0	1.95		0.1	8.19
10.0	0.98		0.1	32.41
11.0	0.49		0.05	64.81

The dispersed sample is placed into a graduated cylinder containing 1000-ml dispersant solution. The suspension is stirred vigorously by means of either end-over-end shaking or using a manually operated stirring device that travels from bottom to top of the water column and ensures all sediment is in suspension. The column should be maintained at 20°C. As soon as stirring ceases, a timer should be started and a pipette slowly and gently inserted into the standing fluid. A 20-ml sample is withdrawn (slowly and gently), the timing and depth of the sample are determined by the times taken for a particle of given diameter to fall through 0.20, 0.10 and 0.05 m (Table 6-1). The sample is then discharged into a beaker and the pipette flushed with water into the same beaker to capture all the material. The samples are dried overnight at 100°C, cooled in a desiccator and the mass recorded (three decimal places). The mass of the dispersant must be subtracted to give the final mass of the sediment fraction. Clifton et al. (1999) found that turbulence continued after stirring ceased and recommended that timing commence after 20 seconds have elapsed.

In calculating the grain-size distribution, the fraction by mass of material finer than the nominal diameter for a time t is expressed as:

$$F_t = \frac{M_t}{V_t} \div \frac{M_o}{V_o}$$

Where M_0 is the mass of the entire sample plus dispersant, V_0 is the original volume (i.e., M_0/V_0 is the original concentration), and M_t is the mass of sediment and dispersant in volume V_t taken at time t (M_t/V_t is the aliquot concentration).

5.4   Hydrometer Method

The hydrometer method is another, now less frequently used, gravimetric technique. However, here particle size is inferred from the measurement of the specific gravity of the suspension, which is dependent on the weight of the particles suspended in it. A hydrometer is inserted into the settling cylinder, and variation in the density of the suspension is reflected in the depth at which the hydrometer bulb floats below the fluid surface. Details of the hydrometer method and calculation of particle size are given in Gee and Orr (2002). The hydrometer method has some advantages over the pipette method in that no samples are withdrawn, and suspensions do not need to be dried and weighed, reducing the time required for analysis. However, results may not be very precise due to the difficulties in accurately measuring the depth of the hydrometer bulb below the meniscus. The hydrometer method is based on the same assumptions as the pipette method in terms of particle density, sphericity, absence of interparticle reactions and a constant laminar flow (Goossens, 2008). Consequently, as for the pipette method, particle shape, density and hence mineralogy can have a significant impact on the size distribution determined (Lu et al., 2000; Wen et al., 2002). In addition, if the hydrometer is left in the fluid for the entire duration of the experiment, settling material may adhere to it, introducing accuracies. It is possible to remove the hydrometer and restart the sedimentation after each measurement, but this increases the time involved.

5.5.   Laser diffraction

In laser granulometry, particle size is inferred from the diffraction of incident light by a particle, where the angle of diffraction is inversely proportional to the particle diameter, and the intensity of the diffracted light is proportional to the number of particles within the cross-sectional area of the optical path (Agrawal et al., 2007). Its use is relatively recent, and there is a wealth of literature comparing its application and reliability to other more established techniques, in particular sieving and gravimetric settling (McCave et al., 1986; Loizeau et al., 1994; Konert and Vandenberghe, 1997; Beuselinck et al., 1998; Wen et al., 2002; Eshel et al., 2004; Rodriguez and Uriarte, 2009; Di Stefano et al., 2010).

Detailed explanations of the theory of laser granulometry and instrumentation are provided in Agrawal et al. (2007). The sample (usually dispersed and pretreated to remove organic matter) is suspended in deionised water and introduced to the particle size analyser. Ultrasonification is applied for up to 3 min to the keep particles in suspension during analysis (Di Stefano et al., 2010). Monochromatic light is then passed through a cell containing this sample suspension and focused onto a bank of detectors. To calculate the particle diameter from the diffraction angle, two diffraction theories, the Fraunhofer and the Mie theories, are commonly applied. Both theories assume that particles are spherical, but the Fraunhofer theory assumes that only diffraction takes place, whereas the Mie theory acknowledges that some light may be transmitted through particles and hence includes an absorption coefficient (Beuselink et al., 1998; Eshel et al., 2004; Di Stefano et al., 2010). The Fraunhofer theory is the most commonly used, but becomes inapplicable when the particle diameter approaches the wavelength of light. Hence, when a significant proportion of the sediment is < 10 μm in size, the Mie theory is preferred (Pye and Blott, 2004). The Mie theory requires an estimation of refractive index for the sediment analysed; it is therefore less robust for heterogeneous sediments. As a consequence, a combination of both theories is used to calculate the particle size distribution of sediments comprising clays, silts and sands. During analysis, gas bubbles can be produced within the sample, producing erroneous measurements, and samples should only be stirred gently, and degassed water should be used. To maintain data quality, background measurements should be made, and the instrument should be regularly cleaned to prevent cross-contamination between samples.

Laser granulometry is widely used as it offers rapid (5–10 min per sample), auto-mated, accurate and reliable analysis for small samples (approx. 50 mg; Beuselink et al., 1998; Pye and Blott, 2004), frequently without the need for an additional sieving step. It operates within the range of 0.1–2000 μm, depending on the instrument used, but detection size can be reduced to 0.04 μm by the addition of a detector that measures the polarization intensity differential scattering of light – the PIDS system (Pye and Blott, 2004). The particle diameter calculated is determined based on the two-dimensional (2-D) shape of the particle, and there is an assumption that particles are spheres, and that the orientation of particles is random (Konert and Vandenberghe, 1997). Therefore, laser granulometry, particularly where the Fraunhofer theory has been applied, can overestimate the clay-sized fraction compared with gravimetric techniques, and this is attributed to the properties (mineralogy, size and shape) of the clay fraction (e.g., McCave et al., 1986; Loizeau et al., 1994; Konert and Vandenberghe, 1997; Beuselinck et al., 1998; Eshel et al., 2004). Significant differences have also been noted between laser diffraction and sieving results for sand (e.g., Blott and Pye, 2006; Rodriguez and Uriarte, 2009).

5.6   Electro-Resistance Size Analysis Using a Coulter Counter

Electro-resistance size analysis is based on the measurement of electrical resistivity between two electrodes. Particles are suspended in an electrolyte solution, which is then drawn through an aperture of known diameter. Two electrodes sit on either side of this aperture, and they measure the change in electrical impedance that occurs when a particle displaces the electrolyte solution. The difference in impedance is proportional to the volume of electrolyte displaced and hence the volume of the particle, which can then be calibrated against spheres of known volume. Thus, these instruments are particle counters, but can also measure the volume and concentration of particles in the electrolyte solution. Further details of the theory of electro-resistance size analysis can be found in Milligan and Kranck (2007).

Details of the operating instructions and conditions can be found in the manufactur-ers' handbooks and have changed little from well-established earlier texts (e.g., Shide-ler, 1976) and hence will not be covered in detail here. Sediments are first treated to remove organic matter, dried, weighed and then added to an electrolyte and dispersed using ultrasonification. A range of electrolytes are used including seawater, NaCl and more viscous solutions for coarser sands (Milligan and Kranck, 2007). Particles can only be sized if they are between 2 and 50 per cent of the aperture diameter, and therefore multiple aperture tubes with overlapping diameters are required (Poppe, Eliason, Fredericks, et al., 2003; Milligan and Krank, 2007). As a result, instruments cannot resolve clays that are < 0.5 µm in diameter, which is a significant disadvantage in comparison with laser diffraction techniques. Care must be taken to ensure that the sample does not contain particles larger than the aperture in order to avoid blockage, and this may require an additional screening step. In addition, the concentration of suspended particles must be such that two particles are not sized together because they overlap. This is termed 'coincidence' (Milligan and Kranck, 2007) and can result in an underestimation of the finest particles.

Generally, the accuracy of electro-resistance techniques is high, although reproducibility may be poor because only a very small number of particles are actually counted. This can become problematic in the larger size classes, where the number of particles may be extremely low if coincidence is avoided, but can be overcome by increasing count times (up to 5 min) and electrolyte volumes (Goossens, 2008). This technique has been used widely since the 1970s, and the eponymous Coulter Counter is perhaps the best-known of the instruments available. The instruments can be automated, operate over the range of 0.5–1000 µm, and only small volumes of material (0.05–0.5 g) are required. Hence, for some time this has been an attractive technique for the grain-size analysis of estuarine sediments.

5.7 Other Techniques

There is a range of additional techniques that are perhaps now less frequently used for the analysis of grain size in the estuarine environment and will not be covered here in detail. The SediGraph and Atterberg Cylinder (a modification of the pipette method) are both based on gravimetric settling and assume that particles settle out of suspension according to Stokes' Law. For the SediGraph, X-ray transmittance through a sediment suspension is related to turbidity, and by carefully controlling the position of the X-ray sensor with depth and time, analysis times can be reduced very significantly compared with the pipette technique (Coakley and Syvitski, 2007). The SediGraph produces data that are comparable with other sedimentation techniques and shows similar issues when compared with laser granulometry, i.e., underestimation of the clay-sized fraction (Molinaroli et al., 2011). Image analysis, which uses software to analyse the geometric properties of particles and particle images, can provide data on both particle size and shape, although its use tends to be restricted to sand-sized particles (e.g., Urbanski et al., 2011).

6 Grain-Size Distribution

The wide range of techniques discussed earlier uses many different parameters as a descriptor of particle size, including the diameter of an equivalent sphere, the volume of an equivalent sphere and the sieve aperture dimension. The data generated by different techniques rarely agree, and other particle characteristics such as shape, density and mineralogy clearly have a strong influence on results. To ease presentation and interpretation, grain-size data are first subdivided into size classes, with five first-order

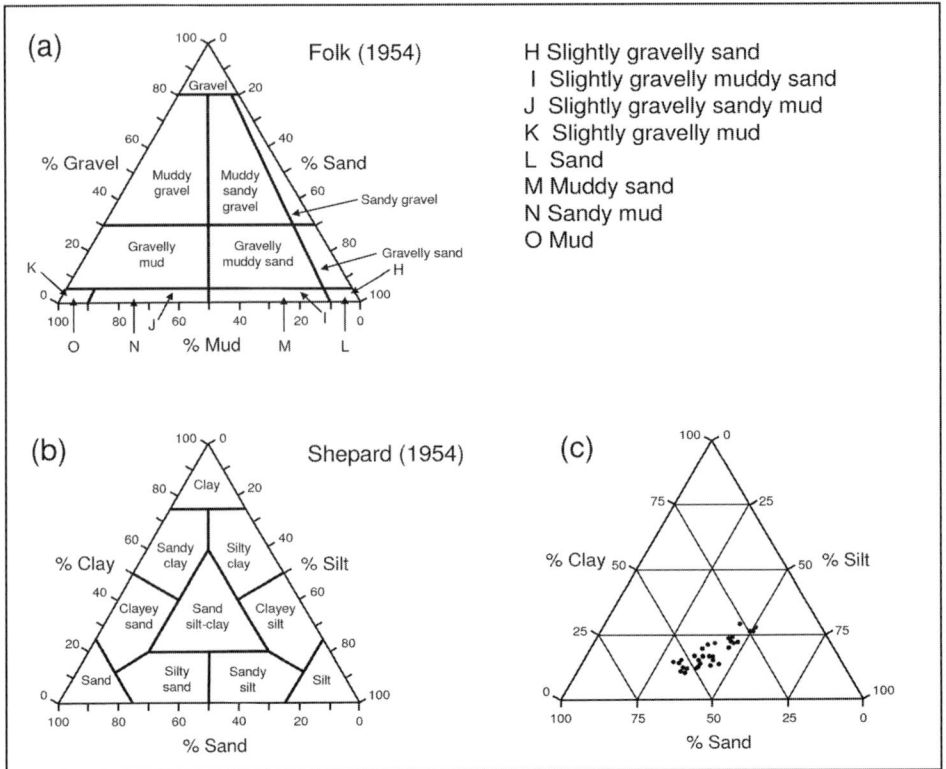

Figure 6-4 Trigons showing sediment classification schemes based on the relative percentages of: (a) gravel, sand and clay (reproduced with minor modifications from Folk, 1954; with permission); (b) sand, silt and clay (reproduced with minor modifications from Shepard, 1954; with permission); and (c) classification of sediment from a saltmarsh in the Thames Estuary, UK (Courtesy of Dr Francis O'Shea, 2016; personal communication; with permission).

classes (boulder, gravel, sand, silt and clay) and variable numbers of subclasses. The size class boundaries most frequently used (as in Figure 6-3) are perhaps those of Wentworth (1922), although the relative merits of other schemes proposed by, e.g., Atterberg and the US Department of Agriculture have been discussed (Blott and Pye, 2012). The boundaries between these class sizes are also frequently expressed in phi (φ) values, where $\varphi = -\log_2 d$ and d is the particle diameter (in units of mm). Data can then be recorded as percentage mass or percentage volume in each size class.

Grain size is perhaps most simply presented using ternary (trigon or triplot) diagrams, which display the relative abundance of three size fractions; gravel, sand and fines (silt and clay combined); or sand, silt and clay (Figure 6-4 a, b and c). These schemes were developed to describe texture using laboratory (Shepard, 1954) and field (Folk, 1954) data, and whilst they provide a standardised approach, results are dependent on the size class boundaries used. More recent modifications have sought to use this simple approach to provide more information on sediment genesis, for example, distinguishing between various hydrodynamic regimes (Flemming, 2000), and graphical representations can now be rapidly achieved using appropriate software packages (Poppe, Eliason, and Hastings, 2003; Blott and Pye, 2012).

Grain-size distribution is usually described in terms of its deviation from a normal (Gaussian) distribution, and a geometric scaling (log-normal) is used to allow equal weighting to both small differences between fine particles and larger differences between coarse particles (Blott and Pye, 2001). The parameters most commonly used to describe and interpret these distributions are the mean, sorting (spread around the mean), symmetry or 'skew', and peakedness or 'kurtosis'. These parameters can be determined using the mathematical 'method of moments' (Krumbein, 1938) and can be determined graphically by plotting a cumulative frequency curve for the grain-size data. More recent and widely available statistical packages, such as 'Gradistat' (Blott and Pye, 2001), which is based on Microsoft Excel, make this much quicker and simpler to accomplish. In addition to these simple descriptive statistics, many workers have used multivariate statistical analysis, for example, to classify sediments using cluster analysis and interpret depositional environments using factor analysis; such approaches are reviewed in Syvitski (2007).

7   Shape

Sediment is highly variable in composition, and, depending on source, weathering and transport history, particle shape will also vary, ranging from plate-like clay minerals to near-spherical quartz grains. Particle shape provides information on the transport and settling characteristics of sediment, and whilst most sediment transport models require information regarding sediment grain size and density, some determination of shape would provide significant improvements. Shape also significantly influences grain-size analysis, specifically those techniques that rely on gravimetric settling, with most techniques using a single dimension as a size descriptor (Matthews, 2007).

Shape is perhaps a rather vague term and generally includes an assessment of form, sphericity, roundness (or angularity) and irregularity. The use of these terms and approaches to classifying shape were reviewed thoroughly by Blott and Pye (2008). Perhaps the simplest characterisation of shape is the measurement of the three dimensions that best describe the particle (form), and the definitions of shape class originally developed by Zingg (1935) and modified by Sneed and Folk (1958) are still widely used today. These measurements can easily be acquired for larger particles (> sand sized), by direct measurement using a tape, calipers or measuring frame, although this is time consuming, and it is difficult to obtain large data sets. Examining the shape of finer particles (sands and to a lesser extent silts) requires the acquisition of high-resolution digital images and the use of image analysis software (Chapter 8) to allow rapid, automated assessment of a wide number of shape characteristics (e.g., Graham and Midgely, 2000; Urbanski et al., 2011; Lira and Pina, 2009).

8   Sediment Structure, Porosity and Density

There is a wide range of additional physical properties that can be used to characterise sediments, including moisture content, porosity, permeability and bulk density. Where sediment cores have been collected and care taken not to disturb their physical characteristics (see Section 2.2), observations of the spatial distribution and arrangement of these physical properties can provide information on sediment structure. Such information is frequently of use in the study of deep marine and estuarine cores because it can provide information on stratigraphy, allow correlation between multiple cores, and provide information on sediment accumulation trends. However, information on porosity and density can also be used to characterise sediments in terms of their engineering and hydraulic characteristics (e.g., their erodibility) and to understand the movement of solutes through sediment, which may be of use to those who manage the estuarine environment and operate within it, such as dredging companies, port authorities and berth operators. In order to examine physical characteristics such as density and porosity, sediments can be returned to the laboratory following collection and analysed ex situ. However, there is also a wide range of continuous, nondestructive analytical techniques that typically have been used to log marine sediment cores.

Dry bulk density is the mass of dry solids in a given bulk volume (Brady, 1984) and is perhaps the simplest physical characteristic to obtain. Sediment is extracted using a core or bulk density rings of known size. To obtain dry bulk density, sediments are either dried at 100°C overnight or freeze-dried (Bartholdy et al., 2010), and dry bulk density is then calculated using:

$$\rho_b = M_s / V_t$$

In which ρ_b is dry bulk density (kg m^{-3}), M_s is mass of dry solids (kg) and V_t is total volume of wet solids (m^3).

Estuarine sediments may frequently have high water and organic matter contents, and porosity is a measure of the volume that is not occupied by solids (Chapter 9). Porosity is calculated from the bulk dry density and particle density, which is usually taken as 2650 kg m^{-3}, the density of quartz, and assumes that silica is the main mineral component of the sediment. Therefore, porosity can be simply expressed as:

$$P = \left(1 - \rho_b / \rho_p\right) \times 100$$

in which P is the % porosity and ρ_p is the particle density. When the organic matter content is known, this can also be included in the calculation; for example, Bartholdy et al. (2010) assume a bulk density for organic carbon of 1140 kg m^{-3}.

Continuous nondestructive logging provides a rapid and convenient way to analyse sediments collected offshore, where a return to the laboratory may not be practical and/or cost effective. Physical sediment parameters such as density and porosity can be obtained using a variety of techniques, including natural gamma radiation, gamma density, p-wave velocity, magnetic susceptibility, electrical resistivity and colour reflectance. Data are obtained over centimetre scales and can be interpreted to provide detail regarding sediment structure and stratigraphy (e.g., St-Onge et al., 2007).

The use of X-rays combined with digital image analysis can provide information on sediment structure. X-rays are attenuated as they pass through sediment, and the degree of attenuation is largely due to the density of the material, although water content, mineralogy and compaction are also important. Initial studies were carried out using medical computed tomography (CT) scanners, and these provided 2-D images that showed bulk sediment structural features such as laminations, gas bubbles and burrows (Amos et al., 1996; Solan and Kennedy, 2002). However, with improved resolution and data processing power, the development of micro-CT could enable the 3-D observation and quantification of benthic processes (Mazik et al., 2008).

9   Final Remarks

This chapter has discussed the sampling and analysis of estuarine deposited sediments and some of their important properties, including particle shape, sediment structure, porosity and density. Several methods of sediment collection have been described, ranging from scoops or lightweight grab samplers for very shallow-water work, to the use of heavy grab samplers from ships in deeper waters. Core samplers range from handheld tubes that are pushed into the sediment, which are often used on saltmarshes during low-water exposure, to gravity and piston corers, vibrocorers and percussion corers and box corers. Some mention has been made of the techniques used to sample sediment within the interface region between water column and bed sediment. Survey design and the frequency of sediment sampling, as well as the storage, preparation and pretreatment of samples and particle sizing have been described. Pretreatment may include washing, biogenic calcite and organic matter removal, flocculent-dispersal and drying. The coarse-grained sediment fraction sizes are often determined by sieving, whereas the fine-grained sizes are usually measured via sieving, gravimetric techniques, or methods such as laser diffraction and electro-resistance, using the Coulter Counter.

References

Adachi, H., Yamano, H., Miyajima, T., Nakaoka, M., 2010. A simple and robust procedure for coring unconsolidated sediment in shallow water. *Journal of Oceanography* 66, 865–872.

Agrawal, Y. C., McCave, I. N., Riley, J. B., 2007. Laser diffraction size analysis. In: Syvitski, J. P. M. (ed.), *Principles, Methods and Application of Particle Size Analysis* (3rd ed.). Cambridge: Cambridge University Press, pp. 119–128.

Allen, T. A., 1990. *Particle Size Measurement* (4th edition). London: Chapman and Hall.

Allen, J. R. L., Thornley, D. M., 2004. Laser granulometry of Holocene estuarine silts: Effects of hydrogen peroxide treatment. *Holocene* 14, 290–295.

Amos, C. L., Sutherland, T. F., Radzijewski, B., Doucette, M., 1996. A rapid technique to determine bulk density of fine-grained sediments by X-ray computed tomography. *Journal of Sedimentary Research. Section A, Sedimentary Petrology and Processes* 66, 1023–1025.

Apitz, S. E., 2010. Waste or resource? Classifying and scoring dredged material management strategies in terms of the waste hierarchy. *Journal of Soils and Sediments* 10, 1657–1668.

Apitz, S. E., 2012. Conceptualizing the role of sediment in sustaining ecosystem services: Sediment-ecosystem regional assessment (SEcoRA). *Science of the Total Environment* 415, 9–30.

Barnett, P. R. O., Watson, J., Connelly, D., 1984. A multiple corer for taking virtually undisturbed samples from shelf, bathyal and abyssal sediments. *Oceanologica Acta* 7, 399–408.

Bartholdy, J., Pedersen, J. B. T., Bartholdy, A. T., 2010. Autocompaction of shallow silty salt marsh clay. *Sedimentary Geology* 223, 310–319.

Beuselinck, L., Govers, G., Poesen, J., Degraer, G., 1998. Grain-size analysis by laser diffracto-metry: Comparison with the sieve-pipette method. *Catena* 32, 193–208.

Birch, G. F., Chang, C. H., Lee, J. H., Churchill, L. J., 2013. The use of vintage surficial sediment data and sedimentary cores to determine past and future trends in estuarine metal contamination (Sydney estuary, Australia). *Science of the Total Environment* 454, 542–561.

Blomqvist, S., 1991. Quantitative sampling of soft-bottom sediments – problems and solutions. *Marine Ecology Progress Series* 72, 295–304.

Blott, S. J., Pye, K., 2001. GRADISTAT: A grain size distribution and statistics package for the analysis of unconsolidated sediments. *Earth Surface Processes and Landforms* 26, 1237–1248.

Blott, S. J., Pye, K., 2006. Particle size distribution analysis of sand-sized particles by laser diffraction: an experimental investigation of instrument sensitivity and the effects of particle shape. *Sedimentology* 53, 671–685.

Blott, S. J., Pye, K., 2008. Particle shape: A review and new methods of characterization and classification. *Sedimentology* 55, 31–63.

Blott, S. J., Pye, K., 2012. Particle size scales and classification of sediment types based on particle size distributions: Review and recommended procedures. *Sedimentology* 59, 2071–2096.

Brady, N. C., 1984. *The Nature and Properties of Soils, 9.* New York: Macmillan Publishing Co.

Burningham, H., French, J., 2006. Morphodynamic behaviour of a mixed sand-gravel ebb-tidal delta: Deben estuary, Suffolk, UK. *Marine Geology* 225, 23–44.

Caeiro, S., Painho, M., Goovaerts, P., Costa, H., Sousa, S., 2003. Spatial sampling design for sediment quality assessment in estuaries. *Environmental Modelling and Software* 18, 853–859.

Clifton, J., McDonald, P., Plater, A., Oldfield, F., 1999. An investigation into the efficiency of particle size separation using Stokes' Law. *Earth Surface Processes and Landforms* 24, 725–730.

Coakley, J. P., Syvitski, J. M. P., 2007. SediGraph Technique. In: Syvitski, J. P. M., 2007. *Principles, Methods and Application of Particle Size Analysis.* Cambridge: Cambridge University Press, 129–142.

Di Stefano, C., Ferro, V., Mirabile, S., 2010. Comparison between grain-size analyses using laser diffraction and sedimentation methods. *Biosystems Engineering* 106, 205–215.

Environment Canada, 1994. *Guidance Document on Collection and Preparation of Sediments for Physicochemical Characterisation and Biological Testing.* Ottawa: Environmental Protection Series. Environment Canada.

Eshel, G., Levy, G. J., Mingelgrin, U., Singer, M. J., 2004. Critical evaluation of the use of laser diffraction for particle-size distribution analysis. *Soil Science Society of America Journal* 68, 736–743.

Flemming, B. W., 2000. A revised textural classification of gravel-free muddy sediments on the basis of ternary diagrams. *Continental Shelf Research* 20, 1125–1137.

Folk, R. L., 1954. The distinction between grain size and mineral composition in sedimentary-rock nomenclature. *Journal of Geology* 62, 344–359.

Gee, G. W., Or, D., 2002. Particle-size analysis. In: Dane, J. H., Topp G. C. (eds.), Soil Science Society of America Book Series. Vol. 5, *Methods of Soil Analysis*. Part 4. Physical methods, 255–293. Madison, WI: Soil Science Society of America.

Glew, J. R., 1991. Miniature gravity corer for recovering short sediment cores. *Paleolimnology* 5, 285–287.

Glew, J. R., Smol, J. P., Last, W. M., 2002. Sediment core collection and extrusion. In: Last, W. S., Smol, J. P. (eds.), *Tracking Environmental Change Using Lake Sediments, Basin Analysis, Coring, and Chronological Techniques Developments in Paleoenvironmental Research.* London: Kluwer Academic Publishers. 73–105.

Goossens, D., 2008. Techniques to measure grain-size distributions of loamy sediments: A comparative study of ten instruments for wet analysis. *Sedimentology* 55, 65–96.

Grabowski, R. C., Wharton, G., Davies, G. R., Droppo, I. G., 2012. Spatial and temporal variations in the erosion threshold of fine riverbed sediments. *Journal of Soils and Sediments* 12, 1174–1188. doi:10.1007/s11368-012-0534-9.

Graham, D. J., Midgley, N. G., 2000. Graphical representation of particle shape using triangular diagrams: An Excel spreadsheet method. *Earth Surface Processes and Landforms* 25, 1473–1477.

Gray, A. B., Pasternack, G. B., Watson, E. B., 2010. Hydrogen peroxide treatment effects on the particle size distribution of alluvial and marsh sediments. *Holocene* 20, 293–301.

IAEA, 2003. *Collection and Preparation of Bottom Sediment Samples for Analysis of Radio-Nuclides and Trace Elements.* Vienna: International Atomic Energy Agency.

Inglett, P. W., Viollier, E., Roychoudhury, A. N., Van Cappellen, P., 2004. A new idea in marsh coring: The wedge. *Soil Science Society of America Journal* 68, 705–708.

Jahnke, R. A., Knight, L. H., 1997. A gravity-driven, hydraulically damped multiple piston corer for sampling fine-grained sediments. *Deep Sea Research, Part I* 44, 713–718.

Jones, B., Phimester, K. F., Hunter, I. G., Blanchon, P., 1992. A quick, inexpensive, self-contained coring system for use underwater. *Journal of Sedimentary Petrology* 62, 725–728.

Kersten, M., Smedes, F., 2002. Normalization procedures for sediment contaminants in spatial and temporal trend monitoring. *Journal of Environmental Monitoring* 4, 109–115.

Konert, M., Vandenberghe, J., 1997. Comparison of laser grain size analysis with pipette and sieve analysis: A solution for the underestimation of the clay fraction. *Sedimentology* 44, 523–535.

Kornijów, R., 2013. A new sediment slicer for rapid sectioning of the uppermost sediment cores from marine and freshwater habitats. *Journal of Paleolimnology* 49, 301–304.

Krumbein, W. C., 1938. Size frequency distribution of sediments and the normal phi curve. *Journal of Sedimentary Petrology* 8, 84–90.

Krumbein, W. C., Pettijohn, F. J., 1938. *Manual of Sedimentary Petrography.* New York: Plenum.

Kuenen, P. H., 1968. Settling convection and grain size analysis. *Journal of sedimentary Petrology* 38, 817–831.

Kunze, G. W., Dixon, J. B., 1982. Pre-treatment for mineralogical analysis. In: Klute, A. (ed.), *Methods of Soil Analysis, Part 1. Physical and Mineralogical Methods – Agronomy Monograph no. 9*, Madison, WI: Soil Science Society of America, Inc., pp. 91–100.

Lane, C. M., Taffs, K. H., 2002. The LOG corer – a new device for obtaining short cores in soft lacustrine sediments. *Journal of Paleolimnology* 27, 145–150.

Lewis, D. W., 1984. *Practical Sedimentology*. New York: Hutchinson Ross Publishing Company.

Lira, C., Pina, P., 2009. Automated grain shape measurements applied to beach sands. *Journal of Coastal Research* SI56, 1527–1531.

Loizeau, J. L., Arbouille, D., Santiago, S., Vernet, J. P., 1994. Evaluation of a wide range laser diffraction grain-size analyzer for use with sediments. *Sedimentology* 41, 353–361.

Loring, D. H., 1991. Normalisation of heavy metal data from estuarine and coastal sediments. *ICES Journal of Marine Science* 48, 101–115.

Lu, N., Ristow, G. H., Likos, W. J., 2000. The accuracy of hydrometer analysis for fine-grained clay particles. *Geotechnical Testing Journal* 23, 487–495.

Macreadie, P. I., Hughes, A. R., Kimbro, D. L., 2013. Loss of 'blue carbon' from coastal salt marshes following habitat disturbance. *PLoS ONE* 8, 7, e69244.

Matthews, M. D., 2007. The effect of grain shape and density on size measurement. In: Syvitski, J. P. M. (ed.). *Principles, Methods and Application of Particle Size Analysis*. Cambridge: Cambridge University Press, 22–33.

Mazik, K., Curtis, N., Fagan, M. J., Taft, S., Elliott, M., 2008. Accurate quantification of the influence of benthic macro- and meio-fauna on the geometric properties of estuarine muds by micro computer tomography. *Journal of Experimental Marine Biology and Ecology* 354, 192–201.

McCave, I. N., Bryant, R. J., Cook, H. F., Coughanowr, C. A., 1986. Evaluation of a laser-diffraction analyzer for use with natural sediments. *Journal of Sedimentary Petrology* 56, 561–564.

Milligan, T. G., Kranck, K., 2007. Electroresistance particle size analyzers. In: Syvitski, J. P. M., 2007. *Principles, Methods and Application of Particle Size Analysis*. Cambridge: Cambridge University Press, 109–118.

Milligan, T. G., Law, B. A., 2013. Contaminants at the sediment-water interface: Implications for environmental impact assessment and effects monitoring. *Environmental Science & Technology* 47, 5828–5834.

Molinaroli, E., De Falco, G., Matteucci, G., Guerzoni, S., 2011. Sedimentation and time-of-transition techniques for measuring grain-size distributions in lagoonal flats: Comparability of results. *Sedimentology* 58, 1407–1413.

Mudroch, A., MacKnight, S. D. (eds.), 1994. *Handbook of Techniques for Aquatic Sediments Sampling*. Boca Raton, FL: CRC.

Muschenheim, D. K., Milligan, T. G., 1996. Flocculation and accumulation of fine drilling waste particles on the Scotian Shelf. *Marine Pollution Bulletin* 32, 740–745.

O'Shea, F., 2016. *Personal Communication. School of Geography*, Queen Mary University of London, Mile End, London E1 4NS, UK.

Otvos, E. G., 2000. Beach ridges – definitions and significance. *Geomorphology* 32, 83–108.

Pedersen, T. F., Malcolm, S. J., Sholkovitz, E. R., 1985. A lightweight gravity corer for undisturbed sampling of soft sediments. *Canadian Journal of Earth Sciences* 22, 133–135.

Percival, J. B., Lindsay, P. J., 1997. Measurement of physical properties of sediments. In: Mudroch, A., Azcue, J. M., Mudroch, P. (eds.), *Manual of Physico-Chemical Analysis of Aquatic Sediments*. Boca Raton, FL: CRC Press, 7–38.

Poppe, L. J., Eliason, A. H., Fredericks, J. J., Rendigs, R. R., Blackwood, D., Polloni, C. F., 2003. Grain-size analysis of marine sediments: Methodology and data processing. *U. S. Geological Survey Open-File Report* 00–358.

Poppe, L. J., Eliason, A. H., Hastings, M. E., 2003. A Visual Basic program to classify sediments based on gravel-sand-silt-clay ratios. *Computers and Geosciences* 29, 805–809.

Pye, K., Blott, S. J., 2004. Particle size analysis of sediments, soils and related particulate materials for forensic purposes using laser granulometry. *Forensic Science International* 144, 19–27.

Rodriguez, J. G., Uriarte, A., 2009. Laser diffraction and dry-sieving grain size analyses undertaken on fine- and medium-grained sandy marine sediments: A note. *Journal of Coastal Research* 25, 257–264.

Sansone, F. J., Hollibaugh, J. T., Vink, S. M., Chambers, R. M., Joye, S. B., Popp, B. N., 1994. Diver-operated piston corer for nearshore use. *Estuaries* 17, 716–720.

Schuenemeyer, J. H., Drew, L. J., 2011. *Statistics for Earth and Environmental Scientists.* Hoboken, NJ: John Wiley and Sons. doi:10.1002/9780470650707.

Shepard, F. P., 1954. Nomenclature based on sand-silt-clay ratios. *Journal of Sedimentary Petrology* 24, 151–158.

Shideler, G. L., 1976. Comparison of electronic particle counting and pipette techniques in routine mud analysis. *Journal of Sedimentary Petrology* 46, 1017–1025.

Simpson, S. L., Batley, G. E., Chariton A. A., Stauber J. L., King C. K., Chapman J. C., Hyne R. V., Gale S. A., Roach A. C., Maher W. A., 2005. *Handbook for Sediment Quality Assessment.* Bangor: CSIRO.

Sneed, E .D., Folk, R. L., 1958. Pebbles in the Lower Colorado River, Texas: A study in particle morphogenesis. *The Journal of Geology* 66, 114–150.

Solan, M., Kennedy, R., 2002. Observation and quantification of in situ animal-sediment relations using time-lapse sediment profile imagery (t-SPI). *Marine Ecology Progress Series* 228, 179–191.

St-Onge, G., Mulder, T., Francus, P., Long, B., 2007. *Continuous Physical Properties of Cored Marine Sediments, Developments in Marine Geology, Volume 1.* Amsterdam: Elsevier, 63–98.

Stoodley, J., 1998. A monolith sampler for saltmarsh sediments. *Journal of Sedimentary Research* 68, 1046–1047.

Syvitski, J. P. M., 2007. *Principles, Methods and Application of Particle Size Analysis.* Cambridge: Cambridge University Press.

Teasdale, P. A., Collins, P. E. F., Firth, C. R., Cundy, A. B., 2011. Recent estuarine sedimentation rates from shallow intertidal environments in western Scotland: Implications for future sea-level trends and coastal wetland development. *Quaternary Science Reviews* 30, 109–129.

Turner, R. E., Milan, C. S., Swenson, E. M., 2006. Recent volumetric changes in salt marsh soils. *Estuarine Coastal and Shelf Science* 69, 352–359.

Urbanski, J., Wochna, A., Herman, A., 2011. Automated granulometric analysis and grain-shape estimation of beach sediments using object-based image analysis. *Journal of Coastal Research* SI64, 1745–1749.

U.S. EPA, 2001. *Methods for Collection, Storage and Manipulation of Sediments for Chemical and Toxicological Analyses: Technical Manual. EPA 823-B-01-002.* Washington, DC: U.S. Environmental Protection Agency, Office of Water.

U.S. EPA, 2005. *Collection of Undisturbed Surface Sediments: Sampler Design and Initial Evaluation Testing. EPA/600/R-05/076.* Las Vegas, NV: U.S. Environmental Protection Agency.

Weaver, P. P. E., Schultheiss, P. J., 1990. Current methods for obtaining, logging and splitting marine sediment cores. *Marine Geophysical Research* 12, 85–100.

Webster, R., 1999. Sampling, estimating and understanding soil pollution. In: Gómez-Hernández, J., Soares, A., Froidevaux, R. (eds.), *GeoEnvII 98—Geostatistics for Environmental Applications. Quantitative Geology and Geostatistics.* Dordrecht: Kluwer Academic Publishers.

Wen, B. P., Aydin, A., Duzgoren-Aydin, N. S., 2002. A comparative study of particle size analyses by sieve-hydrometer and laser diffraction methods. *Geotechnical Testing Journal* 25, 434–442.

Wentworth, C. K., 1922. A scale of grade and class terms for clastic sediments. *Journal of Geology* 30, 377–392.

Winterwerp, J. C., van Kesteren, W. G. M., 2004. *Introduction to the Physics of Cohesive Sediment in the Marine Environment.* Vol. 56, *Developments in Sedimentology*. Oxford: Elsevier.

Wright, H. E., 1991. Coring tips. *Journal of Paleolimnology* 6, 37–50.

Wright, H. E., 1993. Core compression. *Limnology and Oceanography* 38, 699–701.

Xu, J. R., Wang, Y. S., Yin, J. P., Lin, J. P., 2011. New series of corers for taking undisturbed vertical samples of soft bottom sediments. *Marine Environmental Research* 71, 312–316.

Ysebaert, T., Herman, P. M. J., Meire, P., Craeymeersch, J., Verbeek, H., Heip, C. H. R., 2003. Large-scale spatial patterns in estuaries: Estuarine macrobenthic communities in the Schelde estuary, NW Europe. *Estuarine Coastal and Shelf Science* 57, 335–355.

Zingg, T., 1935. Beitrag zur schotteranalyse. *Swiss Bulletin of Minerology and Petrology* 15, 39–140.

7 Suspended Particulate Matter: Sampling and Analysis

S. B. Mitchell, R. J. Uncles, J. A. Stephens

1 Introduction

There are many reasons to sample and measure suspended particulate matter (SPM) concentrations in estuaries and coastal waters. Perhaps foremost among these, from an engineering viewpoint, is the need to determine rates of sediment transport, erosion and deposition, in order to inform those responsible for the management of harbours, ports and navigation channels and the various stakeholders in these facilities. Although numerical sediment-transport models have advanced considerably in recent years, their application to a particular estuary or coastal area, harbour or port, still requires a good understanding of all the numerous factors related to sediments and their behaviour. Another reason for sampling SPM concentrations, from an ecological viewpoint, is the importance of light availability for photosynthesis and its reduction due to water-column SPM. For example, algal productivity and biomass are functions of light availability; therefore, SPM and its associated turbidity also contribute to the inhibition of chlorophyll-*a* production, even when there is a plentiful supply of nutrients (Cloern, 1987). In addition, the transport of fine-grained SPM in estuaries is often linked to the passage or storage of chemical species, due to their adsorption onto the surfaces of suspended fine particles (e.g., Millward et al., 1990; Turner et al., 1994).

Numerous techniques are available for analysing SPM (which includes suspended inorganic minerogenic sediment, Chapter 6) and depending on the focus of study and the location, these methods may be more or less relevant to the sampling team; however, the concentration of material suspended in the water column is invariably of interest. The aim of this chapter is to give some details of the techniques available for estimating a grams-per-litre concentration of SPM, in order to build up a picture of the sediment regime at or in a particular site or system, and then to consider the analysis of some other important SPM properties. These include the SPM mineralogy, its size distribution, specific surface area and in situ settling velocity, as well as the fraction of SPM resulting from loss-on-ignition and the fractions that can be attributed to the organic, chlorophyll-*a*, and extracellular-polymeric-substances' contents of SPM. The techniques commonly used for boat-based, mooring-based and land-based surveys of SPM in estuaries and coastal waters are discussed. We also touch on the topic of remote sensing of SPM concentrations using satellite and aircraft imagery (Chapter 11), although the main focus is on the need to 'ground-truth' data obtained in this way.

In the following sections we discuss some of the issues involved in SPM sampling from moored equipment and from boat-based and land-based surveys. The procedures used for analysing samples are then considered, together with the ways in which optical and acoustic data can be calibrated to give more useful grams-per-litre values, thereby allowing for, e.g., the possibility of calculating an SPM flux in kg m^{-2} s^{-1} (Chapter 9). We then briefly discuss some of the research that is directed toward making measurements of SPM properties using satellites and aircraft (Chapter 11). Finally, we consider the analysis of some of the SPM properties that are particularly important to quantify for sediment transport and ecological studies.

2   SPM Sampling

The measurement of SPM can be carried out either directly, using a sample of known volume that is then filtered, thus allowing the suspended material to be weighed and the concentration to be calculated, or more commonly using some form of optical or acoustic instrument, which can be used to gain an indirect measurement of the SPM concentration. The former method (gravimetric analysis) is generally accurate but time consuming; the latter methods save a great deal of time, especially if some form of logger is used as well, but instrument calibration is clearly necessary to relate the optical or acoustic measurement to the grams-per-litre values. This calibration is usually highly site specific and may also be time specific, relating generally to the nature of the sediment and to the recent history of the system concerned. Nevertheless, the vast majority of studies use some form of sensor/logger that is generally left in place for periods that typically range from about a day to several months or even longer. A number of suppliers have developed instruments to perform this function (e.g., YSI (2016) – an instrument in preparation for deployment is shown in Figure 7-1e). The collection of samples involves careful consideration of the available fieldwork infrastructure, especially sampling platforms.

2.1   Sampling Platforms

Boat-based surveys generally allow the best results, given that these provide a degree of flexibility both in where the samples are collected (or where the water properties are measured, e.g., Figure 7-1k, l, m, n, o) and in allowing the effects of fixed objects (bridge piers and channel edges, for example) to be minimised. However, a boat may not be available, or there may be constraints on the availability of personnel or equipment. In these cases, samples may be collected using some sort of floating platform (e.g., Figure 7-1g), or simply from the edge of the channel (e.g., Figure 7-1a), but then it is important to consider the effects of the boundary on the flow regime. Provided the water near the channel edge is sufficiently deep, it may be possible to obtain a representative sample, but in many cases the proximity to the sides, or the shallowness of the sampling area, profoundly affect the water velocity and hence the representativeness of the sampling. As with all field data collection campaigns, safety is

Figure 7-1 Types of survey and SPM sampling: (a) Deployment of instruments from a jetty, utilising land-based access to the deep channel with improvised cranes and winches (an Owen Tube is rigged for deployment from the boat-trailer crane); (b) Time-series profiling from a bridge dolphin (ship-impact protective structure); (c) Collecting surface water samples with a bucket from Maisemore Bridge in the upper Severn, UK, for SPM analysis; (d) Utilising the water-access steps of a jetty as a platform to pump water samples from depth; (e) A simple mooring comprising a pole driven deeply into the sand holding a YSI instrument deployed at LW of a spring tide; (f) A bottom-mounted frame with logging instrumentation carried into shallow water at LW of a spring tide; (g) Collecting samples and profiling using a floating platform comprising a twin-hulled frame, tower and winch, shown with a sea truck alongside; (h) Using a research vessel to deploy a bed frame with an ADV and logger in the mid-estuarine channel; (i) Using a research vessel to deploy a bed frame with OBS and other instrumentation and logger near the mouth of an estuary; (j) Using a rigid inflatable to attach self-recording instrumentation to an upright structure in an estuary close to the LW spring-tide line; (k) Preparing an inflatable for a longitudinal survey of a small, sheltered, weakly tidal estuary; (l) Using a rigid inflatable at anchor to obtain a short (few hours) time-series of vertical profile data; (m) A modified sea-truck shown at anchor and in use for tidal-cycle profiling in a relatively sheltered estuary; (n) A larger, conventional research vessel at anchor and in use for tidal-cycle profiling in the Humber Estuary, UK, a large, exposed estuary; and (o) A research ship preparing for work over a few weeks in the North Sea and Humber Estuary plume region. Photographs: R. J. Uncles.

paramount, and this consideration, together with the cost of personnel and equipment, often dictate the method to be used.

2.2   Pumped Sampling

SPM concentrations can be determined directly via the gravimetric analysis of samples (i.e., filtering and weighing, Section 4.1). In highly turbid systems, where data interpretation can be affected by uncertainties about the reliability or 'saturation' of optical and acoustic devices, it is still possible to obtain valuable survey data for SPM concentrations using this method (e.g., Mitchell et al., 1998; Uncles and Stephens, 1999).

Surface water samples can be collected either by means of a bucket (e.g., deployed from a bridge, Figure 7-1c) or other receptacle, or in a more controlled way, and at depth, via a small submersible pump, such as a whale pump or similar (e.g., Whale, 2016). A simple pumping arrangement of bucket, tubing, pump and battery (shown in Figure 7-1d) is easily constructed. In collecting a pumped sample, it is important that a precalculated volume of the pumped water is rejected before it is collected in the sample storage bottle, in order to be sure that the collected water is from the required depth, and that the sample storage bottle is rinsed before use, to reduce the risk of cross-contamination from earlier samples. Sampling at depth can also be achieved using shallow-water sample-collection bottles (e.g., Kelly et al., 2000) or, in deeper water, Van Dorn or Niskin bottles (McCave, 1979).

At each stage during water collection and during any subsequent subsampling, it is important to ensure that the sample is well mixed and that any settling effects are carefully considered, particularly for high concentrations that might contain heavier fractions or large, fast-settling flocs (Chapter 8). There is no particular requirement to use one type of sample storage bottle or another for gravimetric analysis; generally, 250-ml plastic bottles will be adequate for most physically based determinations in many estuaries, although multiple or larger bottles may be needed in low-turbidity systems.

3   Types of Survey

It is possible to track a particular pool of sediment or tracer that is transported by the current (the Lagrangian survey technique, e.g., Huhn et al., 2012), but we will focus on surveys at fixed stations (the Eulerian approach, e.g., Green and Hancock, 2012). Survey work in estuarine and coastal waters is generally best achieved by teams of researchers working toward an agreed set of goals. It is always preferable to undertake surveys in good weather and light conditions if at all possible. The selection of a suitable sampling site is important, if this is not already specified. For example, it is usually advantageous to be in the centre of a straight stretch of channel or in an area where the currents and sediments are as representative as possible of the conditions present in a particular reach of estuary or coastal site. Considerations related to the influence of any estuarine turbidity maximum at or near the sampling point (e.g., Mitchell et al., 2012) should also inform the timing and the nature of any sampling regime. The site should be as far as possible from obstacles and away from shipping lanes, other craft, and other users of the water. The timetables of ferries and other scheduled vessels could be helpful. For safety and welfare reasons, boat-based surveys should be near an access point, and there should be a sufficient number of personnel to allow members of the survey team to take breaks or to provide assistance from the land. Good lines of visibility (small-boat workers should be clearly visible from the land) are highly desirable. In most cases it is necessary to inform and request permission from the local harbour or port authorities.

Although this chapter is concerned with SPM data, it is good practice in all deployments of this type to take complementary sets of readings of some other variables at the same time. Salinity, for example, should always be measured if possible, together with temperature, water level above a fixed datum, and depth of measurement, usually by means of a CTD sensor (Chapter 4). To this end, a turbidity sensor (or sensors with different ranges) can be attached to a CTD probe and preferably connected to the same logging system, ensuring synchronicity of readings as far as possible. Local water depth and tide level, possibly from some sort of stage board or from some continuous water-level monitors nearby, should be recorded (Chapter 2), and it is also recommended to take readings of position using a GPS. We will subdivide survey types into land-based, moored platform-based and boat-based surveys, whilst acknowledging that air-based (helicopter and light aircraft) and space-based (satellite imagery, Chapter 11) surveys play an important role in SPM data acquisition.

3.1   Land-Based Surveys

Where there are vertical walls, e.g., at a moored pontoon, jetty (Figure 7-1a), bridge (Figure 7-1b, c) or some sort of sheet-piled bank, or intertidal area such as a mudflat or beach (Figure 7-1e, f), it may be possible to take samples from the water's edge or deploy instruments there during low water (LW) of a spring tide (Figure 7-1e, f), although there are obvious risks in terms of proximity to the water's edge and interference from land-based operations. In addition, account needs to be taken of the fact that, in the absence of waves, flow velocities and suspended material concentrations are likely to be lower than in the main channel. Often, it is a requirement that the sampled water column should be as representative as possible of the estuarine channel section as a whole, and it is then preferable to take samples and use instrumentation at a position where the flow is minimally impacted by the effects of the water boundary.

Whenever a bridge, pontoon, jetty or wall are used for sampling and profiling (Figure 7-1a, b, c) there are risks of accidents and slips, and there are potential problems resulting from interference between the flow and the fixed boundary. Samples should be taken (or measurements made) from as far away from the edge of the water as possible, using some sort of crane (i.e., a winch device with a horizontal arm) jutting out toward the main flow (Figure 7-1a and utilised, but not shown, in Figure 7-1b). In general terms, land-based surveys of this type have the advantage that it can be much easier to collect data, requiring smaller teams, less time to organise, and less equipment. Access to the water is often more straightforward, and equipment can sometimes be left unattended.

3.2   Moored Platform-Based Surveys

An often essential alternative to land-based and boat-based surveys is the use of moored platforms (e.g., the twin-hulled frame, tower and winch shown in Figure 7-1g) or other floating arrangements on the water surface (e.g., Cook et al., 2007), or a frame-based structure deployed on the bed of the estuary or coastal zone (Figure 7-1h, i) or beach (e.g., Palanques et al., 2002), which could simply be a pole driven deeply into the sand or intertidal mud at LW springs (Figure 7-1e), or attached to an upright structure close to the LW spring-tide line (Figure 7-1j). Such approaches offer the possibility of reducing the need for personnel but can still allow the user to gain a representative set of readings in mid channel or in the deeper areas away from the edge of the water, either throughout the tide or at higher water levels, or during subsequent smaller tides, provided the instrumentation remains submerged.

In the case of surface moorings, instruments can be deployed or accessed at any time of slower current speeds for maintenance or downloading of data, providing there is a weather and daylight 'window'; however, instrumented frames installed on the bed should additionally be timed for installation and servicing during a time of slack currents, preferably close to low water in an estuary if sufficient time is available, ensuring that access is safe.

Bottom-mounted frames can be constructed using scaffold tubes or similar materials, affording them the versatility of holding any kind of sensor array in more or less any arrangement. Care must be taken to ensure that the different sensors do not interfere with one another or act as 'traps' for seaweed, or as substrates for the growth of large quantities of algae. Biofouling (as discussed in Chapter 4 and Chapter 10) is a hazard with any sensor that is likely to be submerged for long periods of time.

3.3   Boat-Based Surveys

For shorter deployments, such as a longitudinal, vertical-profiling survey of an estuary, or time-series profiling of up to several hours at a fixed station, it is often advantageous to use a small boat. Often, an inflatable (Figure 7-1k), rigid inflatable (Figure 7-1l), or other small craft with an outboard motor is used, in view of the manoeuvrability and fuel economy of this sort of vessel. The smaller inflatable is suitable only for work in small, sheltered estuaries, whereas the rigid inflatable can be used in larger, more exposed estuarine and (in some circumstances) coastal regions. The modified sea-truck is an ideal vessel for survey work or single tide tidal-cycle work in relatively sheltered estuaries because of its covered laboratory space, its capability of grounding on intertidal mudflats or sand flats should the survey require this, and its shallow hull, which minimises the influence of the boat on profiling measurements (Figure 7-1m). Larger, conventional research vessels are required for tidal-cycle surveys or instrumented bed-frame deployments in the larger, more exposed estuaries (e.g., the 16-m-long vessel in Figure 7-1n), and research ships are used for long-term coastal work in the estuarine plume regions (e.g., the 55 m-long vessel in Figure 7-1o). Whatever the survey type, appropriate safety precautions must always be taken, including the use of life jackets, warm/waterproof clothing, communications protocols, and permissions as appropriate.

3.4   Tidal Cycle Stations

There is merit in spending an entire tidal cycle at one measurement station to interpret sediment transport processes over the whole cycle, to compare these processes with those in other systems, to carry out a net tidal-cycle SPM flux calculation, and to improve understanding of the sources and sinks of sediment involved. However, if a boat is used then the difficulties of holding the craft in place should not be underestimated. Even when fore and aft anchoring is used (as in Figure 7-1m), there is likely to be some small movement that is associated with a degree of slack in the anchor ropes following changes in tidal current direction and winds. In some cases a coxswain can hold the craft in place for short periods using the engine, while the anchor lines are being adjusted.

If large numbers of personnel are available for such work, then several teams can operate simultaneously at different stations. In many cases where whole tidal-cycle surveys are being planned, it is worth keeping in mind that for some of the tide the velocity, salinity and SPM concentration might vary only very slowly over time and depth. An example is the upper reaches of a strongly tidal estuary, where conditions can be essentially fluvial for several hours during the ebb, water levels fall very slowly, SPM concentrations are very low and salinity is negligible. The 'interesting' observations as far as SPM concentrations and transport are concerned generally occur over the fast flood, high-water slack water and early-to-mid ebb periods, so that the time-interval between 'less interesting' observations can be much longer and the effort required much less.

3.5   Moored Instruments

Although we have so far emphasised the tidal aspects of SPM surveys, moored instruments (e.g., Figure 7-1e, f, h, i, j) are ideal when personnel are limited in number and for studying additional aspects, such as wind-induced and wave-induced SPM concentrations, or rain-induced resuspension on mudflats. Continuous monitors (sensors connected to a logger) can be left unattended for long periods at locations such as moorings, floating platforms, sheet-piled banks, and bottom-based frames, with only intermittent servicing and calibration requirements. The Thames (UK) water quality monitors, for example, are generally left to log readings of SPM concentrations every 15 min for long periods, allowing a picture to be built up of the long-term effects of tides and flow rates (Mitchell et al., 2012).

When interpreting data from a moored instrument it is important to take into account the instrument's position relative to the water's surface, whether it is fixed relative to the bed, as on a bed frame or a structure fixed to the bed (Figure 7-1e, f, h, i, j), or moves with the surface, as on a pontoon or a floating frame (Figure 7-1g). Statutory and regulatory bodies in the UK (e.g., the Environment Agency of England and Wales) and elsewhere commonly use continuous monitoring to assess the impact of changes in environmental conditions or the effects of a pollution incident. A limitation of such techniques is that they monitor at only one or a few fixed depths beneath, and relative to, the surface, although this is rapidly changing with the increasing deployments and capabilities of ADCPs (Chapter 4).

4   Methods for Measuring SPM Concentration

There are several ways to determine SPM concentrations in the water column. These can be summarised as direct gravimetric analysis, optical measurements, which include remote sensing by satellite and aircraft (Chapter 11), and acoustical measurements. The gravimetric method is exact, i.e., it produces a true concentration of SPM in, e.g., $kg\ m^{-3}$ or, equivalently, $g\ l^{-1}$ within a sampled portion of the water column, whereas all optical and acoustical methods are dependent, respectively, on the responses of the water and SPM to electromagnetic and acoustic wave propagation and scattering. As such, these surrogate methods are highly dependent on the in situ calibrations of the measuring instruments, which in turn are dependent on gravimetric analysis.

Currently, the vast majority of SPM measurements utilise sensors that are either optical or acoustic in nature and are generally set to read at intervals of a few minutes or less, storing the measured data on a logger or sent instantaneously to a hard drive on a remote computer (Chapter 10). We first describe the procedure for the gravimetric analysis of water samples and then discuss these optical and acoustic instruments.

4.1   Gravimetric Analysis

Gravimetric determination of SPM yields the weight of solids per unit volume of a water-sample suspension, either made as a calibration standard or taken in situ from the field. The result is expressed as a weight per unit volume of suspension (e.g., $g\ l^{-1}$) or as weight-for-weight (such as parts per million, ppm, which is approximately equivalent to $mg\ l^{-1}$).

A volume of suspension (water sample) can be collected using any suitable airtight receptacle that will prevent evaporation of the water during the storage period before sample processing. Whenever possible, samples should be kept cool and in the dark to prevent the growth of organisms between the times of collection and sample analysis. The volume of sample required to obtain an accurate result depends on the SPM concentration and, therefore, location. For example, a sample near the ETM (estuarine turbidity maximum) of a highly turbid estuary such as the Humber-Ouse (Humber Estuary, UK) would require only a few millilitres, whereas a sample from the freshwater inflow to an estuary during low runoff conditions might require several litres.

Having obtained a sample of suspension and returned it to the laboratory, it is resuspended fully by shaking end-over-end before subsampling a known volume using a measuring cylinder for low concentrations, or a pipette for very high concentrations. The subsample is then filtered through a dried, preweighed filter paper (Whatman GF/C, 47-mm filter, weighed to five decimal places) using a vacuum pump, taking care to rinse the sides of the filtration cup to collect any material deposited on the glass or shoulders that may have settled there during filtration (using a wash-bottle with Milli-Q water). A water trap should be used between the filter and the pump to prevent saline water reaching the pump. When the water has been drawn off, the filtration cup must be removed and the filter paper washed gently with a small amount of deionised water, which is particularly important for high-salinity samples. If this practice is not observed, the residual salt can weigh more than the SPM in low-concentration, high-salinity samples. Individual filters must be numbered using a permanent marker before drying and weighing, to allow their identification when retrieving them after final drying.

The filter paper containing the SPM must be placed in, or on, a suitable, clean, nonstick receptacle (a made-up tray of aluminium foil is usually effective) that is then placed in a drying oven at 70–80°C until constant weight has been achieved (24 h is generally sufficient). Whilst it is advisable to weigh to five decimal places to obtain as accurate a reading as possible, cruder bulk results can be obtained using a balance with an accuracy of four decimal places. Drying overnight is usually long enough, but in case of doubt the paper should be weighed, placed in the drying oven for a further 2–3 hours, then reweighed. The filters should be allowed to cool in a dessicator before weighing to avoid absorption of moisture from the atmosphere, which would yield a false reading. The weight of the dry SPM is then determined by subtraction of the filter's weight from the final combined weight of the filter and SPM. The SPM concentration is expressed as the dry weight of SPM per volume filtered (e.g., mg l^{-1}) or as ppm on a weight-for-weight basis. It is important to know the repeatability of the results, and it is necessary to carry out duplicate determinations within the same batch and assess the similarity between them.

A manual, site-based filtering arrangement may be possible when large numbers of samples are to be analysed in systems where the SPM concentration can exceed 1 g l^{-1} (e.g., muddy estuaries). Using dry, numbered, preweighed filter papers, known volumes of samples can be passed through a simple filter manifold attached to a manual hand pump. The filters can then be stored for later drying and weighing under laboratory conditions. This method allows reuse of sample storage bottles during a survey and reduces the volume and weight of the samples that need to be transported back to the laboratory for subsequent drying and weighing. Care must be taken to ensure that none of the filtered material is lost from the surface of the filter papers.

4.2   Optical Measurements of SPM Concentration

In situ optical devices fall into two main categories: those that use an optical backscatter (OBS) sensor, such as those supplied by Seapoint (2016), and those that act as transmissometers (e.g., Partech, 2016), which detect the attenuation of light that passes through a known path-length of water. Generally, optical devices return an output voltage that can be measured by any standard logging device. It is sometimes necessary to select a particular level of gain associated with the sensor, which is a function of the likely range of values of SPM concentration; setting the correct voltage gain is largely a trial-and-error procedure.

A third instrument, the Secchi Disc, is largely of historical interest; however, it is briefly included here because of the relatively recent use of archived Secchi Disc data to describe turbidity variability over a period of four decades in the western Wadden Sea (Philippart et al., 2013).

4.2.1   The Secchi Disc

The instrument is a white disc, usually 0.2–0.3 m in diameter, which is lowered into the surface waters to a depth at which it just disappears from view (McCave, 1979). The depth of disappearance is related to near-surface turbidity. However, the instrument can only be calibrated by taking sufficient simultaneous water samples for SPM concentration. Several corrections can be applied, including those for varying solar altitude and wave-roughened sea surfaces (Philippart et al., 2013).

4.2.2   Transmissometer Instruments

A transmissometer measures turbidity in terms of the percentage of light that is transmitted during its traversal over a fixed path length. A review of these older instruments is given by McCave (1979). Double-beam transmissometers split the transmitted beam and send it along two paths of different lengths, which is a procedure that enables the attenuation coefficient to be determined from the two values of light transmittance (McCave, 1979). A twin-gap instrument utilises a common light source that transmits its light along two, oppositely aligned paths with different lengths (a nominally $0–1$ g l^{-1} instrument is shown in Figure 7-2a). The main feature of this sensor is the displacement of two photocells at differing gap lengths and the use of lenses placed either side of the lamp to convey focused light beams that provide, within the sensor electronics, a bridge balance in the absence of light attenuation. These instruments can be affected by ambient sunlight in shallow, low-turbidity systems, and care must be taken to ensure that they are suitably shielded in these environments (the shield is shown in an open position in Figure 7-2a).

Single-gap transmissometers have been used at higher concentrations, where errors that demand a twin-gap principle become much smaller: e.g., window biofouling and effects due to temperature changes (a nominally $0–10$ g l^{-1} instrument is shown in Figure 7-2b). Single-gap transmissometers that utilise infrared light can measure higher SPM concentrations and currently are used in several applications (Partech, 2016; a nominally $0–30$ g l^{-1} instrument, although it can be nonlinearly calibrated to approx. 40 g l^{-1} in the ETM of a muddy estuary [Uncles et al., 2000a], is shown in Figure 7-2c).

The LISST-100X (laser in situ scattering and transmissometry) instrument is a multiparameter system for in situ observations of particle-size distribution and particle-volume concentration. It also records the optical transmission (Sequoia, 2016). LISST transmissometer data and LISST volume concentrations have been

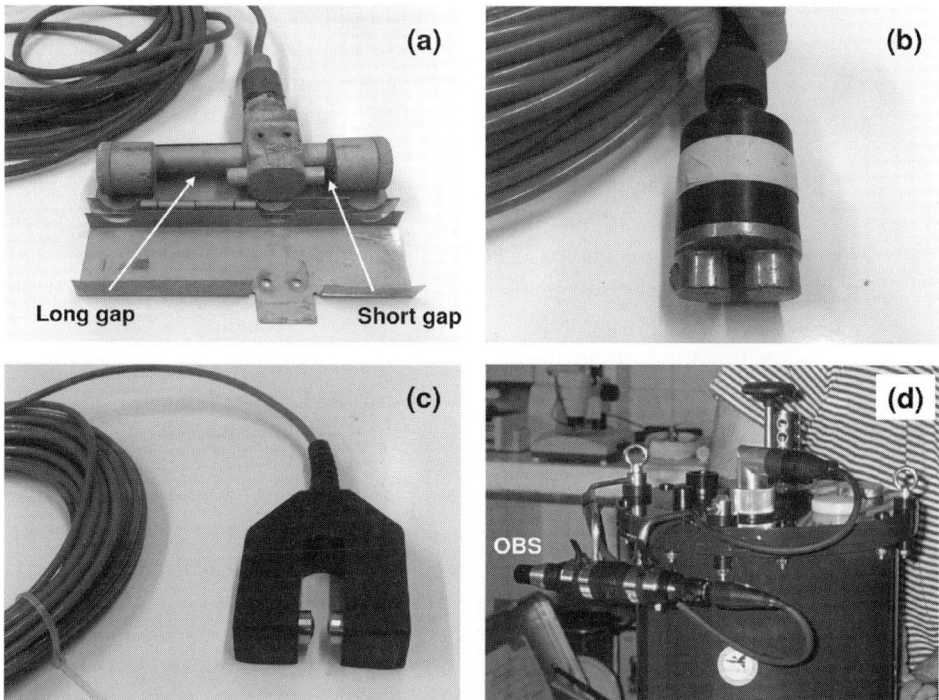

Figure 7-2 Examples of transmissometers and an OBS suitable for estuarine work: (a) A twin-gap transmissometer with its light shield shown in an open position (a nominally 0–1 g l^{-1} instrument); (b) A single-gap transmissometer (a nominally 0–10 g l^{-1} instrument); (c) A single-gap transmissometer that utilises infrared light (a nominally 0–30 g l^{-1} instrument); (d) An optical backscatter (OBS) sensor in preparation for deployment with a wave sensor and logging system. Photographs: R. J. Uncles.

compared with measured estuarine SPM concentrations (which were < 70 mg l^{-1}) by Fugate and Friedrichs (2002) using the now superseded LISST-100.

Transmissometers were frequently used on moorings in the past, and the transmittances often decayed with time as the optical windows become biofouled, necessitating either the abandonment of the record in severe cases or otherwise some form of numerical corrective processing (Chapter 10). Jago and Bull (2000) discuss some of the errors associated with results from transmissometer deployments in coastal waters.

4.2.3   OBS Instruments

The optical backscatter (OBS) sensor is an instrument for measuring turbidity and, via calibration, SPM concentration (Figure 7-2d). Within its detector cone it measures infrared light scattered from SPM in the water column. The sensor consists of an infrared-emitting diode, a detector (comprising photodiodes) and a temperature transducer. Measurements in shallow water are not affected by ambient sunlight because the infrared light utilised by these sensors is strongly absorbed in SPM-free water.

The OBS-3+ and the OBS-300 instruments (Campbell, 2016) are stated to measure turbidity up to 4000 NTU (nephelometric turbidity units) and maximum concentrations of mud and sand suspensions in the approximate ranges 5–10 g l^{-1} and 50–100 g l^{-1}, respectively, with a sampling frequency of up to 10 Hz. These instruments are approx. 0.14 m in length and 0.025 m in diameter. The Seapoint Turbidity Meter (Seapoint, 2016) is stated to measure up to 4000 FTU (formazin turbidity units, where formazin is the polymer reaction product of hydrazine sulphate and hexamethylenetetramine) and are approx. 0.07 m in length, excluding connectors, and 0.025 m in diameter. OBS sensors exhibit greater sensitivity to coarser SPM than transmissometers and suffer less from biofouling (USGS, 2007). Downing (2006) gives a review of the advantages and disadvantages of OBS technology for use in measuring coastal suspended sediments.

4.2.4   Remote Sensing

There is an increasing trend in the use of data buoys and other unmanned platforms for the measurement of turbidity and other variables (Chapter 10). This is especially true of remotely sensed data from satellites (Chapter 11) and from aircraft; e.g., Compact Airborne Spectrographic Imager (CASI) data derived from aircraft overflights (Robinson et al., 1998). The inherent optical properties of a stretch of water can be related to the nature and concentration of its SPM, as well as other variables (Chapter 11), as described by Astoreca et al. (2012) for the Southern North Sea. Doxaran et al. (2009) used satellite and in-water survey data to study the ETM in the Gironde Estuary, France, and Uncles et al. (2001) used CASI data, in conjunction with synchronous ground-truth data from gravimetrically derived SPM concentrations, to investigate the spatial complexity of the ETM in the surface waters of the highly turbid Humber-Ouse Estuary, UK, where observed maximum surface concentrations at that time were 13 g l^{-1}.

Unfortunately, these satellite and aircraft remote sensing methods can only be used to make an assessment of the SPM concentrations near the water surface. Where there are significant vertical gradients in SPM concentration, this additional information will not be available. Therefore, to aid data interpretation, some understanding must first be gained of the magnitude of vertical SPM gradients over space and their variation with time; e.g., Robinson et al. (1998) modelled depth variations of velocity and SPM concentrations, based on CASI-derived surface SPM data, in order to derive SPM transport rates through an estuarine mouth region, and Fettweis et al. (2007) stressed the importance, in their study, of taking into account variations due to tide and weather conditions. Therefore, these techniques can be used to good effect in conjunction with more detailed knowledge of the processes occurring in the water column.

4.3   Acoustic Measurements of SPM Concentration

Acoustic devices make use of sound waves to measure the amplitude change and frequency shift of the signal emitted by a transmitter (Chapter 4). There are acoustic point sensors such as acoustic backscatter (ABS) sensors (Vincent et al., 1998), acoustic Doppler velocimeters (ADVs; Ha et al., 2009), and acoustic profilers such as the

Acoustic Doppler Current Profiler (ADCP; Gartner, 2004; Defendi et al., 2010). Detailed technical information on ADCPs and similar devices can be found at Teledyne RD Instruments (RDI, 2016) and Nortek (2016).

4.3.1   Acoustic Backscatter (ABS) Sensor

The ABS sensor utilises a sonar transducer to emit short pulses of acoustic energy with frequencies that typically are in the approx. range 1–5 MHz (a three transducer instrument is shown in Figure 7-3a). The ABS insonifies SPM within the water column; the SPM, in turn, scatters and reflects some of the acoustic energy back toward the instrument. The magnitude (intensity) of the backscattered signal can be related to the SPM concentration and the constituent particle size via calibration (e.g., Thorne and Hanes, 2002).

A typical ABS deployment could entail mounting the instrument facedown, pointing toward the bed, on a frame (such as that shown in Figure 7-3b) that would also hold

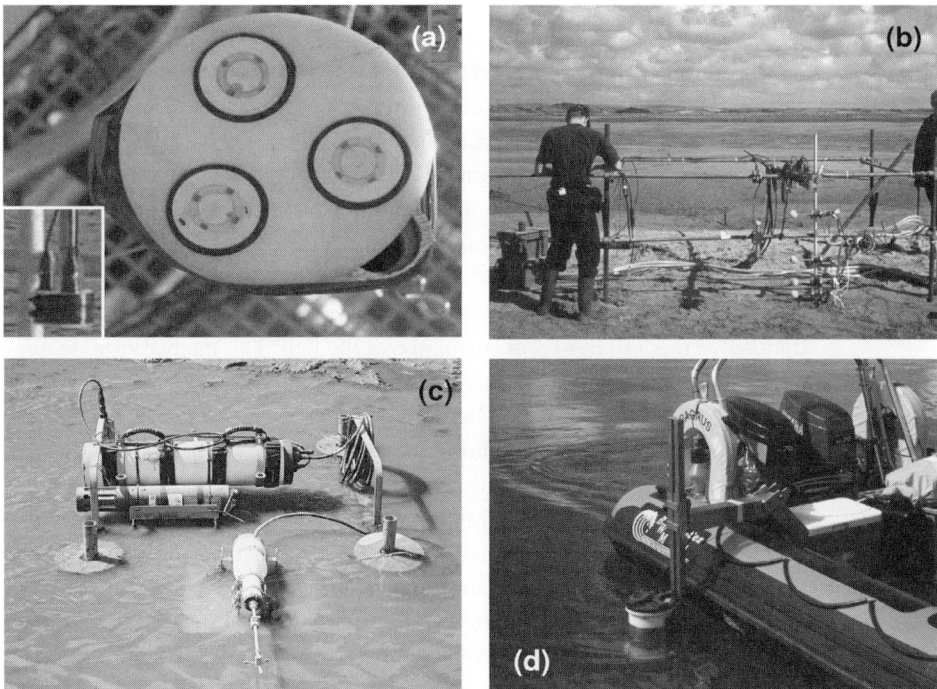

Figure 7-3 Examples of acoustic instruments currently used to measure SPM concentrations: (a) A three-transducer ABS instrument mounted to face toward the sea bed (redrawn, with modifications, from figure 23 of USGS Open-File Report 2007–1194; ref: GSFYXBN3); (b) A bed frame designed to hold downward-facing ABS instruments and salinity, temperature and velocity sensors, together with pumping equipment, under construction on the intertidal sand flats of a large estuary during LW springs; (c) Deployment of an ADV and logger, together with a YSI instrument for recording salinity, temperature and OBS data on the intertidal mudflats of a small subestuary; (d) A downward-facing ADCP mounted on a rigid inflatable, deployed in the Tamar Estuary, UK. Photographs: R. J. Uncles, except (a), USGS – with permission.

instruments to measure salinity, temperature and velocity, in addition to holding pumping equipment that collects water samples for gravimetric analysis and subsequent calibration. Such a configuration would additionally enable small-scale seabed morphology to be investigated. Burst-sampled data would be acquired at high frequency (e.g., 64 Hz) for a prescribed period within a data-collection cycle, followed by averaging (e.g., at 2 Hz) and storage, allowing high-resolution measurements in time and space. Examples of such deployments within a shallow estuary are given by Thorne and Hardcastle (1997) and Thorne and Hanes (2002).

As an example of a commercial instrument, the AQUAscat 1000S (Aquatec, 2016) is a profiling, multifrequency ABS (0.5, 1, 2 and 4 MHz) that utilises four fixed transducers with a maximum transmission pulse rate of 128 Hz for each frequency. The use of multiple frequencies allows an interpretation of the backscattered data in terms of different SPM size ranges. It is sensitive to a wide range of sizes and measures SPM concentrations that typically cover 10 mg l^{-1}–20 g l^{-1} for range-lengths >1 m (comprising 256 cells, where 10 mm is a standard cell length) and greater for shorter range lengths, although the calibration is frequency and particle-size dependent.

4.3.2   Acoustic Doppler Velocimeter (ADV)

The ADV is designed to measure hydrodynamic variables and has been much used for this purpose in shallow estuarine and saltmarsh regions (e.g., Figure 7-3c) as well as in deeper coastal and shelf waters. In addition, Fugate and Friedrichs (2002) report measurements of SPM concentrations in the lower Chesapeake Bay using a Sontek ADV (Sontek, 2016). They showed that acoustic backscatter from the ADV was the best estimator of SPM mass concentration in the lower Bay (compared with OBS and LISST measurements), where the observed, gravimetrically determined SPM concentrations were < 70 mg l^{-1} at 0.1 m above the bed. The intensity (in decibels, dB) of the backscattered sound pulses received at the ADV from 0.03 and 0.23 m above the bed, averaged over 15-min burst-sample determinations, exhibited strong correlations with the logarithm of SPM concentrations measured at 0.10 and 0.25 m above the bed, respectively (Fugate and Friedrichs, 2002).

Ho Kyung Ha (2008) found that ADV backscatter strength could be used as a proxy for SPM concentration within limited concentration ranges. A 5-MHz ADV was used that was found to have an operational range of up to 1 and 4 g l^{-1} for 'Clay Bank' sediment (bimodal, with approx. 1 μm as the first mode and approx. 90 μm as the second) and kaolinite, respectively. Backscattered signals would sometimes be too noisy to follow instantaneous changes of SSC due to the high amplification setting, high sampling rate (e.g., >10 Hz), and the small sampling volume associated with ADVs. It was concluded that caution should be exercised when a measuring site has a significant change of grain size distribution with time; this is because the backscattered signal strength is primarily controlled by the acoustic frequency and the sediment properties of particle size, flocculation status and floc structure.

Sottolichio et al. (2011) used Nortek ADVs operating at 6 MHz and sampling at 32 Hz in the Gironde Estuary, France; applying an 'attenuation compensation' procedure to the acoustic intensities at high concentrations, they showed that SPM concentrations were very similar to those recorded by OBS sensors in the range 0–15 g l^{-1}. Salehi and Strom (2011) used a 6-MHz Nortek Vector Velocimeter in laboratory tests to show that a region of linearity existed between the logarithm of SPM concentration and the instrument's time-averaged signal-to-noise ratio for all muddy sediment types tested when SPM concentrations were <1.5 g l^{-1}.

Although there are issues that need to be resolved with the application of ADV sensors to SPM concentration determinations, overall the results from these various studies are encouraging.

4.3.3   Acoustic Doppler Current Profiler (ADCP)

The ADCP is designed to measure water velocity over the depth, either by looking down through the water column toward the bed whilst mounted on a mooring, fixed platform or research vessel (an ADCP is shown mounted on an estuarine research boat in Figure 7-3d), or by being mounted on the bed and looking up into the water column. The velocity of the water along three or four acoustic beams is measured utilising the Doppler Effect (see Chapter 4 for a description of ADCP operating principles). Despite the fact that ADCPs operate with pulses of the order of 1 ms and frequencies of the order of 1 MHz, as opposed to typically 10 μs and a few MHz for ABS sensors, the intensity of the return echo (acoustic echo intensity in dB) is again related to the concentration of backscattering particles present in the water and their properties.

An ADCP logs the acoustic backscatter intensity in bins, which are cone-shaped, and all SPM within a bin contributes to the backscattered signal. The theoretical relationship between the backscattered intensity and the SPM concentration is based on the sound-scattering equations described by, e.g., Thorne and Hanes (2002) and are functions of distance from the sensor, the attenuation by the water, and the total SPM concentration and its scattering properties. ADCPs have been used in the laboratory to make estimates of grain size in SPM-laden suspensions (Guerrero et al., 2012).

An early application of ADCP data to the determination of SPM concentration in a turbid estuarine environment (the Mersey Estuary, UK) was given by Holdaway et al. (1999). The backscattered signals were calibrated with reference to gravimetrically calibrated transmissometer measurements that were coincident with an ADCP range cell. A reasonable agreement between ADCP and gravimetrically determined SPM concentrations over the observed range (< 200 mg l^{-1}) was demonstrated. Souza et al. (2004) applied a similar methodology to ADCP data collected in the upper Gulf of California. The backscattered signal was corrected for sound absorption and geometric spreading in the water (Deines, 1999). Ten-minute averaged backscatter intensity at the ADCP (in dB) was averaged over the four ADCP beams and the results calibrated with gravimetrically calibrated OBS measurements. The linear regression between backscatter intensity (dB) and the logarithm of SPM concentration explained 86 per cent of the variance; SPM concentrations estimated from the ADCP were < 40 mg l^{-1} (Souza et al., 2004).

An ambitious use of ADCPs for determining SPM concentrations has been reported by Nauw et al. (2014). Long-term measurements between 2003 and 2005 with a hull-mounted ADCP positioned under the public ferry that repeatedly crosses the Marsdiep

Inlet, The Netherlands, were used to estimate profiles of SPM concentration from acoustic backscatter intensity, in which the shift between the low and high turbulent regimes was taken into account (Merckelbach, 2006; Merckelbach and Ridderinkhof, 2006). Calibration was achieved through comparison of ADCP data and gravimetrically calibrated observations over 13-h anchor stations at seven different stations.

5   Calibration of Turbidity Sensors

Calibration is essential in all cases where optical or acoustic sensors are employed to measure SPM concentrations. The relationship between an optical or an acoustic signal and a real, grams-per-litre value for SPM concentration is highly variable and depends on a range of factors, including size, shape, concentration itself, colour, and angularity.

An example of a calibration curve for a nominally $0–10$ g l^{-1} single-gap transmissometer (shown in Figure 7-2b) demonstrates not only the extent of the data scatter to be expected for comparisons of gravimetrically determined, pump-sampled SPM concentrations with in situ transmissometer measurements, but also the overriding linearity that is nevertheless exhibited by these sensors in the presence of such scatter (Figure 7-4a; Uncles et al., 2000a). The single-gap infrared sensor (shown in Figure 7-2c) exhibits considerably greater data scatter for these higher concentrations, as well as a nonlinear (sigmoidal) calibration curve for the highest concentrations, although the calibration is useful up to approximately 45 g l^{-1} (Figure 7-4b). Part of the extra data scatter can be attributed to the very large vertical gradients that exist close to the bed of a muddy, highly turbid estuary and the difficulty of sampling and transmissometer-measuring exactly the same water volume. A laboratory calibration of this instrument, using diluted and mixed suspensions of SPM from the same estuary, lies below the in situ calibration line (Figure 7-4b), so that the instrument has a greater response for the same SPM concentration, illustrating that the transmittance properties of the diluted, well-mixed, and therefore partially deflocculated SPM differs somewhat from the natural, in situ material.

For OBS sensors a large difference in calibration is seen depending on sediment grain size, composition and shape of the suspended particles (e.g., Green and Boon, 1993; Bunt et al., 1999). Battisto et al. (1999) show that, for the same SPM concentration, the OBS response to clay particles of 2-μm diameter is 50 times greater than that for sand of 100-μm diameter. Therefore, OBS calibrations made during field surveys can be expected to vary from one estuary to another (Figure 7-4c); e.g., the OBS response to SPM is greater in the Frome Estuary than the Axe Estuary of southwest England (Figure 7-4c). Uncles et al. (2000b) found that the calibrations of their OBS sensors with very fine sand suspensions led to sensor-measured concentrations (in logger bits) that were approximately seven times less than those for calibrations with estuarine silt and clay for the same milligrams-per-litre SPM concentration (Figure 7-4d). It follows that each sensor has to be calibrated using SPM from its deployment site.

The calibration may therefore be related to seasonal effects via biogeochemical influences, tidal state via resuspension of bed sediments, the weather *via* wind and

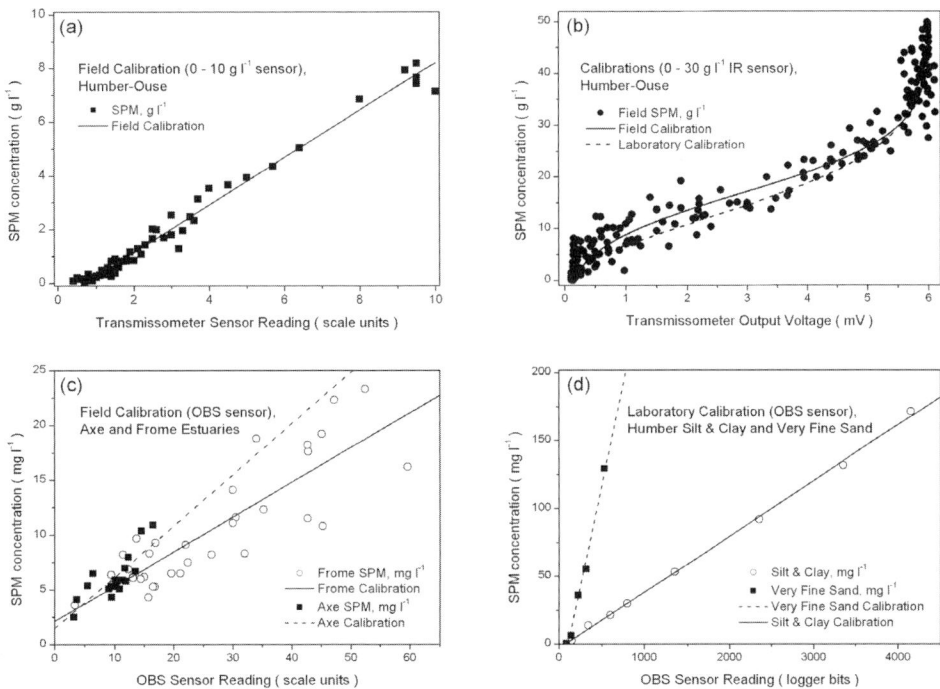

Figure 7-4 Examples are shown of field and laboratory calibrations for transmissometers and OBS sensors: (a) Calibration of a single-gap 0–10 g l^{-1} transmissometer (Figure 7-2b) in the Humber-Ouse Estuary, UK, using in situ pumped samples and simultaneous sensor readings; (b) Calibration of a single-gap, 0–30 g l^{-1} infrared transmissometer (Figure 7-2c) in the Humber-Ouse using in situ pumped samples and simultaneous sensor readings; the sigmoidal calibration curve of least-squares fit is shown as the continuous line. The laboratory calibration, using diluted and mixed suspensions of Humber-Ouse SPM, is shown as the dashed line (the laboratory SPM data are not shown); it lies somewhat below the field calibration; (c) Field calibrations during longitudinal boat surveys of the Devon Axe and Dorset Frome Estuaries, UK, using an OBS sensor (e.g., Figure 7-2d); (d) Laboratory calibrations of an OBS sensor as functions of the sensor output (logger bits) using: (i) very fine sand (filled squares and dashed linear regression calibration line) and (ii) silt and clay (open circles and linear regression calibration line).

freshwater influences on catchment and local sediment erosion, or any local variations in sediment sources that may affect the SPM composition. In taking samples for calibration, it is important to try to cover the range of conditions of interest to the study, covering therefore a range of concentrations, sediment sizes, locations and salinity conditions. It is generally necessary to repeat the calibration process several times during a deployment. Calibrations may also be carried out using a suspension of formazin to gain a non-site-specific calibration of the instrument in terms of the relationship between the output of the sensor and the known FTU or NTU readings, depending on the instrument. Of course, the most important calibration is between the output of the turbidity sensor and the SPM grams-per-litre concentration at the site concerned. Once a reasonable calibration has been established (e.g., via the calibration

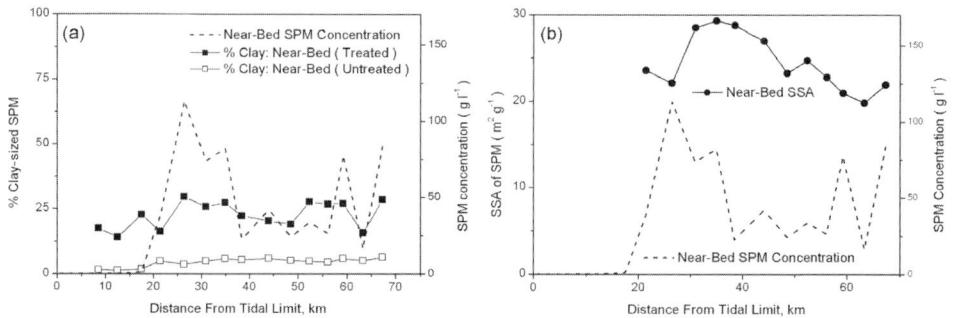

Figure 7-5 Some properties of the SPM: (a) Near-bed SPM concentrations (g l^{-1}) and % clay-sized near-bed SPM for 'treated' and 'untreated' SPM sediment at HW of a neap tide along the Humber-Ouse Estuary, UK; (b) corresponding SSA (m^2 g^{-1}) of near-bed SPM. Subplots (a) and (b) are modified from Uncles et al. (2006), with permission of Springer.

experiments and is thus a measure of their surface area to mass ratio (e.g., Millward et al., 1990). SSA can be measured using Brunauer-Emmett-Teller (BET) analysis (e.g., Lucideon, 2016) after the sediment has been freeze-dried to –50°C and then ground to a powder. The method entails measuring the quantity of gas that is adsorbed as a single layer of molecules (monomolecular layer) onto a sample. The gas mixture generally comprises 30 per cent nitrogen and 70 per cent helium, and the monolayer is established at atmospheric pressure and liquid nitrogen temperature. The liquid nitrogen is placed under the sample, enabling the nitrogen to form a monolayer on it. The liquid nitrogen is then replaced with water at room temperature, so that the adsorbed nitrogen is released back into gas to provide the desorption data.

As an example of SSA data, the SSA of near-bed SPM at HW of a neap tide along the Humber-Ouse Estuary shows a general increase of roughly 20 to 30 m^2 g^{-1} from the mouth of the Ouse to the ETM region before decreasing near the up-estuary limit of the ETM and farther into the upper estuary (Figure 7-5b; Uncles et al., 2006).

6.2   In Situ Settling Velocity of SPM

The LISST instrument has frequently been used to estimate in situ settling velocities of SPM. Measurements by Voulgaris and Meyers (2004) in a tidal creek that drained a saltmarsh recorded mean SPM floc sizes of 25 to 75 μm, settling velocities between 0.02 and 0.2 mm s^{-1} and SPM concentrations that were less than 80 mg l^{-1}. Their effective settling velocity estimates were based on: (a) the mean effective particle density, i.e., the difference between the density of the suspended particles and the density of water, which was derived from the total SPM mass and volume concentrations measured by the OBS and LISST sensors, respectively, and the application of Stokes' Law (e.g., Buller and McManus, 1979) using the mean particle diameter determined by the LISST, and (b) the assumption that gravitational settling of SPM was balanced by turbulent dispersion, using their ADV data to determine the turbulent-mean vertical flux of SPM (kg m^{-2} s^{-1}) and the turbulent-mean SPM concentration (kg m^{-3}), the ratio of which provided an estimate of the settling velocity (m s^{-1}).

In the Pearl River Estuary, China, SPM concentrations typically were less than 100 mg l^{-1}, median floc sizes varied between 10 and 96 μm and settling velocities varied between 0.01 and 0.2 mm s^{-1} (Xia et al., 2004). They used a LISST ST (Sequoia, 2015; now superseded) that measured the size-dependent settling velocity distribution. The optics end of this instrument was enclosed in a settling tube of 0.3-m length. The settling column, which was enclosed within this 0.05-m diameter settling tube, consisted of a 0.05-m x 0.01-m wide x 0.30-m tall rectangular volume. The settling tube had door openings at the top and bottom that were operated by motors and used to draw in water samples for SPM measurement and analysis and expel them afterwards.

In the relatively low turbidity (15 mg l^{-1}) Po River prodelta, Fox et al. (2004) measured sizes that ranged from approximately 60 to 800 μm and settling velocities that were of the order of 1 mm s^{-1}. The instrumentation of Sternberg et al. (1996, 1999) was used to collect size *versus* settling velocity data. The assembly consisted of a baffled settling trap with one transparent side, to which a Sony Hi8 Video Camera in a pressure-housing was attached. Video was recorded for 10 s every 6 h over the course of each tripod deployment. Other, more modern video-based and imaging camera systems are reviewed in Chapter 8.

Settling velocities in the field are frequently measured using settling tubes based on the Owen Tube concept (McCave, 1979; Burt, 1986; Malarkey et al., 2013). The Owen Tube is reviewed in Chapter 8. A modified version of the instrument used by Uncles et al. (2006) comprises a clear, lightweight, tough plastic tube, 0.9 m in length and 0.06 m in diameter, which is lowered into the water and suspended horizontally (shown in a stand and also suspended from an improvised crane in Figure 7-1a). It is initially open at both ends. A water sample is obtained by closing the ends remotely. The instrument is then lifted out of the water, placed vertically, and the SPM settling process quantified by withdrawing samples of known volume from the bottom of the tube at predetermined times until the tube is empty. Use of the Owen Tube in the field is likely to lead to the break-up of at least some of the largest, most fragile flocs within the tube, and several of the instruments reviewed in Chapter 8 more successfully measure the settling velocities of natural flocs that comprise the SPM in situ without breakages.

6.3   Loss-on-Ignition (LOI)

The same filter papers that were used to derive SPM concentrations gravimetrically (Section 4.1) can also be used to provide an indication of POM content (specifically particulate organic content, POC) in the suspension. The gravimetrically determined samples of SPM, on their filter papers, are placed in a muffle furnace at 500°C until constant weight is achieved. The time required may range from 2–6 h, but ashing for periods longer than 6 hours is unlikely to produce any significant increase in weight loss. After removal and cooling in a desiccator, the samples are reweighed, generally with an accuracy of 0.00001 g, to determine the new SPM dry weights. LOI is then determined as a percentage difference (loss) between the two dry-weight values (McCave, 1979; Bale and Kenny, 2005).

The method is quick and straightforward and for this reason is currently still used. However, there are numerous combinations of muffle-furnace temperatures and treatment times presented in the literature, despite the plea for consistency by Heiri et al. (2001) in their analysis of LOI methods for estimating organic and carbonate contents of sediments. For example, Azzoni et al. (2015) determined POM as the LOI at 450°C for 2 h of 0.5 g of dry sediment; Pratt et al. (2015) determined POM as the LOI of dry sediments (dried for 24 h at 60°C) followed by combustion in a furnace at 550°C for 5 h. Whatever the treatment time, it must be sufficiently long that further changes in LOI are negligible with increasing times. To add to the uncertainties, Bale and Kenny (2005) point out in their review that at temperatures $> 400°C$ carbonates (mainly shells) may start to decompose, and structural waters may be driven from clay matrices. For these and other reasons, the percentage of LOI that can be ascribed to POC varies with ignition temperature. As an example of the effect of ignition temperature on LOI determinations, the LOI of 57 SPM samples from several estuaries of southwest England (Uncles et al., 2015) were such that the average ratio of LOI at 350°C to that at 500°C was 0.62 ± 0.01 (R. J. Uncles, unpublished data).

6.4   POC Content of SPM

Measured aliquots of the water samples are filtered through ashed Whatman GF/F filters and stored frozen. On return to the laboratory, samples are freeze-dried prior to particulate carbon and, if required, nitrogen analysis. Total particulate carbon and nitrogen are determined on a subsample of filter and sediment by combustion analysis using, e.g., a Carlo Erba 1500 or Thermo Finnigan Flash analyser. The principles employed by these analysers are outlined in, e.g., Stanford (2015). The organic carbon fraction is measured on a second subsample after sulphurous acid treatment to remove the carbonate.

6.5.   Chlorophyll-a Content of SPM

Chlorophyll-a is a proxy measure for phytoplankton. Commonly, chlorophyll-a and phaeopigments are measured in water samples filtered onto 47-mm Whatman GF/F papers, which are then frozen quickly for transport to the laboratory; in the laboratory (in vitro), pigments are extracted with 90 per cent acetone and analysed according to the fluorometric method of Yentsch and Menzel (1963). An introduction to fluorescence measurements is given in TD (2015) and by Lakowicz (2006).

Some recent determinations of chlorophyll-a are described by Seers and Shears (2015), who used the standard spectrophotometric methods that are detailed in APHA (2005); Hitchcock and Mitrovic (2015) filtered 500 ml of water onto GF/C filters, which were then frozen until subsequent determination by standard methods (APHA, 2005), using the grinding technique and acetone as a solute with correction for phaeophytin. Kimmel et al. (2015) used a modified in vitro fluorescence technique

(Environmental Protection Agency method 445.0, Arar and Collins, 1997) without acidification. Photopigments were separated and quantified using high-performance liquid chromatography (HPLC; Paerl et al., 2003).

6.6   EPS Content of SPM

Many microorganisms secrete extracellular polymeric substances (EPS, e.g., slime and mucus) into their aquatic environments. This material can enhance the stability of intertidal mudflats by increasing their resistance to the mechanical erosion that results from waves and currents (Herman et al., 2000). Microphytobenthos, particularly benthic diatoms, are an important source of EPS in these mudflat systems (Lucas et al., 2003).

Generally, extracellular polymeric substances are measured using the established, colorimetric phenol-sulphuric assay described by Underwood et al. (1995), in which colloidal carbohydrate is measured as a proxy for EPS and the results expressed as glucose equivalents in units of g g^{-1} of dry sediment weight. They showed that the phenol-sulfuric acid assay produced linear relationships between sediment sample size and carbohydrate yield across a range of sediment dry weights from 2 to 1000 mg – weights that are readily obtained from bottle samples of SPM in a wide range of estuaries – although traditionally these measurements are made on samples from the surface layer of intertidal mud, which is the source of the EPS, rather than SPM.

Seasonal measurements in the surface, upper 2 mm of intertidal mudflats in the upper Tamar Estuary, UK, showed that the EPS content of this intertidal sediment (expressed as glucose equivalents in μg g^{-1} of dry sediment) was strongly, positively correlated with the chlorophyll-a fraction of the sediment (in μg g^{-1} of dry sediment, Uncles et al., 2003), illustrating its biological origins:

$$EPS = -54 \pm 22 + (11.1 \pm 0.8) \cdot Chlorophyll - a$$

This equation is applicable when EPS > 0. Values of EPS and chlorophyll-a were typically in the range 50 to 900 and 10 to 80 μg g^{-1} of dry sediment, respectively. Because of this strong EPS relationship, Uncles et al. (2010) hypothesised that the observed occurrence of large SPM aggregates (median floc diameters > 500 μm) at times of faster current speeds in the upper Tamar Estuary, UK, may partly have been due to the binding together of very fine bed particles by sticky EPS coatings, produced by benthic diatoms and by other biologically mediated activity, and their subsequent en masse erosion.

6.7   Mineralogy of SPM

Wave diffraction occurs when a crystalline material is irradiated at an angle by monochromatic X-rays and the distance travelled by the waves that are scattered from successive crystal planes differs by an integer number of wavelengths. The angular positions and intensities of the diffracted peaks produce a characteristic pattern. A laboratory manual for X-ray powder diffraction techniques is available

(Poppe et al., 2001). The mineralogy of SPM can be assessed using X-ray diffraction (XRD); e.g., in the Humber-Ouse, UK, the SPM clay mineralogy within the ETM is dominated by chlorite and illite (Uncles et al., 2006). Every mineral or compound has a characteristic XRD pattern that can be matched against a database; moreover, computer-controlled systems can interpret the diffraction patterns generated both by individual constituents and mixtures. XRD has become an essential technique in the study of clay minerals and is used extensively for studying the mineralogy of crystalline compounds (BGS, 2015).

Several XRD instruments are available; the BGS (2015) currently use a PANalytical X'Pert Pro diffractometer (Panalytical, 2016), whereas the mineralogy of near-surface, intertidal bed sediments in the Tamar Estuary, UK, was analysed using a Philips 1710 diffractometer (Stephens et al., 1992). In this latter study, the goniometer was set to scan from 4°–44° using standard Phillips software. Prior to analysis, a sediment sample was homogenized and treated with sodium hexametaphosphate; a representative droplet was then dried on a slide cover for 20 min. at 30°C, and the sample was covered and stored in a desiccator.

Hillier et al. (2001) identified the composition of SPM in the River Don, UK. SPM concentrations were determined from the dry weight of particulates collected on pre-weighed, 0.45-μm silver membrane filters. The SPM was analysed directly from the silver membrane filters using a Siemens D5000 X-ray diffractometer to generate XRD patterns. Ramaswamy et al. (2007) used XRD to investigate the mineralogy of SPM in the Gulf of Kachchh, India. An aliquot of SPM on a polycarbonate filter was resuspended in distilled water and treated with hydrogen peroxide to remove organic matter. The lithogenic fraction was concentrated by centrifugation, and the slurry at the bottom of the centrifuge tube was smeared on a glass slide, glycolated for 2 h at 150°C, and analysed on a Phillips X-ray diffractometer, using goniometer angles from 3° to 32°.

7   Final Remarks

This chapter has largely been concerned with the determination of SPM concentration, both via gravimetric analysis of sampled waters and the use of acoustic and optical backscatter instruments. The general principles of operation of both types of backscatter sensor (acoustic and optical) are the same. A signal of known amplitude and frequency is transmitted from one part of the instrument and received at another. In both cases the sensors attempt to provide a surrogate measure of sediment concentration, via calibration, which can be related to the intensity and frequency of the received signal that has been scattered back to the sensor by the individual particles or aggregates in suspension.

Both acoustic and optical sensor types have their pros and cons: e.g., Gartner (2004) claimed that acoustic instruments are less likely to suffer the effects of biofouling than optical instruments; on the other hand, they can be sensitive to the effects of particle mineralogy (Moate and Thorne, 2012). Generally, acoustic instruments tend to be favoured for use with larger, sand-sized particles, which are usually associated with the seaward reaches of sandy estuaries and open coastal waters (Green et al., 2000), and optical instruments for the finer SPM found in estuarine environments. A direct comparison between optical and acoustic instruments for the same site was described by Chanson et al (2008), and a comparison between the use of an ADCP and a transmissometer can be found in Holdaway et al. (1999). An example showing the advantages of the complementary use of optical and acoustic devices was given by Hoitink and Hoekstra (2005).

The use of acoustics has been very successful in the study of sediment transport over sandy beds and for measuring predominantly inorganic SPM concentrations in these environments (Thorne and Hurther, 2014; Thorne et al., 2014). However, application of the technology to estuaries and coastal systems in which the SPM is dominated by clays and silts has been more difficult. In particular, interpretation of the sensor data is not straightforward in the presence of particle flocculation, and the application of multifrequency ABS sensors to study these fine-grained environments is much less developed than that for sandy environments (Thorne et al., 2014). Nevertheless, Sottolichio et al. (2011) have made progress in using acoustic sensors to elucidate the processes that occur near the bed in fluid–mud concentrations.

Whatever the method used, in any calculation of transport rates and morphological impact it is important to take into account the variations in SPM concentration over depth and over time. The vertical gradients in SPM can change as the current speeds vary throughout the tide. During fast-moving currents the vertical gradients may be quite small, and the water column may be almost well mixed. At high and low 'slack' water in an estuary the opposite may apply, with much higher concentrations near the bed than near the surface (Mitchell et al., 1998). Furthermore, it has been shown that vertical gradients in salinity caused by differences in density between freshwater and saltwater can sometimes lead to the suppression of turbulence generated at the bed of a channel in an estuary (e.g., Ren and Wu, 2014). In this case it is possible to envisage considerable differences between faster moving surface waters that may, e.g., contain very little SPM, and slower moving layers of very high concentrations near the bed. In some cases these layers of sediment can form fluid mud, a topic of great interest to port operators and those concerned with navigation (e.g., Wan et al., 2014). It follows that salinity and temperature (and thus density) are also important variables to measure and adequately resolve throughout the water column; it is also important to measure meteorological conditions (e.g., for wind-induced current and wave mixing) and hydrological conditions (e.g., for buoyancy inputs due to freshwater flows).

Numerical modellers and others concerned with using data collected from field campaigns are generally interested to know how well those data represent the observed system. Therefore, whatever the methods used to collect data, it is important to record them carefully, together with a note of the ambient conditions, the water state, the tide state, the light levels, and so on. Observed SPM concentrations tend to be highly variable, with modellers relying on a scientific understanding that is only as good as the information provided by the team collecting the data in the first place. Any exceptional events, such as passing vessels or sudden changes in weather, should be carefully noted in any report of the survey. Combined with a suitable numerical model (e.g., Hill et al., 2003), the measured data will enhance the detailed process-understanding of SPM transport in the surveyed waters.

Acknowledgements

RJU thanks Professor Steve de Mora, Chief Executive of the Plymouth Marine Laboratory, for the award of a Senior Research Fellowship.

References

APHA, 2005. *Standard Methods for the Examination of Water and Wastewater*, 21st ed., Washington, DC: American Public Health Association.

Aquatec, 2016. Aquatec Group Ltd, Aquatec House, Stroudley Road, Basingstoke RG24 8FW, UK. www.aquatecgroup.com [accessed June 2016].

Arar, E. J., Collins, G. B., 1997. In Vitro Determination of Chlorophyll a and Pheophytin a in Marine and Freshwater Algae by Fluorescence (Revision 1.2). US Environmental Protection Agency method 445.0, 22pp. Downloadable at: https://cfpub.epa.gov/si/si_public_record_report.cfm?dirEntryId=309417 [accessed June 2016].

Astoreca, R., Doxaran, D., Ruddick, K., Rousseau, V., Lancelot, C., 2012. Influence of suspended particle concentration, composition and size on the variability of inherent optical properties of the southern North Sea. *Continental Shelf Research* 35, 117–128.

Azzoni, R., Nizzoli, D., Bartoli, M., Christian, R. R., Viaroli, P., 2015. Factors controlling benthic biogeochemistry in urbanized coastal systems: an example from Venice (Italy). *Estuaries and Coasts* 38, 1016–1031, doi: 10.1007/s12237-014–9882-6.

Bale, A. J., Kenny, A. J., 2005. Sediment analysis and seabed characterisation. In: Eleftheriou, A., McIntyre, A. D. (eds.), *Methods for the Study of Marine Benthos*. Oxford, UK: Blackwell Science Ltd., 43–86.

Battisto, G. M., Friedrichs, C. T., Miller, H. C., Resio, D. T., 1999. Response of OBS to mixed grain-size suspensions during Sandyduck '97. In: Kraus, N. C., McDougal, W. G. (eds.), *Coastal Sediments '99: Proceedings of the Fourth International Symposium on Coastal Engineering and Science of Coastal Sediment Processes, American Society of Civil Engineers.* Reston, VA: American Society of Civil Engineers.

BGS, 2015. X-ray diffraction (XRD) analysis. Web page at: www.bgs.ac.uk/science Facilities/laboratories/mpb/xrd_tech.html [accessed June 2016].

Binding, C. E., Bowers, D. G., Mitchelson-Jacob, E. G., 2003. An algorithm for the retrieval of suspended sediment concentrations in the Irish Sea from SeaWiFS ocean colour satellite imagery. *International Journal of Remote Sensing* 24, 3791–3806. doi:10.1080/0143116021000024131.

Buller, A. T., McManus, J., 1979. Sediment sampling and analysis. In: Dyer, K. R. (ed.), *Estuarine Hydrography and Sedimentation*. Cambridge, UK: Cambridge University Press, 87–130.

Bunt, J. A. C., Larcombe, P., Jago, C. F., 1999. Quantifying the response of optical backscatter devices and transmissometers to variations in suspended particulate matter. *Continental Shelf Research* 19, 1199–1220.

Burt, N. T., 1986. Field settling velocities of estuary muds. In: Mehta, A. J. (ed.), *Estuarine Cohesive Sediment Dynamics. Lecture Notes on Estuarine and Coastal Studies*. Berlin: Springer-Verlag, 126–150.

Campbell, 2016. Campbell Scientific Inc., 815 W 1800 N, Logan, UT 84321–1784, USA. www.campbellsci.com [accessed June 2016].

Chanson, H., Takeuchi, M., Trevethan, M., 2008. Using turbidity and acoustic backscatter intensity as surrogate measures of suspended sediment concentration in a small subtropical estuary. *Journal of Environmental Management* 88, 1406–1416.

Cloern, J. E., 1987. Turbidity as a control on phytoplankton biomass and productivity in estuaries. *Continental Shelf Research* 7, 1367–1381.

Cook, T. L., Sommerfield, T. K., Wong K.-C., 2007. Observations of tidal and springtime sediment transport in the upper Delaware estuary. *Estuarine, Coastal and Shelf Science* 72, 235–246.

Defendi, V., Kovacevic, V., Arena, F., Zaggia, L., 2010. Estimating sediment transport from acoustic measurements in the Venice Lagoon inlets. *Continental Shelf Research* 30, 883–893.

Deines, K. L., 1999. Backscatter estimation using broadband acoustic Doppler current profilers. In: *Proceedings of the IEEE/OES 6th Working Conference on Current Measurement Technology*, 259–264. Piscataway Township, NJ: The Institute of Electrical and Electronics Engineers. Downloadable from: www.comm-tec.com/Library/Technical_Papers/RDI/echopaper.pdf [accessed June 2016].

Di Stefano, C., Ferro, V., Mirabile, S., 2010. Comparison between grain-size analyses using laser diffraction and sedimentation methods. *Biosystem Engineering* 106, 205–215. http://doi.org/10.1016/j.biosystemseng.2010.03.013.

Doxaran, D., Froidefond, J.-M., Castaing, P., Babin, M., 2009. Dynamics of the turbidity maximum zone in a macrotidal estuary (the Gironde, France): Observations from field and MODIS satellite data. *Estuarine Coastal and Shelf Science* 81, 321–332.

Downing, J., 2006. Twenty-five years with OBS sensors: the good, the bad and the ugly. *Continental Shelf Research* 26, 2299–2318.

Fettweis, M., Nechad, B., van den Eynde, D., 2007. An estimate of the suspended particulate matter (SPM) transport in the southern North Sea using SeaWiFS images, in situ measurements and numerical model results. *Continental Shelf Research* 27, 1568–1583.

Fox, J. M., Hill, P. S., Milligan, T. G., Ogston, A. S., Boldrin, A., 2004. Floc fraction in the waters of the Po River prodelta. *Continental Shelf Research* 24, 1699–1715. http://doi.org/10.1016/j.csr.2004.05.009.

Fugate, D. C., Friedrichs, C. T., 2002. Determining concentration and fall velocity of estuarine particle populations using ADV, OBS and LISST. *Continental Shelf Research* 22 (11–13), 1867–1886. http://doi.org/10.1016/S0278-4343(02)00043-2.

Gartner, J. W., 2004. Estimating suspended solids concentrations from backscatter intensity measured by acoustic Doppler current profiler in San Francisco Bay, California. *Marine Geology* 211, 169–187. http://doi.org/10.1016/j.margeo.2004.07.001.

Green, M. O., Boon, J. D., 1993. The measurement of constituent concentrations in nonhomogeneous sediment suspensions using optical backscatter sensors. *Marine Geology* 110, 73–81.

Green, M. O., Bell, R., Dolphin, T. J., Swales, A., 2000. Silt and sand transport in a deep tidal channel of a large estuary (Manukau Harbour, New Zealand). *Marine Geology* 163, 217–240.

Green, M. O., Hancock, N. J., 2012. Sediment transport through a tidal creek. *Estuarine Coastal and Shelf Science* 109, 116–132.

Guerrero, M., Ruther, M., Szupiany, R. N., 2012. Laboratory validation of Acoustic Doppler Current Profiler (ADCP) techniques for suspended sediment investigations. *Flow Measurement and Instrumentation* 23, 40–48.

Ha, H. K., Hsu, W.-Y., Maa, J. P.-Y., Shao, Y. Y., Holland, C. W., 2009. Using ADV backscatter strength for measuring suspended cohesive sediment concentration. *Continental Shelf Research* 29, 1310–1316.

Heiri, O., Lotter, A. F., Lemcke, G., 2001. Loss on ignition as a method for estimating organic and carbonate content in sediments: reproducibility and comparability of results. *Journal of Paleolimnology* 25, 101–110. Open Access version via Utrecht University Repository.

Herman, P. M. J., Middelburg, J. J., Heip, C. H. R., 2000. Benthic community structure and sediment processes on an intertidal flat: Results from the ECOFLAT project. *Marine Ecology Progress Series* 196, 59–73.

Hill, D. C., Jones, S. E., Prandle, D., 2003. Derivation of sediment resuspension rates from acoustic backscatter time-series in tidal waters. *Continental Shelf Research* 23, 19–40.

Hillier, S., 2001. Particulate composition and origin of suspended sediment in the R. Don, Aberdeenshire, UK. *Science of the Total Environment* 265, 281–293.

Hitchcock, J. N., Mitrovic, S. M., 2015. Highs and lows: The effect of differently sized freshwater inflows on estuarine carbon, nitrogen, phosphorus, bacteria and chlorophyll *a* dynamics. *Estuarine, Coastal and Shelf Science* 156, 71–82.

Ho Kyung Ha, 2008. Acoustic measurements of cohesive sediment transport: Suspension to consolidation. *A Dissertation presented to: The Faculty of the School of Marine Science*, The College of William and Mary, Virginia, USA, in partial fulfilment of the requirements for the Degree of Doctor of Philosophy.

Hoitink, A. J. F., Hoekstra, P., 2005. Observations of suspended sediment from ADCP and OBS measurements in a mud-dominated environment. *Coastal Engineering* 52, 103–118.

Holdaway, G. P, Thorne, P. D., Flatt, D., Jones, S. E., Prandle, D., 1999. Comparison between ADCP and transmissometer measurements of suspended sediment concentration. *Continental Shelf Research* 19, 421–441.

Huhn, F., von Kameke, A., Allen-Perkins, S., Montero, P., Venancio, A., Perez-Munuzuri, V., 2012. Horizontal Lagrangian transport in a tidal-driven estuary – transport barriers attached to prominent coastal boundaries. *Continental Shelf Research* 39–40, 1–13.

Jago, C. F., Bull, C. F. J., 2000. Quantification of errors in transmissometer-derived concentration of suspended particulate matter in the coastal zone: implications for flux determinations. *Marine Geology* 169, 273–286.

Kelly, C. A., Law, R. J., Emerson, H. S., 2000. Methods of analysing hydrocarbons and polycyclic aromatic hydrocarbons (PAH) in marine samples. *Science Series, Aquatic Environment Protection: Analytical Methods, CEFAS, Lowestoft*, 12.

Kimmel, D. G., McGlaughon, B. D., Leonard, J., Paerl, H. W., Taylor, J. C., Cira, E. K., Wetz, M. S., 2015. Mesozooplankton abundance in relation to the chlorophyll maximum in the Neuse River Estuary, North Carolina, USA: Implications for trophic dynamics. *Estuarine, Coastal and Shelf Science* 157, 59–68.

Lakowicz, J. R., 2006. *Principles of Fluorescence Spectroscopy*. New York: Springer US. doi:10.1007/978-0-387-46312-4.

Lucas, C. H., Widdows, J., Wall, L., 2003. Relating spatial and temporal variability in sediment chlorophyll *a* and carbohydrate distribution with erodibility of a tidal flat. *Estuaries* 26, 885–893.

Lucideon, 2016. Lucideon Limited, Queens Road, Penkhull, Stoke-on-Trent, Staffordshire ST4 7LQ, UK. www.lucideon.com/testing-analysis/brunauer-emmett-teller-bet-surface-area-deter mination-test-method [accessed June 2016].

Malarkey, J., Jago, C. F., Hubner, R., Jones, S. E., 2013. A simple method to determine the settling velocity distribution from settling velocity tubes. *Continental Shelf Research* 56, 82–89.

Malvern, 2016. Malvern Instruments Ltd, Enigma Business Park, Grovewood Road, Malvern WR14 1XZ, United Kingdom. www.malvern.com [accessed June 2016].

McCave, I. N., 1979. Suspended sediment. In: Dyer, K.R. (ed.), *Estuarine Hydrography and Sedimentation*. Cambridge, UK: Cambridge University Press, 131–185.

Merckelbach, L. M., 2006. A model for high-frequency acoustic Doppler current profiler backscatter from suspended sediment in strong currents. *Continental Shelf Research* 26, 1316–1335.

Merckelbach, L. M., Ridderinkhof, H., 2006. Estimating suspended sediment concentration using backscatterance from an acoustic Doppler profiling current meter at a site with strong tidal currents. *Ocean Dynamics* 56, 153–168. doi: 10.1007/s10236-005-0036-z.

Millward, G. E., Turner, A., Glasson, D. R., Glegg, G. A., 1990. Intra- and inter-estuarine variability of particle microstructure. *Science of the Total Environment* 97/98, 289–300.

Mitchell, S. B., West, J. R., Arundale, A. M. W., Guymer, I., Couperthwaite, J. S., 1998. Dynamics of the turbidity maxima in the upper Humber estuary system. *Marine Pollution Bulletin* 37, 190–205.

Mitchell, S. B., Uncles, R. J., Akesson, L., 2012. Observations of turbidity in the Thames estuary. *Water and Environment Journal* 26, 511–520.

Moate, B. D., Thorne, P. D., 2012. Interpreting acoustic backscatter from suspended sediments of different and mixed mineralogical composition. *Continental Shelf Research* 46, 67–82.

Nauw, J. J., Merckelbach, L. M., Ridderinkhof, H., van Aken, H. M., 2014. Long-term ferry-based observations of the suspended sediment fluxes through the Marsdiep inlet using acoustic Doppler current profilers. *Journal of Sea Research* 87, 17–29.

Nortek, 2016. NortekUSA, 27 Drydock Avenue, Boston, MA 02210, USA, www .nortekusa.com [accessed June 2016].

Paerl, H. W., Valdes, L. M., Pinckney, J. L., Piehler, M. F., Dyble, J., Moisander, P. H., 2003. Phytoplankton photopigments as indicators of estuarine and coastal eutrophication. *BioScience* 53, 953–964.

Palanques, A., Puig, P., Guillen, J., Jimenez, J., Gracia, V., Sanchez-Arcilla, A., Madsen, O., 2002. Near-bottom suspended sediment fluxes on the micro-tidal low energy Ebro continental shelf (NW Mediterranean). *Continental Shelf Research* 22, 285–303.

Panalytical, 2016. PANalytical Ltd., Environmental Science Centre, Keyworth, Nottingham NG12 5GG, UK. www.panalytical.com [accessed June 2016].

Partech, 2016. Partech Instruments, Rockhill Business Park, Higher Bugle, St Austell, Cornwall PL26 8RA UK, http://www.partech.co.uk [accessed June 2016].

Philippart, C. J. M., Suhyb Salama, M., Kromkamp, J. C., van der Woerd, H. J., Zuur, A. F., Cadée, G. C., 2013. Four decades of variability in turbidity in the western Wadden Sea as derived from corrected Secchi disk readings. *Journal of Sea Research* 82, 67–79.

Poppe, L. J., Paskevich, V. F., Hathaway, J. C., Blackwood, D. S., 2001. A Laboratory Manual for X-Ray Powder Diffraction. US Geological Survey Open-File Report 01–041. Viewable at: http://pubs.usgs.gov/openfile/of01-041/index.htm [accessed June 2016].

Pratt, D. R., Pilditch, C. A., Lohrer, A. M., Thrush, S. F., Kraan, C., 2015. Spatial distributions of grazing activity and microphytobenthos reveal scale-dependent relationships across a sedimentary gradient. *Estuaries and Coasts* 38, 722–734. doi: 10.1007/s12237-014–9857-7.

Ramaswamy, V., Nagender Nath, B., Vethamony, P., Illangovan, D., 2007. Source and dispersal of suspended sediment in the macro-tidal Gulf of Kachchh. *Marine Pollution Bulletin* 54, 708–719.

RDI, 2016. Teledyne RD Instruments, 14020 Stowe Drive, Poway, CA 92064, USA. www.rdinstruments.com [accessed June 2016].

Ren, J., Wu, J., 2014. Sediment trapping by haloclines of a river plume in the Pearl River Estuary. *Continental Shelf Research* 82, 1–8.

Robinson, M.-C., Morris, K. P., Dyer, K. R., 1998. Deriving fluxes of suspended particulate matter in the Humber estuary, UK, using airborne remote sensing. *Marine Pollution Bulletin* 37, 155–163.

Salehi, M., Strom, K., 2011. Using velocimeter signal to noise ratio as a surrogate measure of suspended mud concentration. *Continental Shelf Research* 31, 1020–1032. http://doi.org/10.1016/j.csr.2011.03.008.

Savoye, N., David, V., Morisseau, F., Etcheber, H., Abril, G., Billy, I., Charlier, K., Oggian, G., Derriennic, H., Sautour, B., 2012. Origin and composition of particulate organic matter in a macrotidal turbid estuary: The Gironde Estuary, France. *Estuarine, Coastal and Shelf Science* 108, 16–28.

Seapoint, 2016. Seapoint Sensors, Inc., 142B Front Street, Exeter, NH 03833, USA. www.seapoint.com [accessed June 2016].

Seers, B. M., Shears, N. T., 2015. Spatio-temporal patterns in coastal turbidity – Long-term trends and drivers of variation across an estuarine-open coast gradient. *Estuarine, Coastal and Shelf Science* 154, 137–151.

Sequoia, 2015. LISST-ST Particle Size Analyzer User's Manual, Version 3.0 for LISST-ST based on LISST-100X. Downloadable at: www.sequoiasci.com/wp-content/uploads/2013/06/manual-3.pdf [accessed June 2016].

Sequoia, 2016. Sequoia Scientific, Inc., 2700 Richards Road, Suite 107, Bellevue, WA, USA 98005. www.sequoiasci.com [accessed June 2016].

Sontek, 2016. Sontek, 9940 Summers Ridge Road, San Diego, CA 92121–3091, USA. www.sontek.com [accessed June 2016].

Sottolichio, A., Hurther, D., Gratiot, N., Bretel, P., 2011. Acoustic turbulence measurements of near-bed suspended sediment dynamics in highly turbid waters of a macrotidal estuary. *Continental Shelf Research* 31, S36–S49.

Souza, A. J., Alvarez, L. G., Dickey, T. D., 2004. Tidally induced turbulence and suspended sediment. *Geophysical Research Letters* 31, L20309, doi:10.1029/2004GL021186.

Stanford, 2015. Stanford University, Environmental Measurements Facility, CA 94305, http://em1.stanford.edu [accessed June 2016].

Stephens, J. A., Uncles, R. J., Barton, M. L., Fitzpatrick, F., 1992. Bulk properties of intertidal sediments in a muddy, macrotidal estuary. *Marine Geology* 103, 445–460.

Sternberg, R. W., Ogston, A., Johnson, R., 1996. A video system for in situ measurement of size and settling velocity of suspended particulates. *Journal of Sea Research* 36, 127–130.

Sternberg, R. W., Berhane, I., Ogston, A. S., 1999. Measurement of size and settling velocity of suspended aggregates on the northern California continental shelf. *Marine Geology* 154, 43–53.

Talke, S. A., de Swart, H. E., de Jonge, V. N., 2009. An idealized model and systematic process study of oxygen depletion in highly turbid estuaries. *Estuaries and Coasts* 32, 602–620.

TD, 2015. Turner Designs: An introduction to fluorescence measurements. Downloadable at: www.turnerdesigns.com/t2/doc/appnotes/998-0050.pdf [accessed June 2016].

Thorne, P. D., Hardcastle, P. J., 1997. Acoustic measurements of suspended sediments in turbulent currents and comparison with in-situ samples. *Journal of the Acoustical Society of America* 101, 2603–2614.

Thorne, P. D., Hanes, D. M., 2002. A review of acoustic measurement of small-scale sediment processes. *Continental Shelf Research* 22, 603–632.

Thorne, P. D., Hurther, D., 2014. An overview on the use of backscattered sound for measuring suspended particle size and concentration profiles in non-cohesive inorganic sediment transport studies. *Continental Shelf Research* 73, 97–118.

Thorne, P. D., MacDonald, I., Vincent, C. E., 2014. Modelling acoustic scattering by suspended flocculating sediments. *Continental Shelf Research* 88, 81–91.

Turner, A., Millward, G. E., Tyler, A. O., 1994. The distribution and chemical composition of particles in a macrotidal estuary. *Estuarine, Coastal and Shelf Science* 38, 1–17.

Uncles, R. J., Joint, I., Stephens, J. A., 1998. Transport and retention of suspended particulate matter and bacteria in the Humber-Ouse Estuary, United Kingdom, and their relationship to hypoxia and anoxia. *Estuaries* 21, 597–612.

Uncles, R. J., Stephens, J. A., 1999. Suspended sediment fluxes in the tidal Ouse, UK. *Hydrological Processes* 13, 1167–1179.

Uncles, R. J., Stephens, J. A., Parker, R., 2000a. The Ouse Estuary – a precursor to variability in the Humber. In: Jones, N. V., Elliott, M. (eds.), *Coastal Zone Topics: Process, Ecology & Management, Vol 4*. The Humber Estuary and adjoining Yorkshire and Lincolnshire Coasts, 33–48.

Uncles, R. J., Stephens, J. A., Bloomer, N. H., 2000b. Salinity, temperature, suspended sediments and waves in the coastal zone of the Tweed Estuary, UK. In: Yanagi, T. (ed.), *Interactions between Estuaries, Coastal Seas and Shelf Seas*, 19–36. Tokyo: Terra Scientific Publishing Company (TERRAPUB).

Uncles, R. J., Lavender, S. J., Stephens, J. A., 2001. Remotely sensed observations of the turbidity maximum in the highly turbid Humber Estuary, UK. *Estuaries* 24, 745–755.

Uncles, R. J., Bale, A. J., Brinsley, M. D., Frickers, P. E., Harris, C., Lewis, R. E., Pope, N. D., Staff, F. J., Stephens, J. A., Turley, C. M., Widdows, J., 2003. Intertidal mudflat properties, currents and sediment erosion in the partially mixed Tamar Estuary, UK. *Ocean Dynamics* 53, 239–251. doi:10.1007/s10236-003–0047-6.

Uncles, R. J., Stephens, J. A., Harris, C., 2006. Properties of suspended sediment in the estuarine turbidity maximum of the highly turbid Humber Estuary system, UK. *Ocean Dynamics* 56, 235–247.

Uncles, R. J., Bale, A. J., Stephens, J. A., Frickers, P. E., Harris, C., 2010. Observations of floc sizes in a muddy estuary. *Estuarine, Coastal and Shelf Science* 87, 186–196.

Uncles, R. J., Stephens, J. A., Harris, C., 2015. Estuaries of southwest England: Salinity, suspended particulate matter, loss-on-ignition and morphology. *Progress in Oceanography* 137, Part B, 385–408. http://dx.doi.org/10.1016/j.pocean.2015.04.030.

Underwood, G. J. C., Paterson, D. M., Parkes, R. J., 1995. The measurement of microbial carbohydrate exopolymers from intertidal sediments. *Limnology and Oceanography* 40, 1243–1253.

USGS, 2007. USGS Open-File Report 2007–1194; ref: GSFYXBN3. Online at http://woodshole .er.usgs.gov/pubs/of2007-1194/html/turbidity.html [Accessed August 2016].

Valentim, J. M., Vaz, N., Silva, H., Duarte, B., Caçador, I., Dias, J. M., 2013. Tagus Estuary and Ria de Aveiro salt marsh dynamics and the impact of sea level rise. *Estuarine, Coastal and Shelf Science* 130, 138–151.

Vincent, C. E., Stolk, A., Porter, C. F. C., 1998. Sand suspension and transport on the Middelkerke Bank (southern North Sea) by storms and tidal currents. *Marine Geology* 150, 113–129. http://doi.org/10.1016/S0025-3227(98)00048-6.

Voulgaris, G., Meyers, S. T., 2004. Temporal variability of hydrodynamics, sediment concentration and sediment settling velocity in a tidal creek. *Continental Shelf Research* 24, 1659–1683.

Wan, Y., Roelvink, D., Li, W., Qi, D., Gu, F., 2014. Observation and modeling of the storm-induced fluid mud dynamics in a muddy-estuarine navigational channel. *Geomorphology* 217, 23–36.

Whale, 2016. Whale, 2 Enterprise Road, Bangor, Co. Down BT19 7TA, Northern Ireland. www.whalepumps.com [accessed June 2016].

Xia, X. M., Li, Y., Yang, H., Wu, C. Y., Sing, T. H., Pong, H. K., 2004. Observations on the size and settling velocity distributions of suspended sediment in the Pearl River Estuary, China. *Continental Shelf Research* 24, 1809–1826.

Yentsch, C. S., Menzel, D. W., 1963. A method for the determination of phytoplankton chlorophyll and phaeophytin by fluorescence. *Deep Sea Research* 10, 221–231.

YSI, 2016. YSI Incorporated, 1700/1725 Brannum Lane, Yellow Springs, OH 45387, USA. www.ysi.com/products/multiparameter-sondes [accessed June 2016].

Zhang, X., Stavn, R. H., Falster, A. U., Gray, D., Gould Jr., R. W., 2014. New insight into particulate mineral and organic matter in coastal ocean waters through optical inversion. *Estuarine, Coastal and Shelf Science* 149, 1–12.

8 Suspended Particulate Matter: The Measurement of Flocs

A. J. Manning, R. J. S. Whitehouse, R. J. Uncles

1 Introduction

The majority of the particulate matter that accumulates within an estuary is commonly referred to as mud. Mud typically is composed of mineral grains that originate from both fluvial and marine sources, together with biological matter - both living and in various stages of decomposition. It is the combination of these features that makes estuarine mud sticky in nature, and for this reason these sediment types are referred to generically as cohesive sediments (Whitehouse et al., 2000; Mehta, 2014).

From a water-quality perspective, cohesive sediments have the propensity to adsorb contaminants (Ackroyd et al., 1986; Stewart and Thomson, 1997). This in turn has a direct effect on water quality and related environmental issues (e.g., Uncles et al., 1998; Dankers and Wintwerp, 2007). Therefore, accurately predicting the movement of muddy sediments in an estuary is highly desirable. In contrast to noncohesive sandy sediments, muddy sediments can flocculate (Winterwerp and van Kesteren, 2004), and this poses a serious complication to modellers of estuarine sediment dynamics. It is the aim of this chapter to provide an overview of flocculation processes, floc sampling and data-processing techniques, and floc modelling.

2 Composition of Sediments

2.1 Overview

The underlying mud layers of a mudflat are often dark in colour and emanate a strong odour that is a result of the anaerobic decomposition of organic matter. The primary mineral component of cohesive mud is clay. The terms *clay* and *mud* are often incorrectly used interchangeably. Mud typically is defined as a mixture of water, clay and silt (i.e., particle sizes < 63 μm) and also includes organic material and sometimes gas (e.g., hydrogen sulphide resulting from organic decomposition). Clays have a plate-like structure and generally have a diameter < 2 μm (Figure 8-1a). Clay particle surfaces have ionic charges, thereby creating forces comparable to or exceeding the gravitational force, which causes the clay particles to interact electrostatically. In the estuarine and coastal environments these sediment particles do not behave as individual, dispersed particles (Figure 8-1b), but instead tend to stick together (Figure 8-1c). This process is

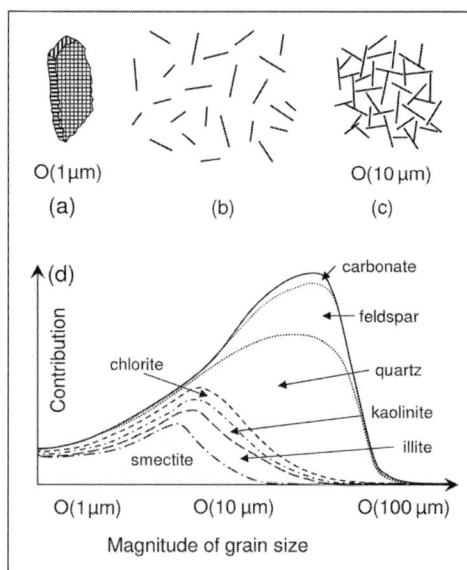

Figure 8-1 This diagram illustrates some fundamental sediment particle properties ('O' indicates 'order of magnitude'). (a) Individual clay particle; (b) dispersed clay particles; (c) individual floc; (d) Qualitative and conceptualised illustration of the distribution of minerals and mineral grains as a function of grain size for an alluvial mud (loosely based on data for Mississippi alluvial mud as illustrated in Potter et al., 1975).

known as flocculation, and the aggregates formed are referred to as flocs, whose size and settling velocity are much greater than those of the individual particles, but whose overall floc density is less (e.g., McDowell and O'Connor, 1977). Recent studies show that natural estuarine floc groups can attain sizes up to several millimetres (much larger than the 200 μm floc illustrated by McDowell and O'Connor, 1977).

Although mud and sand have different physical properties, they commonly occur together in the estuarine and marine environment. Granular classification enables scientists and engineers to assess the nature of soil behaviour: i.e., cohesive, granular or mixed sediments, which has important implications for the mechanical and erosion properties of sediment (Winterwerp and van Kesteren, 2004). Mixed sediments in estuarial regions comprise combinations of clay (diameter, d, such that $d < 2\mu m$), silt ($63\mu m > d > 2\mu m$) and sand grains ($2000\mu m > d > 63\mu m$). Gravel sediments, $d > 2mm$, are rarely found in locations where cohesive sediments occur.

The degree of cohesion increases with the fraction of clay minerals within the sediment and starts to become significant when the sediment contains more than 5–10 per cent of clay by weight (Whitehouse et al, 2000; van Ledden, 2003). Within the clay fraction, a separate subfraction known as the colloidal fraction can be distinguished. Colloidal particles tend to be < 0.1 μm in diameter and are usually removed during gravimetric analysis. Although they are technically a solid, they can pass through a 0.45 μm GFF filter membrane. As a result of their extremely small size, colloids are permanently kept in suspension by Brownian motion. Organic colloids are referred to as dissolved organic carbon (DOC) because they are not retained by the filter paper during filtration.

2.2   Clay Mineralogy

The most common clay minerals that occur within muddy sediments are kaolinite, illite, chlorite, smectite (e.g., montmorillonite) and bentonite (the latter is essentially an impure form of clay, mostly containing montmorillonite). The most important of these are kaolinite, chlorite, montmorillonite and illite, which are termed metal silicates; their chemical structure consists of layers of silica tetrahedra and aluminium hydroxide that are loosely connected by metal ions and water molecules. They are easily broken up by natural degradation processes, which act along cleavage planes (Dyer, 1986). Therefore, clay minerals generally occur as plate-like fragments.

2.3   Examples of Sediment Mineralogy

Whitehouse et al. (2000) tabulate the various clay mineral contributions to mud samples from three estuarine systems and demonstrate the importance of montmorillonite, kaolinite, illite and chlorite. Uncles et al. (2006) investigated the mineralogy of suspended sediment in the turbidity maximum of the Humber-Ouse Estuary, UK. The nonclay mineralogy was formed of quartz, carbonates, mica and feldspar, and the clay mineralogy was dominated by chlorite and illite. Similarly, the clay mineralogy of intertidal mud in the Tamar Estuary, UK, was found to be dominated by chlorite, montmorillonite and illite (Stephens et al., 1992).

Results from a mineralogical analysis by Potter et al. (1975) of alluvial sediments from the Mississippi River, US, are often thought to be illustrative of alluvial mud from other systems (Weaver, 1989). A qualitative, conceptual plot of the distributions of different minerals as a function of grain size illustrates a clay-sized fraction that is primarily composed of the clay minerals: kaolinite, illite, smectite (e.g., montmorillonite) and chlorite (Figure 8-1d). The silt-sized fraction is shown to comprise quartz, feldspar and carbonates, but clay is still present, whereas clay minerals are nonexistent in the sand-sized fraction, so that the distinction between sand and mud is governed by the presence of clay minerals.

2.4   Composition of Biological Material

Most natural mud and mud–sand mixtures will be a combination of sediment minerals and a biological fraction, both living and in various stages of decomposition. In some muddy suspensions the percentage of organic matter can be as high as 30 per cent by dry weight. Biology can influence all aspects of a sediment mixture's physical behaviour, including: deposition (Ginsberg and Lowenstam, 1958), sediment compositional matrix, bed features, consolidation and erodibility (Nowell et al., 1981; Heinzelmann and Wallisch, 1991).

Organic matter in mud exists as particulate or dissolved organic matter (POM or DOC, respectively). The DOC is a polyelectrolyte and does not play a significant role in the natural environment. Organic matter consists mostly of organic polymers that are either transported throughout an estuary or are produced locally by biological activity in the sediment bed or water column (see van Leussen, 1988, 1994). Allochthonous organic matter is degraded during transport and can comprise material such as lignin. Autochthonous organic material is a product of local biological processes, such as bioturbation. The organic substances resulting from these metabolic processes of an individual organism are generically termed extracellular polymeric substances (EPS).

Berner (1980) suggests that the principle organic materials within the structure of natural cohesive sediments comprise three groups: (a) polysaccharides and proteins composed of peptides and amino acids; (b) lipids, hydrocarbons (e.g., cellulose, lignin); and (c) humic acids. Of the three groups, the first represents organic substances that are recognised as flocculants, the second group is neutral, and the third comprises deflocculants (Winterwerp and van Kesteren, 2004).

When organic matter is present in the water column it is oxidised. However, once it is deposited to the bed and begins to undergo consolidation, it starts to reduce its levels of nitrate, manganese, iron and sulphur (Aller, 1982). As these chemical components continue to deplete, the underlying mud layers start to appear dark in colour and emanate a strong odour. This is the result of anaerobic decomposition and can result in the formation of gas within the bed, usually methane, carbon dioxide and the unpleasant odour of hydrogen sulphide. The role of biology on the flocculation process is examined later in this chapter.

3   Mud Flocculation

3.1   Overview

Cohesive sediments have the potential to flocculate into larger aggregates (e.g., Figure 8-2a) called flocs (Winterwerp and van Kesteren, 2004). Floc sizes (D) can range over four orders of magnitude, from individual clay particles to stringer-type floc structures several centimetres in length. An individual floc might comprise up to an order of 10^6 individual particulates, and as it grows in size its effective density (i.e., bulk density minus water density), ρ_e, generally decreases (Tambo and Watanabe, 1979; Klimpel and Hogg, 1986; Droppo et al., 2000). The general trend exhibited by a floc's effective density, as a function of floc size, has been observed by a number of authors (Figure 8-2b).

Settling velocity, in addition to effective floc density, is regarded as a basic variable in determining suspended sediment deposition rates in both still and flowing waters. Much has been documented on noncohesive sediments (coarse silts and larger), and because the only forces involved are gravity and the flow resistance of the particles (e.g., Soulsby, 1997), it is possible to calculate the settling velocity of low-concentration suspensions of these particles from the relative density, size and shape of the particles,

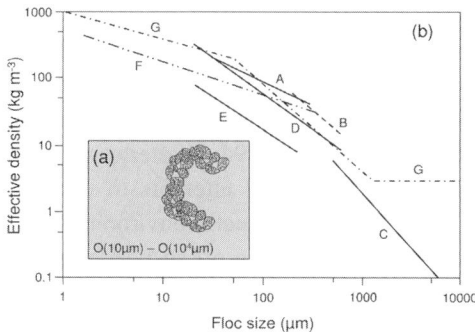

Figure 8-2 Some examples of the shapes and sizes of flocs and their effective densities. (a) Schematic of an individual floc and its potential order of magnitude size range; (b) comparative results of effective density against floc size: A = Manning and Dyer (1999), B = Al Ani et al. (1991), C = Alldredge and Gotschalk (1988), D = Fennessy et al. (1994, a, b), E = Gibbs (1985), F = McCave (1975), and G = McCave (1984); reprinted with minor modifications from Manning and Dyer (1999), with permission from Elsevier.

using well-defined expressions (e.g., Stokes' law; Stokes, 1851). However, the settling velocity of flocculated, cohesive sediments in estuaries is significantly greater than the constituent particles because floc settling velocity rises due to the size-dependent Stokes' law relationship (Mehta and Lott, 1987; Dyer and Manning, 1999). Based on the research of Stolzenbach and Elimelich (1994) and Gregory (1978), Winterwerp and van Kesteren (2004) concluded that although flocs are porous in composition, they can be treated as impermeable entities when considering their settling velocities. For a constant floc settling velocity there is a wide range of D and ρ_e. Similarly, for a constant D there is a large spread in both settling velocity and ρ_e. As with floc sizes, settling velocities typically can range over four orders of magnitude, from an order of 0.01 mms^{-1} up to several cm s^{-1} (Lick, 1994). As a result, the sizes and settling velocities of flocs are key parameters when modelling cohesive sediment transport in near-shore waters (e.g., Mehta and Lott, 1987; Geyer et al., 2000; Cheviet et al., 2002).

Manning (2001) defined microflocs and macroflocs as flocs with diameters such that $D < 160\mu m$ and $D > 160\mu m$, respectively. Flocculation is a dynamically active process that readily reacts to changes in turbulent hydrodynamic conditions (e.g., Krone, 1962; Parker et al., 1972; McCave, 1984; van Leussen, 1994; Winterwerp, 1998; Manning, 2004a). For example, during spring-tide tidal conditions in the Tamar Estuary (UK), Uncles et al. (2010) showed that median-diameter floc sizes could exceed 700 μm, and Manning et al. (2006) showed that macroflocs typically can reach 1–2 mm in diameter with settling velocities up to 20 mm s^{-1}; however, the effective densities of these large 1–2 mm macroflocs generally are <50 kg m^{-3}, which means that they are prone to break up when settling through a region of high turbulent shear.

The degree of flocculation, often referred to as the stability (van Leussen, 1994), is highly dependent on a number of variables, including: mineralogy (Winterwerp and van Kesteren, 2004); electrolytic levels, which tend to be altered because of salinity

less than can be tolerated by the imposed turbulence-induced forces, the floc will fracture. Also, aggregate breakup can occur as a result of high-impact particle collisions during very turbulent events. Floc breakup by three-particle collisions tends to be the most effective (Burban et al., 1989). Images of estuarine flocs from different turbulent and turbid environments illustrate their widely differing shapes and sizes (Figure 8-3b–e).

Turbulent energy is transferred from larger eddies to decreasingly smaller eddies, and this energy is dissipated by viscosity (van Leussen, 1997). These small eddies are defined by the Kolmogorov microscale of turbulence (Kolmogorov, 1941a, b). Kolmogorov proposed a cascade of turbulent energy from energy-containing eddies down to energy-dissipating eddies. McCave (1984) found that turbulence determines the maximum floc size in tidally dominated estuaries. Tambo and Hozumi (1979) showed that an aggregate would break up when the floc diameter was larger than the length-scale of the energy dissipating eddies. Similarly, Eisma (1986) observed a general agreement between the maximum floc size and the smallest turbulent eddies, as categorised by Kolmogorov (1941a, b). Fettweis et al. (2006) and van der Lee et al. (2009) both showed that floc size and the Kolmogorov microscale vary in a similar way with the root mean square of the gradient in the turbulent velocity fluctuations. Puls et al. (1988) and Kranck and Milligan (1992) have hypothesised that both concentration and turbulence are thought to have an effect on the maximum floc size and the resulting size spectra.

Velocity gradients are largest in the lowest 10–20 per cent of the water column, and approximately 80 per cent of the turbulent energy generated by the flow occurs within this zone. It is here that the strongest lift and shear forces occur, and Mehta and Partheniades (1975) have suggested that it is these forces that control the maximum size of the flocs in suspension. Thus, floc measurements in this turbulent, near-bed region are extremely important. On a smaller scale, both Argaman and Kaufman (1970) and Parker et al. (1972) suggested that flocs might decrease in size by gradual breakup through surface erosion of the floc by turbulent drag. The rate at which this takes place is proportional to the floc surface area and the surface shearing stress. These processes of floc breakup are more significant at low SPM concentrations.

3.3   Floc Particle Bonding

3.3.1   Salt Flocculation

Fine-grained sediment particle surfaces, in particular clay surfaces, have ionic charges that create forces comparable with, or exceeding, the gravitational force, and these cause the clay particles to interact electrostatically. The cohesive forces exerted between two clay particles depend both on the mineralogy of the clay and the electrochemical nature of the suspending medium. Most of the individual clay particles, comprising the common clay minerals, have a negative charge on the face of each platelet, mainly due to the exposed oxygen atoms in the broken bonds of their crystal lattices. The mutual forces experienced by two or more clay particles in close proximity are the result of the relative strengths of the repulsive and attractive forces (e.g., van Olphen, 1977; Manning, 2001).

In freshwater suspensions, which contain very few positive ions or possess low electrolyte concentration, the repulsive forces between negatively charged particles dominate, and the particles repel each other. The attractive forces dominate in saline waters due to the abundance of sodium ions that form a cloud of positive ions (cations in a high electrolyte concentration) around the negatively charged clay particles, resulting in the formation of flocs (e.g., Krone, 1962). Consequently, the sediment particles do not behave as individual particles but tend to stick together and form flocs.

Krone (1963) found that flocculation quickly reaches equilibrium at a salinity of about 5–10, which is much less than that for sea water (approx. 35). The potential for fine particles to flocculate is partly governed by their cohesion, and this can vary with mineralogy and the electrolytic level of the suspending fluid. Inevitably, flocculation is controlled by a series of interrelated kinetics that tend to be site specific in nature (Mikeš and Manning, 2010). For example, physical cohesion has been demonstrated to have an important effect on ripple and dune formation in clay–sand mixed environments (Schindler et al., 2015), and this influences the floc formation of locally resuspended sediments. In terms of gauging the importance of salt flocculation, engineering practice (as a simple rule-of-thumb) categorises this behaviour in terms of NaCl concentration. Critical salinities for coagulation of three common clays, expressed both as salinity and milli-equivalents per litre, are (Winterwerp and van Kesteren, 2004):

- Kaolinite: 0.6 or 10 mEq l^{-1}
- Illite: 1.1 or 19 mEq l^{-1}
- Smectite (or Montmorillonite): 2.4 or 36 mEq l^{-1}.

In sea water it might be expected that these critical values of salinity are greatly exceeded. On that basis, the role of salt flocculation should not be one that induces a clay mineral dependency, whereas in brackish environments it could lead to slight dependency of mineral type.

3.3.2 Flocculation Due to Organic Effects

It is increasingly recognised that there is a strong mediation of the physical behaviour of particles and flocs by the biological components of the system. Mineral cohesion effects are further enhanced by the presence of EPS (e.g., Tolhurst et al., 2002), such as mucopolysaccharides produced by microphytobenthos. In estuarine environments, where sediments are predominantly mud and silt, benthic microphytobenthos contribute up to half the total autotrophic production (Cahoon, 1999; Underwood and Kromkamp, 1999). For example, epipelic diatoms (e.g., Paterson and Hagerthey, 2001) secrete long-chain molecule EPS as they move within the sediments. EPS is regarded as a highly effective biostabiliser of muddy sediments (e.g., Uncles et al., 2003; Underwood and Paterson, 2003; de Brouwer et al. 2005; Gerbersdorf et al. 2009; Grabowski et al., 2011) and can significantly enhance interparticle cohesion. Flocculation and deflocculation can result from adsorbed polymers (Figure 8-4a, b). In general, flocs held together by polymers are stronger than those held together solely by electrostatic, London–van der Waals forces (Figure 8-4c and d; Kitchener, 1972).

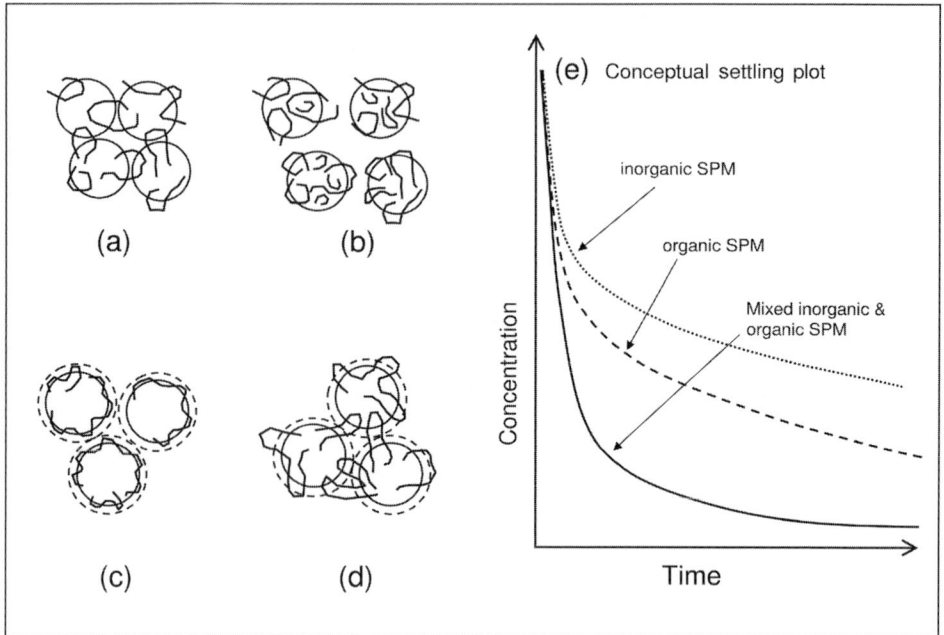

Figure 8-4 Some schematic examples of the effects of organic matter on sediment particle properties are given: (a, b) Flocculation and deflocculation by adsorbed polymers; (c, d) schematic diagram of polymer bridging and the repulsive energy barrier (dotted line) at the particle surface for: (c) low ionic strength (bridging prevented by electrical repulsion); and (d) high ionic strength (bridging across the effective repulsion distance); (e) the conceptual relationships between concentration and time for settling suspensions consisting of inorganic (single mineral) particles, organic matter, and flocs consisting of a mixture of organic matter and mineral particles.

Edzwald and O'Melia (1975) conducted experiments with pure kaolinite and found that the flocculation efficiency was less than 10 per cent. Experiments by Kranck (1984) have shown that the flocculation of mineral particles that contained some organic matter greatly enhanced the settling velocity of the resultant aggregates (shown conceptually in Figure 8-4e). This result was also obtained by Gratiot and Manning (2004) using a grid oscillation tank and video-capture techniques.

Both Cadee (1985) and Kranck and Milligan (1988) reported enhanced flocculation following diatom and coccolith blooms at the entrance of tidal inlets; coccolithophorids may also act as nuclei around which flocs are created. Jackson (1990) modelled flocculation during a diatom bloom by considering both aggregation processes and algal growth. It was concluded that the important parameters were turbulent shear, the algal (particle) concentration and the size and stickiness of the algae.

3.4   Floc Microstructure

Information on the microstructure of flocs can be highly desirable because it provides a detailed insight on how a particular sediment composition flocculates at the floc-floc and primary particle levels. To examine the floc internal microstructure (matrix) at a submicron level (1–2 nm; Buffle and Leppard, 1995), use of a transmission electron microscopy (TEM) has been employed in a series of experiments by Spencer et al. (2010). These TEM images have been a valuable resource to visually identify the presence both of clay minerals and quartz mineral fragments within natural microfloc structures (Spencer et al., 2010). The results suggest that sand particles favour the microfloc fraction when mixed sediments flocculate; this is logical because microflocs tend to have the stronger bonding potential due to the closeness of their bonds. Uptake of individual sand particles will probably be much less in the macroflocs. This is consistent with the order of aggregation theory (Krone, 1962; Eisma, 1986), which states that microflocs will flocculate into macroflocs when the ambient conditions are favourable, and provides a more efficient mechanism (and pathway) for the fine sand grains to move into the macrofloc fractions.

4   Mixed Sediment Flocculation

When modelling sediment transport, it is common practice to assume a single, representative sediment type, such as noncohesive sand or cohesive mud. Typically, modelling single sediment types would be a precursor to more complicated modelling of mixed sediment types and fractions. This is due to the well-documented transport formulae that have already been developed for solely muddy or sandy sediments.

Sediment mixtures may either behave in a segregated way or may interact through flocculation. The phenomenon of mud–sand segregation considers the mud and sand to operate as two independent suspensions (van Ledden, 2002). When a segregated regime dominates there is very little bonding, and flocculation interaction between the fine fraction and the larger, noncohesive sediment fraction is nonexistent. Mixed sediment experiments have shown that fine sediment particles and sand grains, which behave in a segregated manner, settle simultaneously (but at different speeds) at the bed–water interface, thereby forming two well-sorted layers (Migniot, 1968; Ockenden and Delo, 1988; Williamson and Ockenden, 1993; Torfs et al., 1996). However, where the fine fraction and the larger noncohesive sediment coexist as a single mixture (Mitchener et al., 1996), this creates the potential for these two fractions to combine and exhibit some degree of interactive flocculation (Manning et al., 2010, 2013). Whitehouse et al. (2000) describe a process whereby cohesive sediments that are mixed into a predominately cohesionless sandy region can create a 'cage-like' structure that can fully encompass the sand grains, thereby trapping the sand within a clay floc envelope. Within a mixed sediment environment, the degree of cohesion between the various sediment fractions tends to increase with the content of fine clay minerals within the sediment and starts to become significant when the sediment contains more than 5–10 per cent of clay by weight (Dyer, 1986; Raudkivi, 1998; Whitehouse et al., 2000, van Ledden, 2003).

4.1   Biological Effects on Mixed Sediment Flocculation

In addition to physical processes, biological activity, more commonly associated with cohesive sediments, has been highlighted to play an important role in the cohesion of mixed sediments (e.g., Paterson and Hagerthey, 2001) and is an important component that makes mixed sediment flocculation possible. The influence of biology on sand has been reported to a much lesser extent; however, sand grains that are exposed to long-term biological activity may also develop a cohesive bio-coating, which could increase the particle collision efficiency when they are entrained. Hickman and Round (1970) reported that sand particles can be joined by epipsammic (i.e., attached to or moving through sand particles) diatoms that attach to sand grains. Epipsammic macro-algae either adnate (i.e., grow strongly) to the grain surface or attach to sand grains using their mucilage stalks. Epipsammic diatoms that are attached to sand grains demonstrate strong adhesive properties to the grain surface (Harper and Harper, 1967).

When fine sand and biology are combined into a single matrix they can form microbial mats, and the binding strength of these mats can be extremely high. Little (2000) states that because these types of algal threads are sticky with EPS, they can efficiently trap fine sand grains. These sticky bio-coatings can increase the collision efficiency (Edzwald and O'Melia, 1975) of particles when they are entrained into suspension, thus allowing fine sand grains to adhere with the clay fraction. Using microscopic photography, Wolanski (2007) reports the formation of large, muddy flocs formed by mud that creates a sticky membrane around large, noncohesive silt particles. Recent laboratory experiments that examined the effects of EPS biostabilisers on both ripple (Malarkey et al., 2015) and dune (Parsons et al., 2016) evolution in sand–mud sedimentary environments found that just 0.1 per cent EPS was sufficient to prevent the formation of significant bedforms.

The process of bioturbation (i.e., the reworking of bed sediments by living organisms) can also potentially enhance the mixing of bed sediment particles prior to resuspension (e.g., Nowell et al., 1981; Paterson et al., 1990; Widdows et al., 2004). Thus a bed that is initially deposited as a discretely segregated layering of mud and sand may be transformed into a quasi-homogeneous mixture as a result of bioturbation.

4.2   Typical Mixed Sediment Flocculation Behaviour

Different ratios of mud and sand can vary the level of cohesion and influence the resultant level of flocculation. However, due to the wide-ranging variability of mixed-sediment compositional properties, it is extremely difficult to quantitatively describe such a complex sedimentary matrix in a fundamental manner; this is primarily the result of a lack of verification data. Therefore, in order to quantify mixed-sediment flocculation characteristics, a series of laboratory studies were conducted on a range of mud–sand mixtures that possessed different SPM concentrations and were sheared by different levels of turbulent mixing (Manning et al., 2007, 2010). The results revealed the following general, mixed-sediment floc properties:

- $W_{s,micro}$ increases in response to a rising sand content.
- $W_{s,macro}$ decreases with rising sand content.
- The greater the sand content of a mixed suspension, the higher the total mass settling flux (MSF).

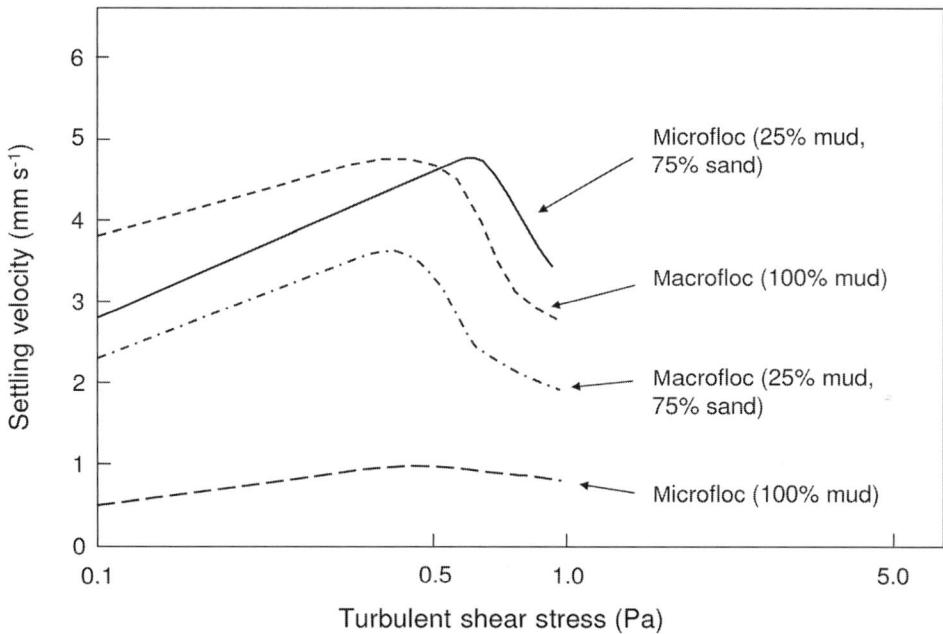

Figure 8-5 Illustration of the settling velocities of macroflocs and microflocs, plotted against shear stress, for a mixed sediment suspension comprising a ratio of 25 per cent mud to 75 per cent sand and a pure mud suspension, all for a total SPM concentration of 5 gl^{-1} (modified from figure 14 of Manning et al., 2013, published March 13, 2013 under CC BY 3.0 license. © The Authors).

- The SPM_{ratio} (defined as the ratio of macrofloc to microfloc SPM concentrations) steadily rises as overall sediment concentration increases, which is similar to a fully cohesive suspension. However, the SPM_{ratio} generally decreases across the suspended particulate matter.

These factors have been included in a set of empirically derived algorithms that describe mixed-sediment flocculation (Figure 8-5; Manning et al, 2011a). It should be noted that any mixed-sediment flocculation effects and sediment–fraction interactions can only be demonstrated empirically through rigorous laboratory settling experiments. A review of mixed sediment flocculation is provided by Manning et al. (2013).

5   Floc Measurements – Data Collection Devices

The fragility of large flocs has tended to preclude the direct measurement of their settling and mass characteristics due to instrumentation limitations (Kranck, 1993; Eisma et al., 1997). Because accurate information on floc sizes and settling velocities is of key importance to the modelling of cohesive sediment transport (e.g., Geyer et al.,

2000; Cheviet et al., 2002), controlled laboratory experiments have been widely used to investigate specific flocculation phenomena in greater detail (e.g., Gibbs, 1985; Tsai et al., 1987; Lick et al., 1993; Ockenden, 1993; Bale et al., 2002). However, in situ data, if collected correctly, are always far more desirable – especially from a modeller's perspective. The influences of floc density variations are required for accurate settling-flux determination. Therefore, a key to rectifying this problem is to use a floc sampling system that directly measures (in situ) both the simultaneous size and settling velocity of the larger and more fragile flocs. This section provides an overview of the various devices and approaches used to measure floc size and settling velocity values, including some of their limitations and advantages.

5.1   Gravimetric-Based Floc Data

Large, fast-settling, fragile macroflocs are easily broken up during sampling (Gibbs and Konwar, 1983). Floc-disruptive devices include field settling tubes (FST) such as the Owen tube (Owen, 1976) and other versions (van Leussen, 1988), BIGDAN (Puls and Kühl, 1996), and the Field Pipette Withdrawal Tube (Cornelisse, 1996).

The Owen tube is the best known FST (Owen, 1971, 1976). It usually comprises a 1-m-long, open tube constructed of Perspex with an internal diameter of approx. 0.05 m. Larger versions, such as the QUISSET – Quasi-In-Situ Settling Tube (Jones and Jago, 1996), have been tested. The Owen tube is deployed in a horizontal orientation, and when positioned at the desired depth, a messenger is sent down the deployment cable that closes the end caps, thus trapping the suspension within the tube. On retrieval, the tube is placed in a vertical position (usually in a frame, see Figure 7-1a). Collected water samples are extracted from the bottom of the tube at preselected time intervals (usually logarithmic time steps), and the settling velocity is inferred from gravimetric analysis (Vanoni, 1975; Chapter 7). There are no direct observations of individual floc properties with FSTs during gravimetric sampling.

The settling velocity values measured using the Owen tube appear to be quite small when compared with more recent, nongravimetric analysis, which probably can be attributed to the disruptive nature of the analysis and the breaking-up of fragile, fast settling macroflocs, thereby skewing the results toward those for the more durable, but slower settling, microflocs (e.g., Eisma et al., 1997). However, Puls et al. (1988) found that the Odd (1988) settling tube measurements sometimes tended to overestimate the settling velocity of suspended mud by a factor of up to approx. 4–5. This was primarily a result of flocs growing quickly by aggregation while sinking in the calm environment of the closed tube (compared with the continuously turbulent state of the waters in the estuary). Therefore, there are ambiguities of interpretation with this technique.

Other gravimetrically based instruments that have been used to determine floc properties, such as the Niskin bottle (Gibbs and Konwar, 1983) or pipettes (e.g., the Andreason Pipette technique, described later), as well as the Owen tube (Owen, 1971), are all very floc-disruptive (Gibbs, 1985; van Leussen, 1988).

5.2   Direct Floc Size Measurements

The presence of large estuarine macroflocs was initially observed in situ using underwater photography (Eisma et al., 1990). However, floc breakage occurs during sampling in response to the additional shear created by the instrumentation (Eisma et al., 1997). To overcome this problem, less-invasive techniques for measuring floc properties in situ have been developed. Usually, these can be divided into devices that solely measure floc size and those that can provide measurements both of floc size and settling velocity.

5.2.1   The Lasentec

This was a commercially available Par-tec 100 (Lasentec Inc., Redmond, WA 98052) laser reflectance instrument, which was modified for underwater use (Figure 8-6a) at the Plymouth Marine Laboratory and the University of Wales at Bangor, UK. Using a

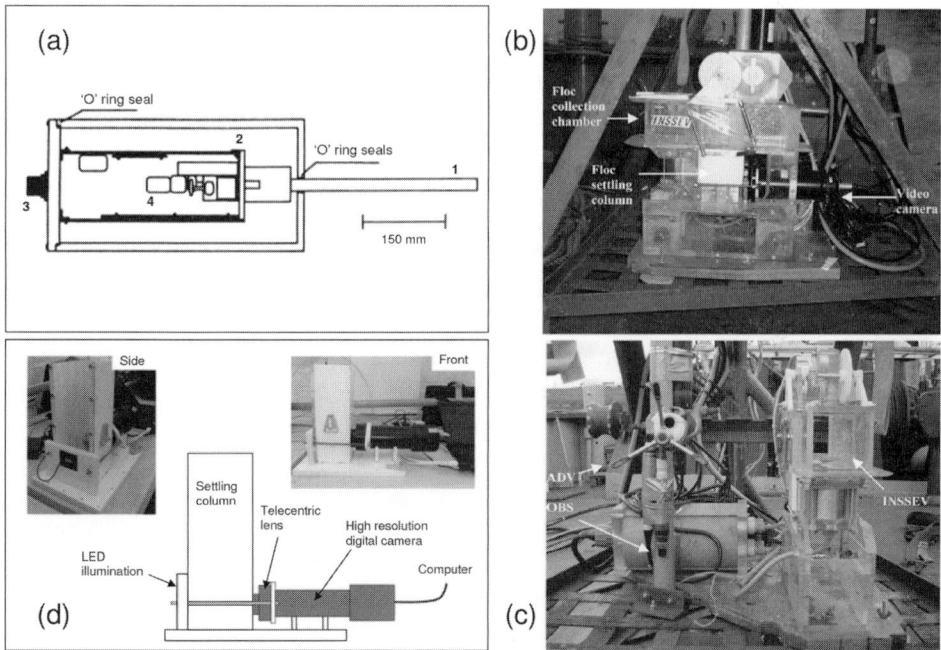

Figure 8-6 Some examples are given of floc measuring instrumentation. (a) Schematic of the adapted Lasentec Par-tec 100 probe unit showing: 1. the light guide, scanning mechanism and focusing lens, 2. the PVC cylinder, 3. watertight cable termination, and 4. electronic circuitry and power supplies (reprinted with minor modifications from Law et al., 1997; with permission from Elsevier). (b, c) The INSSEV instrument (reproduced from figure 7a and b of Manning et al., 2011b; published April 26, 2011 under CC BY-NC-SA 3.0 license. © The Authors); (b) side view of INSSEV mounted on a metal deployment frame; (c) front view of INSSEV (right side of image), together with optical backscatter (OBS) sensors and an acoustic Doppler velocimeter (ADV) positioned on a vertical pole (left side of image). The ADV provides high frequency turbulence data that can be directly related to the floc populations. (d) Views of a LabSFLOC-2 together with a schematic illustration of the instrument (lower centre, after Manning, 2016; with permission).

rapidly scanning laser light, focused to an area of size $0.2\mu m \times 0.8\mu m$ (i.e., smaller than the flocs under observation), the instrument measures the period of the backscattered pulse generated each time the focal point scans a floc in suspension. This sensor arrangement allows the Lasentec P100 to operate in turbid estuarine waters, whilst creating minimal physical disturbance to fragile macroflocs. A minimum of 1000 counts is required for a measurement, and only flocs with a sufficiently coherent reflected signal that can be separated from background noise are registered. A correction algorithm is applied to produce a spherically equivalent diameter. The device is capable of sensing particulate matter within the size range 2–1000 μm, which it bins in 38, approximately logarithmic-sized, intervals. The software generates size distributions as either percentages of the total number of flocs or of the total volume. More details on the Lasentec P100 are given by Law et al. (1997).

5.2.2   The LISST and LISST-Holo

Floc size data can be obtained through particle size data and volume concentration using a LISST (Laser In-Situ Scattering and Transmissometry) instrument. This is an autonomous instrument that measures laser diffraction from suspended particles and assigns each particle to one of 32 logarithmically spaced size classes (1–250 μm; Agrawal and Pottsmith, 1994). It estimates the volume of a floc using the small-angle scattering of the laser beam and applying Mie theory (Mie, 1908; Xu, 2000), which is an analytical solution of Maxwell's equations for the scattering of electromagnetic radiation by small spheres (e.g., Stratton, 1941). The output from LISST is the total volume concentration of a particulate suspension ($\mu l\ l^{-1}$) in each size class (Agrawal and Pottsmith, 2000). This instrument has the advantage that it can provide vertical profiles of suspended sediment sizes through the water column. However, due to the operational nature of these devices, their use is generally limited to low-turbidity environments, compared with the moderate to high turbidity conditions typically found within muddy, mesotidal and macrotidal estuaries (terms defined in Davies, 1964).

The LISST-Holo (laser in situ scattering and transmissivity holographic system) is a submersible digital holographic particle imaging system. Graham and Nimmo Smith (2010) provide an overview of the use of holography for measuring flocculation in low-turbidity environments. The interpretation of complex floc image data produced by a combined LISST and holographic imaging system is discussed by Graham et al. (2012).

5.2.3   The InSiPid

Benson and French (2007) describe a low-cost particle sizing system InSiPID – In situ Particle Imaging Device – that can be operated within shallow water environments where turbidity may reach several hundred mg l^{-1} (higher than most particle sizers can achieve). The InSiPID combines twin CCD (charged coupled device) video cameras, providing an operational imaging range of about 4 to 3000 μm. The instrument has a fully automated digital image acquisition system, together with processing algorithms for the extraction of floc size and floc shape statistics.

5.3   Simultaneous Measurements of Floc Size and Settling Velocity

The development of video-based floc devices has provided tools with which floc sizes and floc settling velocities can be measured simultaneously within a natural flow, whilst creating minimal interference to the aggregates. Furthermore, unlike many standard particle-size instruments, these permit an estimate of an individual floc's effective density, ρ_e, defined as the bulk density minus the water density, using a modified Stokes' law. Therefore, these measurements make possible the computation of floc-mass distribution across a range of sizes (Fennessy et al., 1997; Manning, 2004,a, b) and can provide an insight into the interaction of flocs within the lower layers of the flow, where the suspended concentration gradients are greatest. Such site-specific information on floc settling velocity spectra is a prerequisite for the accurate parameterization of aggregation processes, especially for implementation into sediment transport models (Manning, 2004b, 2008; Baugh and Manning, 2007).

5.3.1   The VIS

A Lagrangian system, VIS, was developed jointly by Rijkswaterstaat and Delft Hydraulics (van Leussen and Cornelisse, 1994), and systems utilising a Braystoke tube to collect the flocs, such as the HR Wallingford video camera system, have also been used during field sampling campaigns (Dearnaley, 1996, and VIL – Video In Lab, Defossez, 1996).

5.3.2   The INSSEV

The most successful of these video-based floc camera systems, when applied to a wide range of aquatic environments, is the low-intrusive INSSEV: IN-Situ SEttling Velocity instrument, which was developed at the University of Plymouth, UK (Fennessy et al, 1994a). A detailed review of the INSSEV instrument and its operation is provided by Manning and Dyer (2002a).

The sampling apparatus comprises a twin-chamber device in which a volume of estuarine water is captured in the upper chamber (Figure 8-6b and c). After a period of 10–20 s has elapsed (to allow the turbulence in the water to decay), a selection of flocs are permitted to pass into a lower settling column through a computer-controlled trap door. An integral underwater backlit analogue video camera views the flocs as they settle within the lower settling chamber. Floc images are recorded using an S-VHS video suite, which has a practical lower size detection limit of 20 μm. The INSSEV device has the important advantage of permitting the simultaneous, in situ measurement of both individual floc size and settling velocity, from which the effective density of flocs can be calculated.

Complete floc population samples are acquired at a rate of one sample every 10–20 min during a typical estuarine sampling run, the interval being dependent on the floc settling velocities and SPM concentration. Usually, the INSSEV apparatus is mounted within a metal frame, with floc measurements made at a nominal height of 0.5 m above the bed. This near-bed region typically contains approximately 80 per cent of the turbulent energy present within an estuarine water column, which is significant both for floc formation and aggregate breakup.

5.3.3   The LabSFLOC

The LabSFLOC – Laboratory Spectral Flocculation Characteristics – instrument (Manning, 2006, 2016) is a laboratory derivation of the in situ INSSEV instrument that enables floc properties to be measured in laboratory studies using suspended particulate matter (SPM) concentrations of several gl^{-1}. LabSFLOC utilises a low-intrusive, high-resolution video camera to observe flocs as they settle in a settling column constructed of Perspex. The video camera utilises a back-illumination system, whereby floc images are manifested as silhouettes; i.e., particles appear to be dark on a light background. This reduces image smearing and renders the floc structure more visible.

A floc sample is extracted from its original environment (e.g., a flume) and is immediately transferred to the column using a modified pipette. The video camera, the centre of which is positioned 0.075 m above the base of the column, views all particles in the centre of the column as they settle from within a predetermined sampling volume. All of the flocs viewed by the LabSFLOC video camera, for each sample, are measured both for floc size and settling velocity. Clear, grey-scale, 2-D optical images of the flocs are recorded by the video suite (analogue-to-digital for version 1 and fully digital for version 2). LabSFLOC-2 (Manning, 2016; Figure 8-6d) can measure floc sizes of 8 mm in diameter and settling velocities approaching 45 mm s^{-1}, providing the flexibility to measure both pure mud and mud–sand mixed sediment floc dynamics. By implementing a sequence of image-analysis algorithms, the floc porosity, fractal dimensions, floc dry mass and the mass settling flux of a floc population can be computed. The calculated dry mass also can be compared with the measured SPM concentration, thus providing an estimate of the efficiency and reliability of each sampling procedure. It follows that the data obtained from LabSFLOC is both qualitative and quantitative.

5.3.4   The INSSEV-LF

The portability of the LabSFLOC instrument has led to the development of the INSSEV-LF: IN-Situ Settling Velocity instrument. The LF (LabSFLOC) version of INSSEV is a hybrid system that combines two key components: (1), the low-intrusive LabSFLOC system and (2), an in situ estuarine floc sampling acquisition unit (to initially obtain the suspension sample). For the latter, a 2.2 l van Dorn horizontal sampling tube with a 10–14 kg weight suspended from the underside of the tube has been used to collect a water sample at a nominal height above the estuary bed. Manning and Schoellhamer (2013) have deployed the INSSEV-LF throughout San Francisco Bay and the Northern Californian River Delta, USA.

5.3.5   The PICS

The Particle Imaging Camera System (PICS) was developed by Smith and Friedrichs (2011) to produce high-quality, in situ floc image sequences within dredge plumes. The PICS consists of a 1-m-long, 0.05-m inner-diameter settling column with a mega-pixel digital video camera and strobed LED lighting. The settling column is equipped with two pneumatically controlled ball valves at the column ends, which permit sample capture, and a third pneumatic actuator for rotating the column from a horizontal to a vertical orientation for image acquisition (Smith, 2010). The associated image-processing routines have been outlined by Smith and Friedrichs (2015).

Smith and Friedrichs (2011) have applied the PICS to dredge-plume monitoring operations. Data collection with PICS proceeds by positioning the profiling frame at the desired depth, capturing a sample of suspended particles by closing the ball valves at the ends of the settling column, rotating the column to a vertical position, permitting turbulence within the column to dissipate (approx. 30–60 s), and collecting, typically, 30 s of video data (Smith and Friedrichs, 2011).

5.4   Other Approaches to the Measurement of Floc Size and Settling Velocity

Optical devices to measure concentration profiles that have been utilised by Spinrad et al. (1989), Kineke et al. (1989), and McCave and Gross (1991) have sought to quantify the rate of water clearance, but they were unable, like all earlier instrumentation, to measure floc size and settling velocity spectra directly. In contrast, sampling devices that directly observe floc size and settling velocity can provide an insight into the interaction of flocs both with turbulent eddies and SPM concentration variations during a tidal cycle, particularly within the lower layers of the flow where the turbulent shearing is at its greatest (Mehta and Partheniades, 1975).

Deploying floc samplers in conjunction with high-frequency velocimeters provides a means of accurately acquiring time series of the spectral distribution of floc dry mass and settling velocity, together with information on the turbulence fluctuations, directly from within a turbulent estuarine water column.

6   Floc Measurements – Data Analysis and Processing

A range of procedures can be used to generate and assess various types of floc data.

6.1   Owen Tube Data

Owen tube measurements generally show an exponentially increasing relationship between median floc settling velocity, $W_{s,50}$, and SPM concentration, C, for concentrations <10 gl^{-1}, in the form of:

$$W_{s,50} = k \cdot C^m \qquad (1)$$

This relationship has been reported for numerous estuaries (Odd, 1988; van Leussen, 1988; Delo and Ockenden, 1992; Figure 8-7a). An exponent, m (Eq. 1), of 1.05 was measured for the Severn Estuary and 0.69 for the neighbouring Parrett Estuary (Figure 8-7a). Variations in m between estuaries are, probably, a result of floc density variations and differences in ambient hydrodynamic conditions.

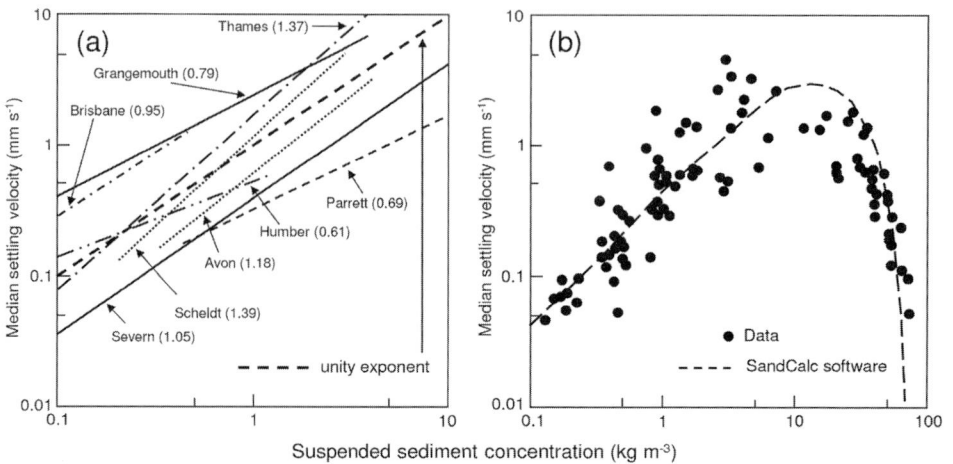

Figure 8-7 Some examples of floc settling velocity measurements are shown. (a) Owen tube determinations of median settling velocity as a function of suspended sediment concentration for different estuaries; the bold dashed line represents an exponent of unity (reproduced with minor modifications from Delo and Ockenden, 1992; with permission). (b) Median settling velocity of Severn Estuary mud as a function of SPM concentration; the Owen-tube data are taken from Odd and Roger (1986), and the dashed line represents results based on the SandCalc software sediment-transport computational algorithm, which incorporates the hindered settling effect at high concentrations (reproduced with minor modifications from Soulsby, 2000; with permission) .

Once the SPM concentration reaches a level of 5–10 g l^{-1} the flocs tend to interact hydrodynamically, and floc growth can be inhibited. A further rise in concentration to >10 g l^{-1} causes the particle interactions to modify properties of the suspension, causing floc settling velocities to gradually slow, which results in a hindered settling flux (e.g., Dankers, 2002; Winterwerp, 2002; Winterwerp and van Kesteren, 2004). The rheological behaviour of these high-concentration suspensions becomes non-Newtonian. For hindered settling conditions, Maude and Whitmore (1958) found that the simple exponential relationship (Eq. 1) had to be modified. A modification algorithm to mimic the hindered settling effect has been integrated into commercially available sediment transport software (SandCalc; HR Wallingford, 1998). Odd and Rogers (1986) illustrate how the median floc settling velocity can vary as a function of suspended sediment concentration using Owen tube data from a number of different estuaries (Figure 8-7b). The data indicate that when the SPM concentration is 65 g l^{-1}, the median settling velocity is approx. 0.1 mm s^{-1}, which is the same settling velocity as a relatively dilute 250-mg l^{-1} suspension.

6.2   The Andreason Pipette – Size- and Weight-Distribution Data

Due to the compositional nature and flocculation potential of fine, cohesive, muddy sediments, it is often impractical to use the sieve analysis techniques that typically are applied to noncohesive sediments (> 63 μm in diameter) in order to determine their sizes. The Andreason Pipette is a simple gravimetric laboratory approach to indirectly determine the particle size and cumulative weight distributions of a cohesive sediment suspension, together with its settling velocity characteristics.

The following procedure is one analytical method that may be applied to Andreason Pipette data (see also Chapter 6); the overarching aim of this experiment is to assess the size distribution of a sample of fine-grained sediment and delineate how the observed results relate to sediment characteristics and particle-settling behaviour:

1. Weigh and mark eight Petrie dishes.
2. Record the temperature of the suspension (T; units of °C).
3. Stir thoroughly and immediately take a 20 or 25 ml sample with the pipette. Decant this sample into a preweighed evaporating dish. When dried and reweighed, this sample will allow calculation of the initial ($t = 0$) concentration of the suspension (Chapter 7).
4. Stir the suspension again. When the major part of the turbulence has died away and the sediment is beginning to settle uniformly, start the timer. Using the pipette stand, lower the pipette 0.10 m into the suspension (it may be helpful to mark the pipette at 0.10 m from the bottom). Take a pipette sample at 30 s. Decant the sample into a Petrie dish.
5. Stop the clock, stir the sample again and repeat the sampling sequences, taking samples at 1, 3 and 5 min. After 5 min there is no need to stop the clock and restart. Take further samples at 10, 20 and 30 min. Note that stirring after each of the early time increments can sometimes create unwanted, higher suspended concentrations; if this is seen to occur, omit the restirring. Dry and weigh the samples. Record all data.
6. Complete the analysis form (Table 8-1). Each time interval is a measurement of the time that it takes a particle of size d to settle (W_s is the settling velocity) over

Table 8-1 Example spreadsheet for recording Andreason pipette data.

Time (min)	Dish No.	Dish weight without sediment, w_c (g)	Dish + dry sediment weight w_d (g)	Dry sediment weight $w_d - w_c$ (g)	Cumulative weight P finer than d (%)	Particle size d (μm)
0					100	$> d[t = 0.5\,\text{min}]$
0.5						
1						
3						
5						
10						
20						
30						

the sample depth of 0.10 m. The sample that is extracted will contain particles of size d or less. Calculate d at each time increment from the formula:

$$d^2 = W_s \cdot F^2 \tag{2}$$

Where W_s is derived from the depth of sampling (cm) divided by the elapsed time (minutes). Eq. 2 is a simplification of the full Stokes' law formula (e.g., Dyer, 1986). The parameter F generally represents the frictional (i.e., viscosity) and density (i.e., sediment bulk and water) influences and can be derived from the following relationship (Galehouse, 1971; Dyer, 1986):

$$F = -0.148T + 16.55 \tag{3}$$

The percentage cumulative weight finer than d, $P(t)$, is calculated from:

$$P(t) = \frac{(w_d[t] - w_c[t]) \times 100}{(w_d[t = 0] - w_c[t = 0])} \tag{4}$$

Where w_c is the dish weight and w_d the dish plus dry sediment weight; $w_d[t]$ indicates the dish plus dry sediment weight at a specific time step (i.e., time increment, t); $w_d[t = 0]$ indicates the dish plus dry sediment weight at the initial time (i.e., $t = 0$). The cumulative weights (Eq. 4) are straightforward to compute once the data are tabulated in a spreadsheet format, such as Microsoft Excel, and once computed can be plotted as a cumulative curve of sediment grain size distribution, i.e., P vs. D.

Logically, one would expect to see a gradual decrease in $P(\%)$ with decreasing particle size. Gee and Or (2002) note that the asymmetrical structure of clay platelets may demonstrate a 'drift' or 'wobbling' throughout settling and will inevitably have reduced settling velocities relative to a sphere, due to increased flow resistance. Also, despite measuring temperature prior to the analysis, it is difficult to prevent minor temperature fluctuations that may ultimately affect the fluid molecular viscosity (μ) and therefore the particle settling velocity (Lau, 1994). When applied to natural muddy sediments, the test does not account for suspensions that comprise a mixture of clay minerals, silt grains, some sand grains and biological matter, all of which might be flocculated to some degree. Therefore, if a potentially erroneous rise in $P(\%)$ is measured for smaller particle sizes, one solution to improve data quality could be to redo the experiment, but without restirring after the initial $t = 0$ baseline measurement.

6.3   Dilute Suspension and Hindered Settling Tests

The aim of this experimental test is to measure the settling velocity of interfaces formed within a muddy suspension and to analyse and compare the results in relation to the properties of the suspended sediment concentration and its supporting fluid. To start the test, a sediment suspension (with a predetermined concentration) is decanted into a graduated cylinder (usually of 1 litre capacity).

For each cylinder:

1. Record the concentration, C_0, and salinity, S.
2. Record the initial heights of the water column, h_0, and the bed layer - if it is apparent.
3. Shake up each of the solutions, and start the stopwatch. Synchronising this activity will save time.
4. Record the elevation (h) of the interface falling from the surface as well as the one rising from the bed (if present) as a function of time (t). Use 1 min intervals until 15 min, and then use 2-min intervals. Quantify the uncertainties in the measurements and comment on them in the report. Note that these height measurements must be from the bottom of the cylinder.
5. Make notes on the settling behaviour in each cylinder, and observe any individual flocs. If the flocs are large enough (several hundred microns in diameter), it may be possible to estimate the settling velocity of individual flocs. The use of backlighting can improve the visual distinction of flocs and interfaces.
6. The initial settling will usually occur in a linear fashion with time; after a longer time the settling will follow a logarithmic settling regime.
7. Plot interface height (on the y-axis) against time (on the x-axis) for each individual settling column data set. It is possible to estimate the settling velocity of the interface for each column by applying a linear regression analysis (i.e., height versus time, which can be used to provide velocity values from the regression gradient) throughout both the linear and logarithmic settling sections.
8. After a period of time the falling interface will meet the rising bed layer, and the rate of fall (settling velocity) of the interface will become logarithmic. In this logarithmic period the fall rate of the interface can be described by the equation:

$$h = h_\infty k(1/t) + h_\infty \tag{5}$$

Where the height of the bed layer at $t = \infty$ is h_∞ and k is the consolidation rate constant. This equation has the form: $Y = aX + b$; therefore, a graph of h as a function of $(1/t)$ will result in a straight line with a slope of $h_\infty k$ and an intercept of h_∞ at $(1/t) = 0$.

The duration required to conduct these hindered settling experiments will depend on the concentration of sediment within the settling columns. Hindered settling typically starts to occur when the SSC approaches 5–10 gl^{-1}. The initial linear settling phase can usually take 20–30 min, whereas the subsequent logarithmic hindered settling phase may occur over many hours.

6.4   Laboratory Floc Simulation and Floc Production Techniques

Suspended sediment (solids, particulate matter) concentrations (SSC) are required to calibrate and mass-reference raw floc data. The methodology and protocols (usually gravimetric techniques) for obtaining SSC are discussed in Chapter 7.

6.4.1   Jar Testing

This technique is often used as a first approach when assessing and comparing the physicochemical interactions of substances such as flocculated natural mud suspensions (e.g., Bouyer et al. 2005; Mietta et al. 2009). A series of 'jar tests' (EPA, 2002; Phipps and Bird, 2016) can be conducted to assess how potentially flocculating particles behave in turbulent waters and to establish whether they do flocculate.

A sediment suspension (slurry) of predetermined concentration is placed in a glass jar and agitated using a stirring bar (suspended in the fluid) for a sufficiently long time that the flocs reach a quasi-equilibrium state of flocculation. This technique has the advantage of providing a good control volume (i.e., a 1 litre jar) within which to examine typical flocculation behaviour. Usually, each suspension is stirred for approximately 30 min to allow the particles sufficient time to flocculate and attain floc equilibrium, in accordance with the theoretical flocculation time (van Leussen 1994). Although a wide range of shearing stresses can be used, a good choice is to set the level of turbulent shear stress within the jar at approximately 0.3–0.4 Pa (simple calibration protocols are outlined by Hudson, 1981 and Droste, 1997) because this can provide the ideal stimulation for maximum floc growth (Manning 2004a) and will establish whether or not a suspended sediment flocculates.

Mietta et al. (2009) used jars that were 0.085 m in height and 0.125 m in diameter, equipped with four 0.012-m-wide baffles to increase and homogenize the shear rate. The suspension was stirred using a single rectangular paddle that was placed 0.01 m above the bottom of the jar. The paddle was 0.075 m in diameter and 0.025 m high. The average shear rate in the jar for each stirring frequency was determined by measuring the power dissipation of the propeller, P (Nagata, 1975; Bouyer et al., 2005). If V is the volume of the suspension, the power dissipation per unit volume, $P_V = P/V$, can be related to the turbulent energy dissipation parameter, ε. The shear rate for each stirring frequency can then be derived from the dissipation parameter (Tennekes and Lumley, 1972) by substitution into Nagata's (1975) relationship between the 'power number' of a propeller, which Nagata (1975) expresses as a function of the power dissipation, the stirring frequency (in rotations) and the diameter of the propeller. Detailed floc characteristic can then be observed using particle size or video observational techniques, using subsamples extracted from the jar with a pipette.

6.4.2   Annular Flumes

Sediment dynamicists studying cohesive sediments often choose to work with annular flumes (Chapter 9), in which a circulating flow is initiated in the annulus by the rotation of an annular ring because their infinite flow length results in a fully developed boundary layer (e.g., Ockenden and Delo, 1991; Williamson and Ockenden, 1993; Black and Paterson, 1997; Manning and Dyer, 1999; Lau and Droppo, 2000; Bale et al., 2002; Stone and Krishnappan, 2003; Graham and Manning, 2007). In contrast to straight recirculating flumes, annular flumes are more suited to investigations of cohesive sediment dynamics, despite potential secondary circulation effects, because the flocculated suspensions are not disrupted by recirculating pumps and filters.

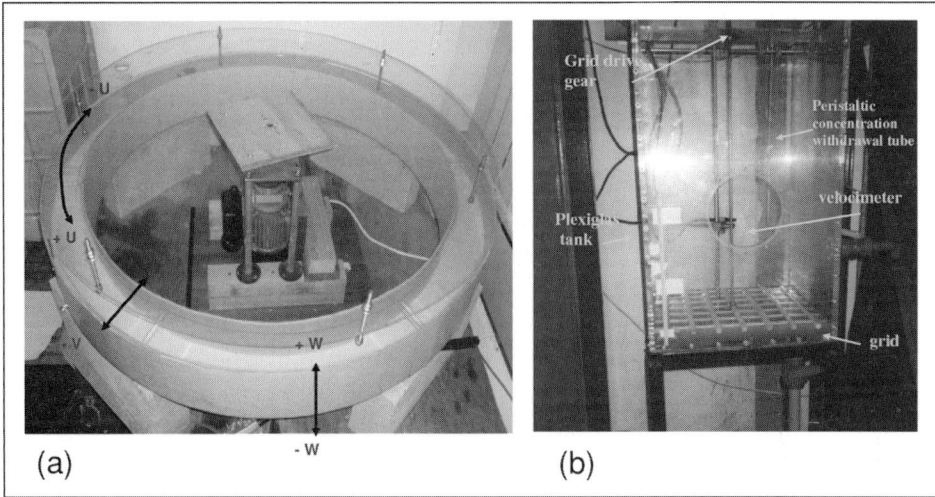

Figure 8-8 Some examples are given of floc measuring instrumentation. (a) Mini-annular flume setup; the sign convention for 3-D flow is also illustrated (reproduced from figure 3a of Manning et al., 2013; published March 13, 2013 under CC BY 3.0 license. © The Authors). (b) LEGI turbulent grid tank; the height-adjustable peristaltic withdrawal tube and velocimeter are located in the centre of the tank.

Whilst jar testing has the advantage of providing a good control volume within which to examine typical flocculation behaviour, the larger capacity of an annular flume is better suited to the generation of a range of controlled turbulent environments within which to observe how flocculating suspensions behave.

6.4.2.1 Annular Flume Flocculation Example

As an example, Manning et al. (2007) reported the findings from a series of collaborative laboratory flocculation studies in which the LabSFLOC-1 system was used to measure the properties of flocs formed in a mini-annular flume (Figure 8-8a). This study examined how the floc properties of a natural estuarine mud from the Medway Estuary (UK) evolved in response to varying levels of suspended sediment concentration and induced turbulent shearing. The flows created in the mini-flume produced average shear stresses at the floc sampling height that ranged from 0.03 Pa to a peak of 0.97 Pa. Nominal suspended particulate matter concentrations of 100, 600 and 2,000 mg l^{-1} were introduced into the flume. The experimental runs produced individual flocs ranging in size from microflocs of diameter 22.2 μm to macroflocs of diameter 583.7 μm. The averaged settling velocities ranged from 0.01 mm s^{-1} to 26.1 mm s^{-1}, whilst the floc effective densities varied from 3.5 kg m^{-3} up to 1550 kg m^{-3}.

The results showed that low concentrations and low shear stresses produced a uniform distribution of floc mass between the macrofloc (>160 μm) and microfloc (<160 μm) fractions. As both concentration and stress rose, the proportion of macrofloc mass increased until it represented over 80 per cent of the suspended matter.

A maximum average macrofloc settling velocity of 3.3 mm s^{-1} was attained at a shear stress of 0.45 Pa. Peak turbulence conditions resulted in deflocculation, limiting the macrofloc fall velocity to only 1.1 mm s^{-1} and placing over 60 per cent of the mass in the microfloc size range. A statistical multiple regression analysis of the data suggests that the combined influence of suspended concentration and turbulent shear controls the settling velocity of the fragile, low density macroflocs (Manning et al., 2007).

6.4.2.2 *Annular Flume Resuspension Tests*

Resuspension experiments conducted in an annular flume (Chapter 9) can provide an indication of how consolidation of the settled bed material can affect the subsequent erosion and re-entrainment of bed sediment with rising shear stress. These tests also provide an indication of the potential critical erosion shear stress thresholds that are exhibited by the bed material after periods of consolidation.

6.4.3 Turbulence Tanks

As an alternative to annular flumes, the generation of a diffusive turbulent flow within a cuboidal tank can facilitate the examination of floc properties at different levels of turbulence and turbidity, particularly within the vertical profile. To generate turbulence within the water column, a horizontal grid can be placed inside the tank and then oscillated vertically with a sinusoidal displacement in time at a selectable frequency and amplitude. The porosity of the grid is defined as the ratio of surface of the holes to the surface of the grid. Work conducted by Hopfinger and Toly (1976) on turbulent mixing across a density interface indicated that grid porosity is an important parameter to consider when assessing the generation of mean secondary flows. For example, a porosity of 64 per cent produces only minimal mean flow inside the tank.

Gratiot and Manning (2007) have assessed floc populations from both dilute suspensions and within concentrated benthic suspension layers, all generated in a diffusive turbulent grid tank (Figure 8-8b) using sediment suspensions (slurries) with concentrations that ranged from 200 mg l^{-1} up to 5 g l^{-1}.

6.5   Floc Video Data

Video-based floc devices tend to record floc images in real time with their cameras linked to a PC; the images are usually stored on a hard drive.

6.5.1   Raw Floc Images

To enable image processing with software such as MATLAB$^{®}$, it is important that the images are noncompressed (i.e., no Codec compression used). The acquisition or frame rate typically can vary from fast collection rates of 50 Hz (50 frames per second) down to lower acquisition rates of 7.5 Hz. Usually, a lower video frame capture rate can be used for cameras with large viewable areas because the flocs are kept in view longer for successive frames. Video floc data can either be streamed as a continual video file in an AVI format, and then still images 'grabbed' as JPEGs at required frame intervals during postprocessing, or they can be recorded directly as JPEGs in real time. For example, the LabSFLOC-1 system has a viewable area of just 3-mm high by 4-mm wide. The analogue recordings are digitized into AVI format files with an acquisition rate of 25 Hz and an image resolution of 640x480 pixels, producing an individual pixel that represents 6.25 μm. The LabSFLOC-1 AVI files are then converted into a series of snapshot images using the MATLAB frame-grabbing routine AVI2JPEG to provide a series of JPEG frames for each floc sample.

The automatic processing of floc images to obtain floc sizes tends to be simpler than obtaining fully automatic settling velocity data. The latter requires sophisticated particle tracking software. Judgement needs to be used when processing floc image data to determine the most practical solution for a particular task. Although a fully automatic software routine may take only a brief time to initially capture floc diameters and settling velocities, it may take further time to manually quality check the outputs; e.g., to assess any multiple counting of images of a settling floc, whereas a more manual approach may take longer but probably would require less rechecking. Automatic software would struggle and potentially become more erroneous as the ambient SSC and floc number increased; a manual approach would be possible, but would take significantly longer because more flocs would need to be counted. An ideal solution would be the middle ground, whereby the software could automatically size the particle images, but an operator could still provide inputs when tracking particles for settling velocity. Therefore, it is advisable to factor in sufficient floc image processing time to a specific project when assessing for floc size and settling velocity distributions. As an illustration, some early analogue video floc devices used a simple calibrated grid on the output display to estimate floc size and fall velocity. DigiFloc software (HR Wallingford, 2016) has been used extensively to semiautomatically process digital floc images (i.e., a series of JPEG frames per floc sample) to obtain floc size and settling spectra for LabSFLOC-1 and LabSFLOC-2 image data.

Some systems have attempted to utilise more automated image processing techniques. Smith and Friedrichs (2011) used a Particle Tracking Velocimetry (PTV) technique to enhance digital imagery, identify and track particles between successive image frames, and thus determine particle characteristics such as size and settling velocity. During the PTV analysis, raw binary images are evaluated through cross-correlation and Kalman filtering methods applied to match particles between adjacent frames in the image sequence. As Smith and Friedrichs (2011) point out, it is always prudent to gauge regularly the performance of any automated particle tracking routine through verification comparisons with a range of manual tracking results, coupled with inspection of the automated particle track sequences.

6.5.1.1 *Floc Size and Settling Velocity*
Representative floc sizes can be measured from the digital image by overlaying an ellipse onto each floc image, which yields both major and minor axis floc dimensions, D_x and D_y, and provides an indication of the floc shape in terms of the height/width aspect ratio. It is also possible to use image analysis routines that can provide a higher level of information regarding the floc shape. If the individual pixel size has been calibrated, it is possible to obtain the real size dimensions. To aid the interpretation of

floc size data, the two orthogonal dimensions can be converted into a spherical equivalent floc diameter, D, using Eq. 6:

$$D = \left(D_x \cdot D_y\right)^{0.5} \tag{6}$$

Settling velocity can be determined by measuring the vertical distance that the centre of each floc travels between two frames, which are separated by a known time step.

6.5.1.2   Floc Effective Density

A modified version of Stokes' law permits an accurate estimate of floc effective density, ρ_e, when both D (Eq. 6) and W_s are measured by the video floc device:

$$\rho_e = \left(\rho_f - \rho_w\right) = \frac{18\mu \cdot W_s}{D^2 g} \tag{7}$$

where μ is the molecular viscosity and g the gravitational acceleration. The effective density, also referred to as the density contrast, is the difference between the floc bulk density, ρ_f, and the water density, ρ_w. The water density can be calculated from measured salinity and water temperature data by applying the International Equation of State of Sea Water, 1980 (Millero and Poisson, 1981). However, Stokes settling only occurs when flocs exhibit low particle Reynolds numbers, $R_e < 0.5$, where:

$$R_e = \frac{\rho_e \cdot W_s \cdot D}{\mu} \tag{8}$$

For situations where $R_e > 0.5$, the Oseen modification (Oseen, 1927; Schlichting, 1968), as advocated by Brun-Cottan (1986) and ten Brinke (1993), is applied to the standard Stokes' equation to account for the inertial drag on settling particles. This has the effect of increasing the calculated ambient effective density values (and subsequent computed parameters) by the following amount:

$$\rho_{e,Oseen} = \rho_e \frac{1}{(1 + 0.1875 R_e)} \tag{9}$$

where $\rho_{e,Oseen}$ is the Oseen modified ambient effective density. After any required corrections are applied to the effective density, the resultant effective densities (ambient or Oseen modified) are all referred to as ρ_e.

As an example, we take $D_x = 61.93\mu m$ and $D_y = 356.27\mu m$, so that the spherical-equivalent floc diameter (Eq. 6) is $D = 148.54\mu m$. If we determine that the floc has settled 260.9 pixels (each pixel of length, say, 6.8 µm) at a frame interval of 240 ms, the settling velocity can be determined as the ratio of settled distance to the time taken to settle that distance, i.e., 7.4 mm s^{-1}. The floc effective density is determined using Stokes' law (Eq. 7), and if, for example, $\rho_w = 1004 kgm^{-3}$, $\mu = 0.001 s\,Pa$, and acceleration due to gravity is $g = 9.81 ms^{-2}$, then ρ_f and ρ_e are 1619 and 615kgm^{-3}.

6.5.1.3   Floc Porosity, Dry Mass and Mass Settling Flux

Using specially derived algorithms, Fennessy et al. (1997) and Manning (2004b) accurately calculated other physical characteristics for each individual floc, including:

porosity, dry mass and mass settling flux. Initially, the volume of a spherical-equivalent floc (V_f) can be calculated from:

$$V_f = \frac{\pi D^3}{6} \tag{10}$$

Thus, the interstitial water volume (V_{iw}) contained within each floc is given by:

$$V_{iw} = \left(1 - \frac{\rho_e}{\rho_{e,np}}\right) \cdot V_f \tag{11}$$

where:

$$\rho_{e,np} = \rho_{mo} - \rho_w$$

With ρ_{mo} the mean dry density of the primary particles and $\rho_{e,np}$ the mean effective density for solid (nonporous) aggregates. When no mineralogical and organic content information is available, one can consider that estuarine flocs generally have a mineral to organic mass ratio of 90:10, respectively, and a mineral to organic volume ratio of 78:22 (Fennessy et al., 1994b). Similarly, if gravimetric analysis of suspended sediment samples is not available, then for estuarine sediments one could consider an average mineral dry density to be 2600 kg m^{-3} and the mean organic matter dry density to be 1030 kg m^{-3}. Given these values, ρ_{mo} equates to an average value of 2256 kg m^{-3}; a more accurate mineral to organic ratio can be obtained from gravimetric loss on ignition tests (Chapter 7).

The floc porosity, P_f, which provides an indication of the level of compactness of the floc, is evaluated by taking the ratio of the volume of the floc interstitial water to the total floc volume:

$$P_f = \frac{V_{iw}}{V_f} \cdot 100 \tag{12}$$

Assuming the interstitial fluid density to be equal to ρ_w, the dry floc mass ($M_{f,dry}$) is calculated from:

$$M_{f,dry} = V_f \cdot \frac{\rho_e \rho_{mo}}{(\rho_{mo} - \rho_w)} \tag{13}$$

If all visible flocs are measured within a controlled volume, it is possible to transform the observed floc population into accurate estimates of SPM concentration and settling flux spectra. For example, for the INSSEV instrument, it is assumed that the measured flocs originated from a constant water-volume sample of 400 mm^3. This comprised a nominal 4 mm screen width (a function of the lens properties), a 1-mm depth of field (a function of the fixed-focus properties of the lens optics), and a 100 mm high decelerator (floc sample capturing) chamber. All of the $M_{f,dry}$ values (Eq. 13) for an individual sample are then summed to produce the total dry mass within the 400 mm^3 sampling volume, which is then converted into an SPM concentration. Typically, this value is referenced to the ambient SPM concentration obtained using gravimetric analysis of the water samples (Chapter 7), collected simultaneously with the floc-sample collection using a Niskin bottle or a van Dorn water sampler (or similar).

The mass settling flux (MSF) is computed as the product of the SPM concentration and settling velocity. Therefore, the mass settling flux can be calculated from floc data by multiplying the SPM concentration associated with each floc by its respective floc settling velocity, so that spectral estimates of mass settling flux can be made. This type of flux computational technique has also been applied successfully by Syvitski et al. (1995), Hill et al. (1998), and Sternberg et al. (1999).

6.5.1.4   Fractal Dimensions

A mathematical approach that has recently gained much interest by numerical modellers is the fractal representation of flocs (e.g., Hill, 1996; Winterwerp, 1999; Merckelbach, 2000). Fractal theory is dependent on the successive aggregation of self-similar flocs, thereby producing a structure that is independent of the scale considered. This is similar to Krone's (1963) order-of-aggregation theory. In practice, the total number of particles in a fractal aggregate (N) is related to the floc size (D), primary particle diameter ($D_p = d$) and the fractal dimension (nf) as follows:

$$N = \left(D/D_p\right)^{nf} \tag{14}$$

with:

$$nf = \ln\left(m_1\right)/\ln\left(m_2\right) \tag{15}$$

where m_1 is the number of primary particles that form a first-order floc and m_2 is a factor by which the floc will increase in size during this hierarchical process. From this, Kranenburg (1994) derived Eq. 16:

$$\left(\rho_f - \rho_w\right) = \left(\rho_{sed} - \rho_w\right)\cdot\left(D_p/D\right)^{3-nf} \tag{16}$$

where ρ_f, ρ_w and ρ_{sed} are the densities of the mud flocs, the (interstitial) water and the sediment (primary particles).

Substituting Eq. 16 into the classic Stokes' Equation, Eq. 7, Winterwerp (1998) obtained the following relationship:

$$W_s = \alpha'\cdot D_p^{3-nf}\left(\rho_{sed} - \rho_w\right)\cdot g/\mu D^{nf-1} \tag{17}$$

where α' is an empirically derived constant (collision efficiency factor). Values for the fractal dimension generally range between 1.0 and 1.4 (representative of fragile aggregates) and 2.5 and 3.0 (representative of more strongly bonded estuarine flocs). However, to make a fractal-based model solvable analytically within a numerical sediment transport simulation, a mean nf of 2 is commonly assumed. As a starting point, an estimation of the fractal dimension can be obtained using Eq. 17 with the assumptions that the primary particle diameter is between 2 and 4 µm and the empirical coefficient α' lies between 0.1 and 0.3 (Gibbs, 1985).

6.5.1.5   Floc Data Parameterization

To aid the interpretation of floc characteristics, each floc population can be segregated into various subgroupings according to floc size. Sample-mean floc values can be computed (i.e., a single value per floc population) to show generalized floc property trends. Similarly, maximum and minimum floc parameters can be computed from the average value obtained from the 0–10 and 90–100 percentile flocs. The MAX4 or MAX6 parameters are illustrative of the average values obtained from the four or six largest flocs in each population, respectively.

A conclusion drawn from an intercomparison experiment of various floc-measuring devices conducted in the Elbe Estuary (Dyer et al., 1996) was that a single mean or median settling velocity did not adequately represent an entire floc spectrum, especially in considerations of a flux to the bed. Dyer et al. (1996) recommended that the best approach for accurately representing the settling characteristics of a floc population was to split a floc distribution into two or more components, each with their own mean settling velocity. Both Eisma (1986) and Manning (2001) concur with this finding by suggesting that a more realistic and accurate generalization of floc behaviour can be derived from the macrofloc and microfloc fractions. These two floc fractions form part of Krone's (1963) classic order-of-aggregation theory and produce two floc property values per floc population. From extensive tests of floc data sets, Manning (2001) defined macroflocs as those flocs with $D > 160 \mu m$ and microflocs as those flocs with $D < 160 \mu m$. A smaller demarcation size of $D = 120 \mu m$ (Eisma, 1986) can also be used as an alternative when required.

Macroflocs are large, highly porous (> 90 per cent), fast-settling aggregates, which typically are the same size as the turbulent Kolmogorov (1941a, b) microscale; they are recognised as the most important subgroup of flocs because their fast-settling velocities tend to have the most influence on the mass settling flux (Mehta and Lott, 1987). Their fragile, low-density structure means that they are sensitive to physical disruption during sampling. Macroflocs are progressively broken down as they pass through regions of higher turbulent shear stress and reduced again to their component microfloc substructures (Glasgow and Lucke, 1980); they rapidly attain equilibrium with the local turbulent environment.

The smaller microflocs generally are considered to be the building blocks from which the macroflocs are comprised. Many field studies (McCave, 1984; Alldredge and Gotschalk, 1988; Fennessy et al., 1994(a, b); and Manning et al., 2007, 2010) have shown that the microfloc class of aggregates tend to display a much wider range of effective densities and settling velocities than the macrofloc fraction. Microflocs are much more resistant to breakup by turbulent shear; they tend to have slower settling velocities, but exhibit a much wider range of effective densities than the larger macroflocs (e.g., McCave, 1975; Alldredge and Gotschalk, 1988; Fennessy et al., 1994a and b).

In terms of flocculation kinetics (Overbeek, 1952), the macroflocs tend to control the fate of purely muddy sediments in an estuary (Mikeš and Manning, 2010); this is because the smaller microflocs generally settle at less than 1 mm s^{-1}, whereas macroflocs settle in the 1–15 mm s^{-1} range, enabling them to deposit to the bed (Pouët, 1997). However, when flocculation of mixed sediments occurs, the microflocs can potentially demonstrate settling velocities comparable to those of the macroflocs (Manning et al., 2010, 2011a).

In order to illustrate the spectral variability of floc properties, each floc population can be divided into various size bands. The band divisions can be chosen to best fit the data collected. For example, Manning and Dyer (2002b) have used twelve size bands (SB) to represent floc data in the turbidity maximum of the Tamar Estuary (UK), with SB1 representing microflocs less than 40 μm in size, whilst SB12 is representative of macroflocs greater than 640 μm in diameter; SB2 to SB6 range from 40 to 240 μm in five steps, each of 40 μm, and SB7 to SB11 range from 240 to 640 μm in five steps of 80 μm. LabSFLOC data for a mud sample from the Medway Estuary, UK, provide a graphical illustration of size banding and the increasing settling velocities associated with larger flocs (Figure 8-9).

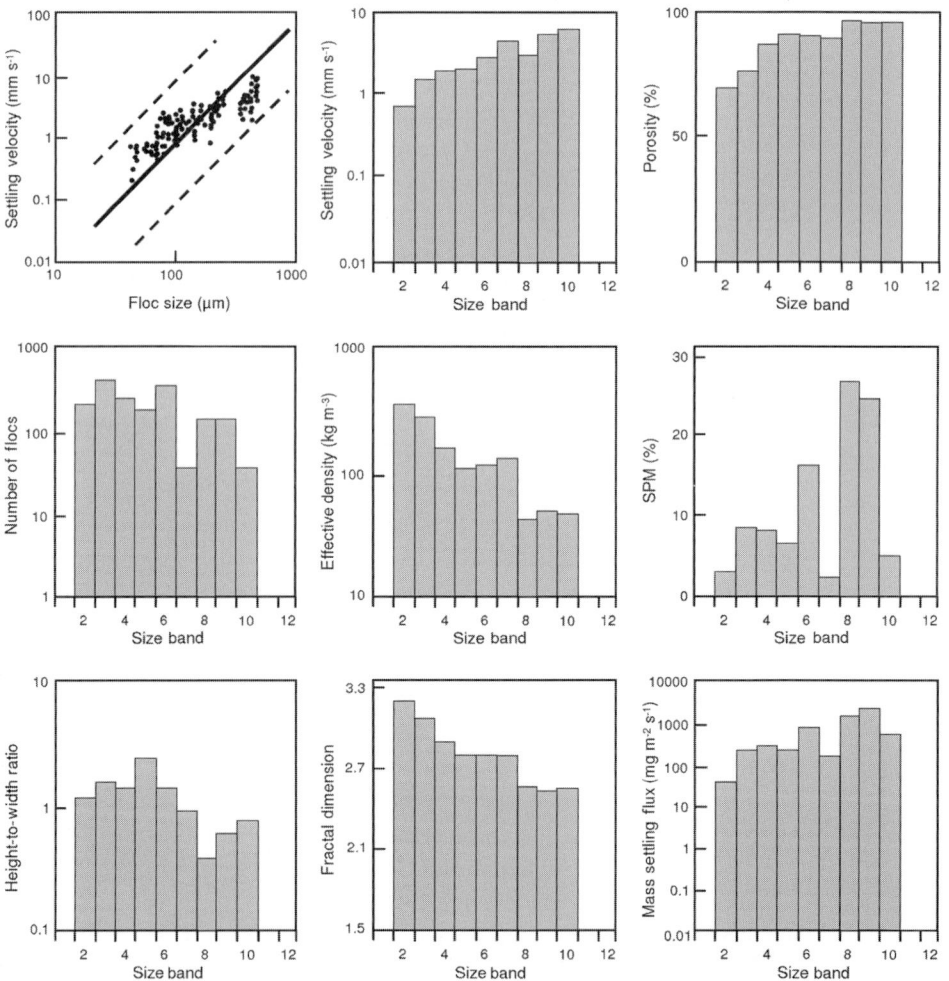

Figure 8-9 An example of various size-banded floc properties for a LabSFLOC SPM sample is given. The complete population of size versus settling velocity data is illustrated in the top-left panel. These data apply to Medway Estuary mud slurry (1.6 gl^{-1}) that has been sheared in the Southampton Oceanography Centre (UK) miniflume at a shear stress of 0.37 Nm^{-2} (data from results in Manning et al., 2007).

7   Some Modelling Approaches

7.1   Overview

Computer simulation models are commonly the chosen tools that estuarine management groups use to predict sediment transport rates for various tasks; e.g., the
requirements for routine maintenance dredging, or estimating the potential impacts
that new port-related construction could have on an existing hydrodynamic regime.
For these numerical models to provide meaningful results, they require a good
scientific understanding of the phenomena under consideration, and these processes
need to be adequately described (i.e., parameterized) by the model coding. In particular, a quantitative understanding of the dynamics of the vertical structure of suspended
cohesive sediments is essential for an accurate estuarine sedimentation model (Kirby,
1986; van Leussen, 1991). This requires an understanding of the physical processes
related to the entrainment, advection and deposition of muddy sediments. One physical process that has caused particular difficulty is the modelling and mathematical
description of the vertical mass settling flux of sediment, which becomes the depositional flux near to slack water. This is the product of the concentration and the settling
velocity. Manning and Bass (2006) found that mass settling fluxes can vary over four
or five orders of magnitude during a tidal cycle in mesotidal and macrotidal estuaries;
therefore, a realistic representation of flux variations is crucial to an accurate
depositional model.

The specification of the flocculation term within numerical models depends on the
sophistication of the model. Until recently, even the conceptual relationship between
floc size, SPM concentration and turbulent shear stress proposed by Dyer (1989,
Figure 8-3a) remained largely unproven. It is useful to briefly introduce some commonly used turbulence concepts (dealt with in greater detail in Chapter 4).

7.2   Turbulent Kinetic Energy (TKE) and the Parameters G and η

The turbulent kinetic energy per unit volume, E, is the result of turbulent fluctuations in
the three components of velocity (u', v', w'):

$$E = 0.5\rho_w \left(u'^2 + v'^2 + w'^2 \right) \qquad (18)$$

Within the wall region, where the flow is directly affected by conditions at the
bottom, the energy production equals the energy dissipation (Chapter 4) so that the
turbulent shear stress is proportional to the turbulent kinetic energy. Using the factor of
proportionality observed over a wide range of flows, Soulsby (1983) suggests that the
turbulent shear stress, τ, can be calculated from:

$$\tau = 0.19E \qquad (19)$$

For intercomparison purposes, the τ values can be recast into two commonly used
alternative measures of turbulent shearing. The first, G, is the root-mean-square of the

gradient in the turbulent velocity fluctuations, with units of s^{-1}. The rate of turbulent energy production per unit mass, P, by turbulent shear stresses at a height z above the bed is:

$$P = \rho_w^{-1} \tau \cdot \partial u / \partial z \qquad (20)$$

The vertical profile of the mean velocity, u, for an unstratified flow in the logarithmic, near-bed, constant shear-stress layer is given by the classic von Karman–Prandtl equation (e.g., Dyer, 1986):

$$u = \frac{u^*}{\kappa} \cdot \ln z / z_0 \qquad (21)$$

where κ is the von Karman constant, which is usually taken as 0.4, although McCave (1979) suggests that it may be smaller for high-sediment concentrations. The height above the bed is z, whilst z_0 is the roughness length, which is dependent on the bed characteristics. The frictional velocity, u^*, is related to the bed shear stress (τ_b) and the water density:

$$\tau_b = \rho_w u_*^2 \qquad (22)$$

Within the logarithmic, constant stress layer, Eq. 22 and Eq. 20 combine to give:

$$P = \frac{u_*^3}{\kappa \cdot z} \qquad (23)$$

Nakagawa et al. (1975) examined the structure of turbulence in open channel flow. They found that the turbulent energy production and turbulent energy dissipation (ε) were nearly in balance with each other for a fully developed turbulent flow. This is a key assumption that is made when adopting the TKE approach for turbulence data analysis. Generally, P is larger than ε within the wall region; however, Ueda and Hinze (1975) suggest that this difference is a result of the transfer of turbulence by diffusion.

Usually, G is defined as:

$$G = (\varepsilon / v)^{0.5} \qquad (24)$$

where v is the kinematic viscosity (molecular viscosity divided by the density of water, μ / ρ_w). Assuming that $P = \varepsilon$, then G can be written:

$$G = \left[\frac{u_*^3}{\kappa \cdot v \cdot z} \right]^{0.5} \qquad (25)$$

An alternative parameter classifies the turbulence level by the size of dissipating eddies in a turbulent flow, as defined by Kolmogorov (1941a, b), and is referred to as the microscale of turbulence, η (units are metres):

$$\eta = \left(v^3 / \varepsilon \right)^{0.25} = (v / G)^{0.5} \qquad (26)$$

The following sections outline various approaches that are used to represent flocculation and settling within numerical models.

7.3   Constant Settling Velocity

Specification of the flocculation term within numerical models depends on the sophistication of the model. The simplest parameterization is a single, floc settling velocity value that remains constant in both time and space (one coefficient). These fixed settling values are usually in the range of 0.5–5 mm s^{-1} and typically are selected on an arbitrary basis and sometimes used as a tuning parameter to match predicted erosion and deposition patterns to observations for an undisturbed estuary.

7.4   Power Law Settling Velocity

The next step is to use gravimetric data provided by field settling-tube experiments to relate floc settling velocity to the instantaneous SPM concentration, using a power law with two coefficients (e.g., Whitehouse et al., 2000). Empirical results have shown a generally exponential relationship between the mean, or median, floc settling velocity and SPM concentrations for concentrations <10 gl^{-1}. This approach sometimes includes hindered settling. However, both the constant settling velocity and the power law parameterization techniques do not include the important and influential effects of turbulence.

7.5   The Van Leussen Parameterization

More recently, a number of authors have proposed simple theoretical formulae interrelating a number of floc characteristics that can then be calibrated using empirical studies. Such an approach has been used by van Leussen (1994), who utilised a formula that modifies the floc settling velocity in still water by a floc growth factor, due to turbulence, and then reduces it by a turbulent floc disruption factor:

$$W_s = W_{s0}\left(\frac{1 + aG}{1 + bG^2}\right) \qquad (27)$$

where a and b are empirically determined constants. The parameter G (units s^{-1}) is obtained from Eq. 24 and Eq. 25. The reference settling velocity (taken at zero turbulent conditions), W_{s0}, is then related to the SPM concentration (C) using Eq. 1:

$$W_{s0} = k \cdot C^m \qquad (28)$$

where k and m are empirical constants. Eq. 27 is a qualitative simplification of a model originally developed for the sanitation industry (Argaman and Kaufman, 1970), with only a limited number of interrelated parameters, and hence does not provide a complete description of floc characteristics within a particular sheared environment.

7.6   The Lick et al. Parameterization

A number of authors have attempted to observe how the floc diameter changes in turbulent environments. In particular, Lick et al (1993) derived an empirical relationship based on laboratory measurements made in a flocculator. They found that the floc diameter varied as a function of the product of the SPM concentration and the turbulence parameter, G:

$$D = c \cdot (C \cdot G)^{-d} \tag{29}$$

where c and d are empirically determined values. However, this formulation provides no information on the important floc settling or floc dry mass properties.

7.7   The Fractal Approach

A relationship between the floc settling velocity and floc properties and fractal dimensions is given by Eq. 17 (Winterwerp, 1998). A fractal approach has been used by Winterwerp (1999) to solve a differential equation that simulates the time-varying representative floc diameter, from which floc density is derived from fractal considerations, and settling velocity obtained from a Stokes-like formula. Winterwerp et al. (2006) also used a simplified fractal model to relate settling velocity to a turbulent shear parameter, the instantaneous concentration, and water depth.

7.8   The Manning and Dyer Parameterization

The Manning Floc Settling Velocity (MFSV) algorithm for settling velocity (Manning and Dyer, 2007) is based entirely on empirical observations made in situ using nonintrusive floc and turbulence data acquisition techniques in a wide range of estuarine conditions. The floc population size and settling velocity spectra were sampled using the video-based INSSEV instrument (Fennessy et al., 1994a; Manning and Dyer, 2002a).

The strength of video-based floc measurements is that they minimise the number of assumptions used during the data processing and interpretation stages. Other types of device, e.g., the laser diffraction device LISST (Agrawal and Pottsmith, 2000), only measure the size component and require additional gross and often incorrect assumptions regarding the relationship between settling velocity, floc size, and floc density. The settling velocity of a floc is a function of both its size and effective density, and both of these floc components can display variations spanning three to four orders of magnitude within any one floc population (ten Brinke, 1994; Fennessy and Dyer, 1996; Manning, 2001). In conclusion, selection of the most appropriate instrumentation is paramount when attempting to parameterize flocculated cohesive sediments.

The Manning algorithms were generated by a parametric multiple regression statistical analysis of key parameters, which were generated from the raw, spectral floc data. Detailed derivations and preliminary testing of the floc-settling algorithms are described by Manning (2004b; 2006). Although the resulting empirical formulae are not presented in a fully dimensionless form, these formulae have the merit of being based on a large dataset of accurate, in situ settling velocity measurements (157 individually observed floc populations), acquired from different estuaries (Tamar, Gironde and Dollard) and different estuarine locations, such as the turbidity maximum and the intertidal zone.

The algorithms (Eqs. 30–32) are based on the segregation of flocs into macroflocs ($D > 160\mu m$) and microflocs ($D < 160\mu m$), which comprise the constituent particles of the macroflocs. This distinction permits the discrete computation of the mass settling flux (MSF) at any point in an estuarine water column:

$$MSF = \left[\left(1 - \frac{1}{1 + SPM_{ratio}}\right) \cdot (SPM \cdot W_{s,macro})\right] + \left[\frac{1}{1 + SPM_{ratio}} \cdot (SPM \cdot W_{s,micro})\right] \quad (30)$$

Equations are given below for: the settling velocity of the macrofloc fraction ($W_{s,macro}$), the settling velocity of microflocs ($W_{s,micro}$) and the ratio of macrofloc mass to microfloc mass in each floc population – termed the SPM_{ratio} (Manning, 2004b). These equations require the input of a turbulent shear stress (τ) and an SPM concentration, SSC (denoted by $SPM(mgl^{-1})$ in Eq. 31(a–c), Eq. 32(a, b) and Eq. 33 for consistency with earlier publications):

$W_{s,macro}(\text{mm s}^{-1})$ is given by:

(a)   for $0.04 < \tau < 0.6Pa$

$$W_{s,macro} = 0.644 + 0.000471 \cdot SPM + 9.36\tau - 13.1\tau^2 \quad (31a)$$

(b)   for $0.6 < \tau < 1.5Pa$

$$W_{s,macro} = 3.96 + 0.000346 \cdot SPM - 4.38\tau + 1.33\tau^2 \quad (31b)$$

(c)   for $1.4 < \tau < 5Pa$

$$W_{s,macro} = 1.18 + 0.000302 \cdot SPM - 0.491\tau + 0.057\tau^2 \quad (31c)$$

$W_{s,micro}(\text{mm s}^{-1})$ is given by:

(a)   for $0.04 < \tau < 0.55Pa$

$$W_{s,micro} = 0.244 + 3.25\tau - 3.71\tau^2 \quad (32a)$$

(b)   for $0.51 < \tau < 10Pa$

$$W_{s,micro} = 0.65\tau^{-0.541} \quad (32b)$$

The SPM_{ratio} (no units) is given by:

$$SPM_{ratio} = 0.815 + 0.00318 \cdot SPM - 0.00000014 \cdot SPM^2 \quad (33)$$

These regression equations (Eqs. 31–33) provide a realistic approximation to the field data. Graphical representations of the equations, together with the data, are presented in Manning and Dyer (2007). The Manning settling algorithm is valid for SPM concentrations in the range 10–8600 mg l^{-1} and shear stress values of $\tau < 2.13Pa$, with extrapolation extending this to 5–10 Pa. The algorithm is a major step forward in establishing a reliable estimate of settling velocity. It has been developed based on a large and reliable dataset, it caters for the spectrum of hydrodynamic conditions that occur during a typical tidal cycle (a feature often lacking in the settling terms of many estuarine sediment transport models) and has been shown to more accurately reproduce the distribution of suspended sediment compared with simpler settling models. An example of this is the implementation of the algorithm in a TELEMAC-3D numerical model of the Thames Estuary, UK (Baugh and Manning, 2007), in which it was shown that the use of the Manning algorithm greatly improved the reproduction of observed distributions of SPM concentrations compared with the other formulations, both in the vertical and horizontal dimensions.

An extension of this model (Eq. 31–33) developed by Manning and Dyer (2007), the Manning–Dyer formulation, gives a good compromise between the representation of physicochemical processes and computational simplicity. Soulsby et al. (2013) has developed a more 'physics-based' version of the empirical model based on the Manning–Dyer formulation, called Soulsby-Manning 2013. It should be noted that for flocculation algorithms and models that include turbulence as a contributing variable, it is vital to ensure that the turbulence data are accurate, otherwise it has significant implications for the accuracy of the calculated floc settling characteristics (Eq. 31 and 32).

7.9   Complex Population Approaches

Lee et al. (2011) applied a time-evolving two-class population balance equation to determine the spatially and temporally changing distribution of fixed-size microflocs and size-varying macroflocs for bimodal floc distributions, with a fractal relationship between floc size and mass to derive the distribution of settling velocities. However, the authors felt that further intensive investigation of the aggregation and breakage kinetics would be required before their model was generally applicable when compared with the simpler approach of Manning and Dyer (2007) and, presumably, Soulsby et al. (2013).

Verney et al. (2010) applied a time-evolving, multifraction model to determine the spatially and temporally changing distribution of the numbers of flocs in each size fraction, with a fractal relationship between floc size and mass to derive the distribution of settling velocities.

8   Final Remarks

Particle flocculation is recognised to be important to the behaviour of fine sediments and mixtures of fine sediments in water. The process of flocculation leads to the formation of larger sized particles (flocs) and floc size and density become important parameters when predicting the transport and settling of fine sediments. The flocculation process is dynamically active and directly affected by environmental conditions, primarily the complex set of interactions between sediment, fluid and flow. This chapter has provided a review of many of the important flocculation processes. In particular, it has provided

- An overview of the composition of suspended sediment in estuarial waters
- An outline of the flocculation process and floc settling behaviour
- A brief review of mixed sediment flocculation and the role of biology
- Flocculation measurement methods and the importance of field data
- Analysis and processing of floc data (including model parameterization)
- A brief review of turbulent shear stresses for application to floc models
- Examples of cohesive sediment depositional model approaches, including ways to parameterize flocculation

Many parameters need to be determined to fully describe a cohesive sediment type and its behaviour. Flocs are multicomponent, being composed of varying proportions and types of inorganic and organic particles, and the packing (i.e., density) of these grains within a floc can significantly affect their resultant size and settling velocity. It is this complexity that means it is not a simple task to describe mathematically the mud flocculation process on a fundamental basis. The principal reason for such a poor understanding of cohesive sediment settling fluxes and deposition rates has been a lack of reliable floc data, although, as this chapter has demonstrated, the situation is now rapidly improving.

Acknowledgements

AJM and RJSW thank their HR Wallingford colleagues (including Jez Spearman, Tim Chesher, Graham Siggers, Giovanni Cuomo and Mike Dearnaley) for their continued support during the production of this book chapter. AJM also thanks his University of Hull colleague, Professor Dan Parsons, for continued support of his sediment transport research. AJM's contribution to the preparation of this chapter was partly funded both by the US Geological Survey Co-operative Agreement Awards (G11AC20352 and G16AC00214) with HR Wallingford (DDS0280), and the HR Wallingford Company Research projects 'Sediment in Transitional Environments – SiTE-2' (DDY0427) and 'Dynamics of Fine-grained Cohesive Sediments at Varying Spatial and Temporal Scales – FineScale' (DDY0523). RJU thanks Professor Steve de Mora, Chief Executive of the Plymouth Marine Laboratory, for the award of a Senior Research Fellowship.

References

Ackroyd, D. R., Bale, A. J., Howland, R. J. M., Knox, S., Millward, G. E., Morris, A. W., 1986. Distributions and behaviour of Cu, Zn and Mn in the Tamar estuary. *Estuarine, Coastal and Shelf Science* 23, 621–640.

Agrawal, Y. C., Pottsmith, H. C., 1994. Laser diffraction particle sizing in STRESS. *Continental Shelf Research* 14, 1101–1121.

Agrawal, Y. C., Pottsmith, H. C., 2000. Instruments for particle size and settling velocity observations in sediment transport. *Marine Geology* 168, 89–114.

Al Ani, S., Dyer, K. R., Huntley, D. A., 1991. Measurement of the influence of salinity on floc density and strength. *Geo-Marine Letters* 11, 154–158.

Alldredge, A. L., Gotschalk, C., 1988. In situ settling behavior of marine snow. *Limnology and Oceanography* 33, 339–351.

Aller, R. C., 1982. The effects of macrobenthos on chemical properties of marine sediment and overlying water. In: McCall, P. L., Tevesz, M. J. S. (eds.), *Animal-Sediment Relations*. New York: Plenum Press, 53–102.

Argaman, Y., Kaufman, W. J., 1970. Turbulence and flocculation. *Journal of the Sanitary Engineering Division* 96, 223–241.

Bale, A. J., Uncles, R. J., Widdows, J., Brinsley, M. D., Barrett, C. D., 2002. Direct observation of the formation and break-up of aggregates in an annular flume using laser reflectance particle sizing. In: Winterwerp, J. C., Kranenburg, C. (eds.), *Fine Sediment Dynamics in the Marine Environment, Proceedings in Marine Science*, Vol. 5. Amsterdam: Elsevier, 189–201.

Baugh, J. V., Manning, A. J., 2007. An assessment of a new settling velocity parameterisation for cohesive sediment transport modelling. *Continental Shelf Research* 27, 1835–1855. (doi:10.1016/j.csr.2007.03.003).

Benson, T., French, J. R., 2007. InSiPID: A new low cost instrument for in situ particle size measurements in estuaries. *Journal of Sea Research* 58, 167–188.

Berner, R. A., 1980. *Early Diagenesis: A Theoretical Approach*. Princeton, NJ: Princeton University Press.

Black, K. S., Paterson, D. M., 1998. LISP-UK Littoral investigation of sediment properties: An introduction. In: Black, K. S., Paterson, D. M., Cramp, A. (eds.), *Sedimentary Processes in the Intertidal Zone*. Geological Society Special Publications, 139, 1–10.

Bouyer, D., Coufort, C., Linè, A., Do-Quang, Z., 2005. Experimental analysis of floc size distributions in a 1-L jar under different hydrodynamics and physico-chemical conditions. *Journal of Colloid Interface Science* 292, 413–428.

Brun-Cottan, J. C., 1986. Vertical transport of particles within the ocean. In: Buat-Ménard, P. (ed.), *The Role of Air Sea Exchange in Geochemical Cycling*. Dordrecht: Reidel, 83–111.

Davies, J. L., 1964. A morphogenic approach to world shore-lines. *Zeitschrift für Geomorphologie* 8, 127–142.

de Brouwer, J. F. C., Wolfstein, K., Ruddy, G. K., 2005. Biogenic stabilization of intertidal sediments: The importance of extracellular polymeric substances produced by benthic diatoms. *Microbial Ecology* 49, 501–512.

Buffle, J., Leppard, G. G., 1995. Characterisation of aquatic colloids and macromolecules. 2. Key role of physical structures on analytical results. *Environmental Science and Technology* 29, 2176–2184.

Burban, P.-Y., Lick, W., Lick, J., 1989. The flocculation of fine-grained sediments in estuarine waters. *Journal of Geophysical Research* 94 (C6), 8323–8330.

Cadee, G. C., 1985. Macroaggregates of *Emiliana huxleyi* in sediment traps. *Marine Ecology Progress Series* 24, 193–196.

Cahoon, L. B., 1999. The role of benthic microalgae in neritic ecosystems. *Oceanography and Marine Biology: An Annual Review* 37, 47–86.

Chassagne, C., Mietta, F., Winterwerp, J. C., 2009. Electrokinetic study on kaolinite suspensions. *Journal of Colloid and Interface Science* 336, 352–359.

Cheviet, C., Violeau, D., Guesmia, M., 2002. Numerical simulation of cohesive sediment transport in the Loire estuary with a three-dimensional model including new parameterisations. In: Winterwerp, J. C., Kranenburg, C. (eds.), *Fine Sediment Dynamics in the Marine Environment – Proc. in Mar. Sci. 5*. Amsterdam: Elsevier, 529–543.

Cornelisse, J. M., 1996. The field pipette withdrawal tube (FIPIWITU). *Journal of Sea Research* 36, 37–39.

Dankers, P. J. T., 2002. *The Behaviour of Fines Released due to Dredging – A Literature Review*, Delft, The Netherlands: Hydraulic Engineering Section, Faculty of Civil Engineering and Geosciences, Delft University.

Dankers, P. J. T., Winterwerp, J. C., 2007. Hindered settling of mud flocs: Theory and validation. *Continental Shelf Research* 27, 1893–1907.

Dearnaley, M. P., 1996. Direct measurements of settling velocities in the Owen Tube: A comparison with gravimetric analysis. *Journal of Sea Research* 36, 41–47.

Defossez, J. P., 1996. *Dynamique des macroflocs au cours de cycles tidaux, Mise au point d'un système d'observation: VIL, Video in Lab*. Mémoire de DEA, Université de Rouen, Rouen, France.

Delo, E. A., Ockenden, M. C., 1992. Estuarine Muds Manual. *HR Wallingford Report, SR* 309.

Droppo, I. G., Walling, D., Ongley, E., 2000. The influence of floc size, density and porosity on sediment and contaminant transport. *Journal of the National Centre for Scientific Research* 4, 141–147.

Droppo, I. G., 2001. Rethinking what constitutes suspended sediments. *Hydrological Processes* 15, 1551–1564.

Droste, R. L. 1997. *Theory and Practice of Water and Wastewater Treatment*. New York, NY: John Wiley and Sons.

Dyer, K. R., 1986. *Coastal and Estuarine Sediment Dynamics*. Chichester: John Wiley and Sons.

Dyer, K. R., 1989. Sediment processes in estuaries: Future research requirements. *Journal of Geophysical Research* 94 (C10), 14327–14339.

Dyer, K. R., Cornelisse, J., Dearnaley, M. P., Fennessy, M. J., Jones, S. E., Kappenberg, J., McCave, I. N., Pejrup, M., Puls, W., Van Leussen, W., Wolfstein, K., 1996. A comparison of in situ techniques for estuarine floc settling velocity measurements. *Journal of Sea Research* 36, 15–29.

Dyer, K. R., Manning, A. J., 1999. Observation of the size, settling velocity and effective density of flocs, and their fractal dimensions. *Journal of Sea Research* 41, 87–95.

Edzwald, J. K., O'Melia, C. R., 1975. Clay distributions in recent estuarine sediments. *Clays and Clay Minerals* 23, 39–44.

Eisma, D., 1986. Flocculation and de-flocculation of suspended matter in estuaries. *Netherlands Journal of Sea Research* 20, 183–199.

Eisma, D., Schuhmacher, T., Boekel, H., Van Heerwaarden, J., Franken, H., Lann, M., Vaars, A., Eijgenraam, F., Kalf, J., 1990. A camera and image analysis system for in situ observation of flocs in natural waters. *Journal of Sea Research* 27, 43–56.

Eisma, D., Dyer, K. R., van Leussen, W., 1997. The in-situ determination of the settling velocities of suspended fine-grained sediment – a review. In: Burt, N., Parker, R., Watts, J. (eds.), *Cohesive Sediments – Proceedings of INTERCOH Conference* (Wallingford, England). Chichester: John Wiley and Sons.

EPA, 2002. Environmental Protection Agency. Water treatment manuals. [online] Available at: www.epa.ie/pubs/advice/drinkingwater/EPA_water_treatment_mgt_coag_flocc_clar2.pdf [accessed March 2016].

Krone, R. B., 1963. *A Study of Rheological Properties of Estuarial Sediments*. Berkeley: Hydraulic Engineering Laboratory and Sanitary Engineering Research Laboratory, University of California, Report No. 63–68.

Lau, Y. L., 1994. Temperature effect on settling velocity and deposition of cohesive sediments, *Journal of Hydraulic Research* 32, 41–51. doi:10.1080/00221689409498788.

Lau, Y. L., Droppo, I. G., 2000. Influence of antecedent conditions on critical shear stress of bed sediments. *Water Research* 34, 663–667.

Law, D. J., Bale, A. J., Jones, S. E., 1997. Adaptation of focused beam reflectance measurement to in-situ particle sizing in estuaries and coastal waters. *Marine Geology* 140, 47–59.

Lee, B. J., Toorman, E., Molz, F. J., Wang, J., 2011. A two-class population balance equation yielding bimodal flocculation of marine or estuarine sediments. *Water Research* 45, 2131–2145.

Lick, W., Huang, H., Jepsen, R., 1993. Flocculation of fine-grained sediments due to differential settling. *Journal of Geophysical Research* 98 (C6), 10279–10288.

Lick, W., 1994. Modelling the transport of sediment and hydrophobic contaminants in surface waters. In: *U. S. / Israel Workshop on Monitoring and Modelling Water Quality*, May 8–13, 1994, Haifa, Israel.

Little, C., 2000. *The Biology of Soft Shores and Estuaries*. Oxford: Oxford University Press.

Malarkey, J., Baas, J. H., Hope, J. A., Aspden, R. J., Parsons, D. R., Peakall, J., Paterson, D. M., Schindler, R. J., Ye, L., Lichtman, I. D., Bass, S. J., Davies, A. G., Manning, A. J., Thorne, P. D., 2015. The pervasive role of biological cohesion in bedform development. *Nature Communications* 6, 6257. doi:10.1038/ncomms7257.

Manning, A. J., Dyer, K. R., 1999. A laboratory examination of floc characteristics with regard to turbulent shearing. *Marine Geology* 160, 147–170.

Manning, A. J., 2001. A study of the effects of turbulence on the properties of flocculated mud. PhD Thesis. Institute of Marine Studies, University of Plymouth.

Manning, A.J., Dyer, K.R., 2002a. A comparison of floc properties observed during neap and spring tidal conditions. In: Winterwerp, J. C., Kranenburg, C. (eds.), *Fine Sediment Dynamics in the Marine Environment – Proceedings in Marine Science 5*, Amsterdam: Elsevier, pp. 233–250.

Manning, A. J., Dyer, K. R., 2002b. The use of optics for the in-situ determination of flocculated mud characteristics. *Journal of Optics A: Pure and Applied Optics, Institute of Physics Publishing* 4, S71–S81.

Manning, A. J., 2004a. Observations of the properties of flocculated cohesive sediment in three western European estuaries. *Journal of Coastal Research* 41, 70–81.

Manning, A. J., 2004b. The observed effects of turbulence on estuarine flocculation. In: Ciavola, P., Collins, M. B. (eds.), *Sediment Transport in European Estuaries, Journal of Coastal Research* 41, 90–104.

Manning, A. J., 2006. LabSFLOC – A Laboratory System to Determine the Spectral Characteristics of Flocculating Cohesive Sediments. *HR Wallingford Technical Report, TR* 156.

Manning, A. J., Bass, S. J., 2006. Variability in cohesive sediment settling fluxes: Observations under different estuarine tidal conditions. *Marine Geology* 235, 177–192.

Manning, A. J., Bass, S. J., Dyer, K. R., 2006. Floc properties in the turbidity maximum of a mesotidal estuary during neap and spring tidal conditions. *Marine Geology* 235, 193–211.

Manning, A. J., Dyer, K. R., 2007. Mass settling flux of fine sediments in Northern European estuaries: Measurements and predictions. *Marine Geology* 245, 107–122. doi:10.1016/j.margeo.2007.07.005.

Manning, A. J., Spearman, J., Whitehouse, R. J. S., 2007. Mud:Sand Transport – Flocculation and Settling Dynamics within Turbulent Flows, Part 1: Analysis of laboratory data. *HR Wallingford Internal Report, IT* 534.

Manning, A. J., 2008. The development of algorithms to parameterise the mass settling flux of flocculated estuarine sediments. In: Kudusa, T., Yamanishi, H., Spearman J., Gailani, J. Z. (eds.), *Sediment and Ecohydraulics – Proc. in Marine Science 9*, Amsterdam: Elsevier, 193–210. ISBN: 978-0-444-53184-1.

Manning, A. J., Baugh, J. V., Spearman, J., Whitehouse, R. J. S., 2010. Flocculation settling characteristics of mud:sand mixtures. *Ocean Dynamics* 60, 237–253. doi:10.1007/s10236-009-0251-0.

Manning, A. J., Baugh, J. V., Spearman, J. R., Pidduck, E. L., Whitehouse, R. J. S., 2011a. The settling dynamics of flocculating mud:sand mixtures: Part 1 – Empirical algorithm development. *Ocean Dynamics*, INTERCOH 2009 special issue. doi:10.1007/s10236-011-0394-7.

Manning, A. J., Baugh, J. V., Soulsby, R. L., Spearman J. R., Whitehouse, R. J. S., 2011b. Cohesive sediment flocculation and the application to settling flux modelling. In: Ginsberg, S.S. (ed.), *Sediment Transport*. Rijeka, Croatia: InTech.

Manning, A. J., Schoellhamer, D.H., 2013. Factors controlling floc settling velocity along a longitudinal estuarine transect. *Marine Geology*, San Francisco Bay special issue. doi.org/10.1016/j.margeo.2013.04.006.

Manning, A. J., Spearman, J. R., Whitehouse, R. J. S., Pidduck, E. L., Baugh, J. V., Spencer, K. L., 2013. Laboratory assessments of the flocculation dynamics of mixed mud-sand suspensions. In: Manning, A.J. (ed.), *Sediment Transport Processes and Their Modelling Applications*. Rijeka, Croatia: InTech, 119–164.

Manning, A. J., 2016. LabSFLOC-2 – the second generation of the laboratory system to determine spectral characteristics of flocculating cohesive and mixed sediments. *HR Wallingford Report*.

Maude, A. D., Whitmore, R. L., 1958. A generalized theory of sedimentation. *British Journal of Applied Physics* 9, 477–482.

McAnally, W., 1999. Aggregation and deposition of estuarial fine sediment. PhD Thesis, University of Florida, FL.

McAnally, W. H., Mehta, A. J., 2001. Collisional aggregation of fine estuarine sediments. In: McAnally, W. H., Mehta, A. J. (eds.), *Coastal and Estuarine Fine Sediment Processes – Proceedings in Marine Science*, 3. Amsterdam: Elsevier, 19–39.

McCave, I. N., 1975. Vertical flux of particles in the ocean. *Deep Sea Research* 22, 491–502.

McCave, I. N., 1979. Suspended sediment. In: Dyer, K.R. (ed.), *Estuarine Hydrography and Sedimentation*. Cambridge: Cambridge University Press, 131–185.

McCave, I. N., 1984. Erosion, transport and deposition of fine-grained marine sediments. In: Stow, D. A. V., Piper, D. J. W. (eds.), *Fine-Grained Sediments: Deep Water Processes and Facies*. Oxford: Blackwell, 35–69.

McCave, I. N., Gross, T. F., 1991. In-situ measurements of particle settling velocity in the deep sea. *Marine Geology* 99, 403–411.

McDowell, D. N., O'Connor, B. A., 1977. *Hydraulic behaviour of estuaries*. London: MacMillan.

Mehta, A. J., Partheniades, E., 1975. An investigation of the depositional properties of flocculated fine sediment. *Journal of Hydraulic Research* 92, 361–381.

Mehta, A. J., Lott, J. W., 1987. *Sorting of fine sediment during deposition. Proceedings of the Conference on Advances in Understanding Coastal Sediment Processes* 1, 348–362.

Mehta, A. J., 2014. *An Introduction to Hydraulics of Fine Sediment Transport. Advanced Series on Ocean Engineering*, Vol. 38. Hackensack, NJ: World Scientific Publishing Co.

Merckelbach, L., 2000. *Consolidation and Strength Evolution of Soft Mud Layers*. Communications on hydraulic and geotechnical eng., report 00-2. Delft, The Netherlands: Delft University of Technology, Faculty of Civil Engineering.

Mie, G., 1908. Beiträge zur Optik trüber Medien, speziell kolloidaler Metallösungen. *Leipzig. Annalen der Physik* 330, 377–445.

Mietta, F., Chassagne, C., Manning, A. J., Winterwerp, J. C., 2009. Influence of shear rate, organic matter content, pH and salinity on mud flocculation. *Ocean Dynamics* 59, 751–763. doi:10.1007/s10236-009-0231-4.

Migniot, C., 1968. Study of the physical properties of various very fine sediments and their behaviour under hydrodynamic action. *La Houille Blanche*, 23 (7). (Translation of French text).

Mikeš, D., Manning, A. J., 2010. An assessment of flocculation kinetics of cohesive sediments from the Seine and Gironde Estuaries, France, through laboratory and field studies. *Journal of Waterway, Port, Coastal, and Ocean Engineering (ASCE)* 136, 306–318. doi: 10.1061/(ASCE) WW.1943-5460.0000053.

Millero, F. J., Poisson, A., 1981. International one-atmosphere equation of state of seawater. *Deep-Sea Research* 28, 625–629.

Mitchener, H. J., Torfs, H., Whitehouse, R. J. S., 1996. Erosion of mud/sand mixtures. *Coastal Engineering* 29, 1–25 [Errata, 1997, 30, 319].

Nagata, S., 1975. *Mixing, Principles and Applications*. New York: John Wiley and Sons.

Nakagawa, H., Nezu, I., Ueda, H., 1975. Turbulence of open channel flow over smooth and rough beds. *Proceedings of Japan Society of Civil Engineers* 241, 151–168.

Nowell, A. R. M., Jumars, P. A., Eckman, J. E., 1981. Effects of biological activities on the entrainment of marine sediments. *Marine Geology* 42, 133–153.

Ockenden, M. C., Delo, E. A., 1988. Consolidation and erosion of estuarine mud and sand mixtures – an experimental study. *HR Wallingford Report, SR* 149.

Ockenden, M. C., Delo, E. A., 1991. Laboratory testing of muds. *Geo-Marine Letters* 11, 138–142.

Ockenden, M. C., 1993. A model for the settling of non-uniform cohesive sediment in a laboratory flume and an estuarine field setting. *Journal of Coastal Research* 9, 1094–1105.

Odd, N. V. M., Roger, J. G., 1986. An analysis of the behaviour of fluid mud in estuaries. *HR Wallingford Report, SR* 84.

Odd, N. V. M., 1988. Mathematical modelling of mud transport in estuaries. In: Dronkers, J., van Leussen, W. (eds.), *Physical Processes of Estuaries*. Berlin: Springer Verlag, 503–531.

Oseen, C. W., 1927. *Neuere Methoden und Ergebnisse in der Hydrodynamik*. Leipzig: Akademische Verlagsgesellschaft.

Overbeek, J. T. G., 1952. Kinetics of flocculation. In: Kruyt, H. R. (ed.), *Colloid Science*. Amsterdam: Elsevier Publishing Company, 278–300.

Owen, M. W., 1971. The effects of turbulence on the settling velocity of silt flocs. *Proceedings of the 14th Congress of the International Association of Hydraulic Research* (Paris), D4–1–D4–6.

Owen, M. W., 1976. Determination of the settling velocities of cohesive muds. *HR Wallingford Report* No. IT 161, 8pp.

Parker, D. S., Kaufman, W. J., Jenkins, D., 1972. Floc break-up in turbulent flocculation processes. *Journal of the Sanitary Engineering Division* 98 (SA1), 79–97.

Parsons, D. R., Schindler, R. J., Hope, J. A., Malarkey, J., Baas, J. H., Peakall, J., Manning, A. J., Ye, L., Simmons, S., Paterson, D. M., Aspden, R. J., Bass, S. J., Davies, A. G., Lichtman, I. D., Thorne, P. D., 2016. The role of biophysical cohesion on subaqueous bed form size. *Geophysical Research Letters*, Early View, 1–8. doi:10.1002/2016GL067667.

Paterson, D. M., Crawford, R. M., Little, C., 1990. Subaerial exposure and changes in the stability of intertidal estuarine sediments. *Estuarine, Coastal and Shelf Science* 30, 541–556.

Paterson, D. M., Hagerthey, S. E., 2001. Microphytobenthos in contrasting coastal ecosystems: Biology and dynamics. In: Reise, K. (ed.), *Ecological Comparisons of Sedimentary Shores*, Ecological Studies, vol. 151. Berlin: Springer, 105–125.

Phipps and Bird, 2016. The jar test. [online] Available at: www.phippsbird.com/pbinc/Water WasteWater/Jartest.aspx [accessed March 2016].

Potter, P. E., Heling, D., Shimp, M. F., van Wie, W., 1975. Clay mineralogy of modern alluvial muds of the Mississippi river basin. *Bulletin du Centre Recherche Pau-SNPA* 9, 353–89.

Pouët, M.-F., 1997. La clarification coagulation—Flocculation. *Traitement de l'eau potable cours, EMA, option Environnement et Systèmes Industriels.*

Puls, W., Kuehl, H., Heymann, K., 1988. Settling velocity of mud flocs: Results of field measurements in the Elbe and the Weser Estuary. In: Dronkers, J., Van Leussen, W. (eds.), *Physical Processes in Estuaries*. Berlin: Springer-Verlag, 404–424.

Puls, W., Kühl, H., 1996. Settling velocity determination using the BIGDAN settling tube and the Owen settling tube. *Journal of Sea Research* 36, 119–125.

Raudkivi, A. J., 1998. *Loose Boundary Hydraulics*. 3rd Edition. Rotterdam: Balkema.

Ross, M. A., 1988. Vertical structure of estuarine fine sediment suspensions. PhD thesis, University of Florida, Gainesville.

Schindler, R. J., Parsons, D. R., Ye, L., Hope, J. A., Baas, J. H., Peakall, J., Manning, A. J., Aspden, R. J., Malarkey, J., Simmons, S., Paterson, D. M., Lichtman, I. D., Davies, A. G., Thorne, P. D., Bass, S. J., 2015. Sticky stuff: Redefining bedform prediction in modern and ancient environments. *Geology* 43, 399–402. doi:10.1130/G36262.1.

Schlichting, H., 1968. *Boundary Layer Theory*. New York: McGraw-Hill.

Sternberg, R. W., Berhane, I., Ogston, A. S., 1999. Measurement of size and settling velocity of suspended aggregates on the northern California continental shelf. *Marine Geology* 154, 43–53.

Smith, S. J., 2010. Fine sediment dynamics in dredge plumes. PhD Thesis, Virginia Institute of Marine Science, College of William and Mary, USA.

Smith, S. J., Friedrichs, C. T., 2011. Size and settling velocities of cohesive flocs and suspended sediment aggregates in a trailing suction hopper dredge plume. *Continental Shelf Research* 31, S50–63. doi:10.1016/j.csr.2010.04.002.

Smith, S. J., Friedrichs, C. T., 2015. Image processing methods for in situ estimation of cohesive sediment floc size, settling velocity, and density. *Limnology and Oceanography Methods* 13, 250–264.

Soulsby, R. L., 1983. The bottom boundary layer of shelf seas. In: Johns, B. (ed.), *Physical Oceanography of Coastal and Shelf Seas*. Amsterdam: Elsevier, 189–266.

Soulsby, R. L., 1997. *Dynamics of Marine Sands*. London: Thomas Telford.

Soulsby, R. L., 2000. Methods for predicting suspensions of mud. *HR Wallingford Report TR* 104.

Soulsby, R. L., Manning, A. J., Spearman, J., Whitehouse, R. J. S., 2013. Settling velocity and mass settling flux of flocculated estuarine sediments. *Marine Geology* 339, 1–12. doi.org/ 10.1016/j.margeo.2013.04.006.

Spencer, K. L., Manning, A. J., Droppo, I. G., Leppard, G. G., Benson, T., 2010. Dynamic interactions between cohesive sediment tracers and natural mud. *Journal of Soils and Sediments* 10, 1401–1414. doi:10.1007/s11368-010-0291-6.

Spinrad, R. W., Bartz, R., Kitchen, J. C., 1989. In-situ measurements of marine particle settling velocity and size distributions using the remote optical settling tube. *Journal of Geophysical Research* 94 (C1), 931–938.

Stephens, J. A., Uncles, R. J., Barton, M. L., Fitzpatrick, F., 1992. Bulk properties of intertidal sediments in a muddy, macrotidal estuary. *Marine Geology* 103, 445–460.

Sternberg, R. W., Berhane, I., Ogston, A. S., 1999. Measurement of size and settling velocity of suspended aggregates on the northern California continental shelf. *Marine Geology* 154, 43–53.

Stewart, C., Thomson, J. A. J., 1997. Vertical distribution of butyltin residues in sediments of British Columbia harbours. *Environmental Technology* 18, 1195–1202.

Stokes, G. G., 1851. On the effect of the internal friction on the motion of pendulums. *Transactions of the Cambridge Philosophical Society* 9, 8–106.

Stolzenbach, K. D., Elimelich, M., 1994. The effect of density on collisions between sinking particles: Implications for particle aggregation in the ocean. *Journal of Deep Sea Research I* 41, 469–483.

Stone, M., Krishnappan, B. G., 2003. Floc morphology and size distributions of cohesive sediment in steady-state flow. *Water Research* 37, 2739–2747.

Stratton, A., 1941. *Electromagnetic Theory*. New York: McGraw-Hill.

Syvitski, J. P. M., Asprey, K. W., Leblanc, K. W. G., 1995. In-situ characteristics of particles settling within a deep-water estuary. *Deep-Sea Research II* 42, 223–256.

Tambo, N., Hozumi, H., 1979. Physical characteristics of flocs – II. Strength of flocs. *Water Research* 13, 441–448.

Tambo, N., Watanabe, Y., 1979. Physical characteristics of flocs – I. The floc density function and aluminium floc. *Water Research* 13, 429–439.

Ten Brinke, W. B. M., 1993. The impact of biological factors on the deposition of fine-grained sediment in the Oosterschelde (The Netherlands). PhD Thesis, Utrecht University.

Ten Brinke, W. B. M., 1994. Settling velocities of mud aggregates in the Oosterschelde tidal basin (The Netherlands), determined by a submersible video system. *Estuarine, Coastal and Shelf Science* 39, 549–564.

Tennekes, H., Lumley, J. L., 1972. *A First Course in Turbulence*. Cambridge, MA: MIT Press.

Tolhurst, T. J., Gust. G., Paterson, D. M., 2002. The influence on an extra-cellular polymeric substance (EPS) on cohesive sediment stability. In: Winterwerp, J.C., Kranenburg, C. (eds.), *Fine Sediment Dynamics in the Marine Environment – Proceedings in Marine Science 5*, Amsterdam: Elsevier, pp. 409–425.

Torfs, H., Mitchener, H. J., Huysentruyt, H., Toorman, E., 1996. Settling and consolidation of mud/sand mixtures. *Coastal Engineering* 29, 27–45.

Tsai, C. H., Iacobellis, S., Lick, W., 1987. Flocculation of fine-grained sediments due to a uniform shear stress. *Journal of Great Lakes Research* 13, 135–146.

Ueda, H., Hinze, J. O., 1975. Fine-structure turbulence in the wall region of a turbulent boundary layer. *Journal of Fluid Mechanics* 67, 125–143.

Uncles, R. J., Stephens, J. A., Harris, C., 1998. Seasonal variability of subtidal and intertidal sediment distributions in a muddy, macrotidal estuary: The Humber-Ouse, UK. In: Black, K. S., Paterson, D. M., Cramp, A. (eds.), *Sedimentary Processes in the Intertidal Zone*. London: Geological Society Special Publications 139, 211–219.

Uncles, R. J., Bale, A. J., Brinsley, M. D., Frickers, P. E., Harris, C., Lewis, R. E., Pope, N. D., Staff, F. J., Stephens, J. A., Turley, C. M., Widdows, J., 2003. Intertidal mudflat properties, currents and sediment erosion in the partially mixed Tamar Estuary, UK. *Ocean Dynamics* 53, 239–251. doi:10.1007/s10236-003-0047-6.

Uncles, R. J., Stephens, J. A., Harris, C., 2006. Properties of suspended sediment in the estuarine turbidity maximum of the highly turbid Humber Estuary system, UK. *Ocean Dynamics* 56, 235–247.

Uncles, R. J., Bale, A. J., Stephens, J. A., Frickers, P. E., Harris, C., 2010. Observations of floc sizes in a muddy estuary. *Estuarine, Coastal and Shelf Science* 87, 186–196.

Underwood, G. J. C., Kromkamp, J., 1999. Primary production by phytoplankton and microphytobenthos in estuaries. *Advances in Ecological Research* 29, 93–153.

Underwood, G. J. C., Paterson, D. M., 2003. *The Importance of Extracellular Carbohydrate Production by marine Epipelic Diatoms*. Advances in Botanical Research (incorporating Advances in Plant Pathology), Vol. 40. Amsterdam: Elsevier, 183–240.

van der Lee, E. M., Bowers, D. G., Kyte, E., 2009. Remote sensing of temporal and spatial patterns of suspended particle size in the Irish Sea in relation to the Kolmogorov microscale. *Continental Shelf Research* 29, 1213–1225.

van de Ven, T. G., Hunter, R. J., 1977. The energy dissipation in sheared coagulated soils. *Rheologica Acta* 16, 534–543.

van Ledden, M., 2002. A process-based sand-mud model. In: Winterwerp, J.C., Kranenburg, C. (eds.), *Fine Sediment Dynamics in the Marine Environment – Proceedings in Marine Science 5*, Amsterdam: Elsevier, 577–594.

van Ledden, M., 2003. Sand-mud segregation in estuaries and tidal basins. PhD Thesis, Delft University of Technology, The Netherlands, Report No. 03–2, ISSN 0169-6548, 217pp.

van Leussen, W., 1988. Aggregation of particles, settling velocity of mud flocs: A review. In: Dronkers, J., van Leussen, W. (eds.), *Physical Processes in Estuaries*. Berlin: Springer-Verlag, 347–403.

van Leussen, W., 1991. Fine sediment transport under tidal action. *Geo-Marine Letters* 11, 119–126.

van Leussen, W., 1994. Estuarine macroflocs and their role in fine-grained sediment transport. PhD Thesis, University of Utrecht, The Netherlands, 488pp.

van Leussen, W., 1997. The Kolmogorov microscale as a limiting value for the floc sizes of suspended fine-grained sediments in estuaries. In: Burt, N., Parker, R., Watts, J. (eds.), *Cohesive Sediments*. New York: Wiley, 45–73.

van Leussen, W., Cornelisse, J. M., 1994. The determination of the sizes and settling velocities of estuarine flocs by an underwater video system. *Journal of Sea Research* 31, 231–241.

van Olphen, H., 1977. *An introduction to Clay Colloid Chemistry. For Clay Technologists, Geologists, and Soil Scientists*, 2nd edition. New York: John Wiley and Sons.

Vanoni, V. A., 1975. *Sedimentation Engineering. Manuals and Reports on Engineering Practice, no. 54*. New York: American Society of Civil Engineers, 481–484.

Verney, R., Lafite, R., Brun-Cottan, J. C., Le Hir, P., 2010. Behaviour of a floc population during a tidal cycle: Laboratory experiments and numerical modelling. *Continental Shelf Research* 31, S64–S83.

Wacholder, E., Sather, N. F., 1974. The hydrodynamic interaction of two unequal spheres moving under gravity through quiescent viscous fluid. *Journal of Fluid Mechanics* 5, 417–437.

Weaver, C. E., 1989, *Clays, Muds, and Shales. Developments in Sedimentology, no. 44*. New York: Elsevier.

Whitehouse, R. J. S., Soulsby, R., Roberts, W., Mitchener, H. J., 2000. *Dynamics of Estuarine Muds*. London: Thomas Telford Publications.

Widdows, J., Blauw, A., Heip, C. H. R., Herman, P. M. J., Lucas, C. H., Middelburg, J. J., Schmidt, S., Brinsley, M. D., Twisk, F., Verbeek, H., 2004. Role of physical and biological processes in sediment dynamics of a tidal flat in Westerschelde Estuary, SW Netherlands. *Marine Ecology Progress Series* 274, 41–56.

Williamson, H. J., Ockenden, M. C., 1993. Laboratory and field investigations of mud and sand mixtures. In: Wang, S. S. Y (ed.), *Advances in Hydro-science and Engineering, Proceedings of the First International Conference on Hydro-science and Engineering*, Washington D.C. (7–11 June 1993), Volume 1. University, MS: Center for Computational Hydroscience and Engineering, the University of Mississippi, 622–629.

Winterwerp, J. C., 1998. A simple model for turbulence induced flocculation of cohesive sediment. *Journal of Hydraulic Engineering* 36, 309–326.

Winterwerp, J. C., 1999. On the dynamics of high-concentrated mud suspensions, PhD thesis, Delft University of Technology, Delft.

Winterwerp, J. C., 2002. On the flocculation and settling velocity of estuarine mud. *Continental Shelf Research* 22, 1339–1360.

Winterwerp, J. C., van Kesteren, W. G. M., 2004. Introduction to the physics of cohesive sediment in the marine environment. In: van Loon, T. (ed.), *Developments in Sedimentology*, 56. Amsterdam: Elsevier.

Winterwerp, J. C., Manning, A. J., Martens, C., de Mulder, T., Vanlede, J., 2006. A heuristic formula for turbulence-induced flocculation of cohesive sediment. *Estuarine, Coastal and Shelf Science* 68, 195–207.

Wolanski, E., 2007. *Estuarine Ecohydrology*. Amsterdam, The Netherlands: Elsevier.

Xu, R., 2000. *Particle Characterization: Light Scattering Methods*. Dordrecht: Kluwer Academic Publishers.

9 Sediment Transport: Instrumentation and Methodologies

K. Black, J. Poleykett, R. J. Uncles, M. R. Wright

1 Introduction

This chapter is concerned with the methods used to examine and quantify suspended particulate matter (SPM) transport and bedload sediment transport in estuarine and coastal environments. In these systems the SPM is generally dominated by suspended sediment (see Chapter 6 for a broad definition of sediment), and the terms *SPM* and *suspended sediment* are used interchangeably here. The movement of sediment in rivers, estuaries and the coastal zone is a fundamental attribute of the Earth surface system and an important component driving sediment cycling. Methods to investigate the nature of suspended and deposited sediment, including the primary and flocculated sizes of suspended particles and aggregates, have been addressed in Chapters 6, 7, and 8. The emphasis of this chapter is on quantifying the transport of sediment as suspended load and bedload.

Both older (but still used) and modern methods are described for the measurement of suspended sediment fluxes and bedload transport rates. A particularly powerful method of tracking sediment movements – particle tracing or tracking – is also described in some detail. Such experiments have been undertaken for many years; e.g., Ingle (1966) tracked the movement of beach sand using fluorescent grains and Marsh et al. (1991, 1993) used fluorescent tracer particles to study sediment movements in lacustrine and turbid estuarine environments. However, this technology has recently been refined and improved on so that it might be applied to the determination of bedload transport rates as well as other related issues, including mixing-depth assessment and bioturbation rates. 'Particle tracking', or as it is also referred to in the geological sciences 'sediment tracing' or 'sediment tracking', offers a unique methodology with which to track the movement through space and time of environmental particulates. It is a relatively straightforward, practical methodology that involves the introduction of particulate tracers into the environment (e.g., the beach face or subtidal waters) labelled with one or more signatures in order that they may be unequivocally identified following release (Black et al., 2007; McLaren et al., 2007).

Also considered in this chapter are the erosion-deposition properties of deposited sediments and the application of benthic flumes to their study. During the 1970s and 1980s a significant volume of work utilised laboratory annular flumes to investigate the erosion and deposition of fine-grained sediments (e.g., Mehta and Partheniades, 1982). Flumes are devices that can be employed to apply a controlled flow stress (or velocity) onto the surface of bed sediments. Whilst this approach produced some of the key developments in the mechanistic understanding of sediment processes, the work was of limited utility because the sediments studied were usually remoulded; i.e., extracted from their environment and transferred or stored in a way that was not consistent with their original placement. Remoulding of sampled, fine (muddy) sediments gave rise to changes in their physical character, and by the mid-1980s this deficiency led to a drive to develop the technology for in situ measurements, which arrived in the form of subsea annular, or benthic, flumes (Black et al., 2002).

2   Suspended Sediment Transport

Suspended sediment (SPM) fluxes can be computed directly via measurements of current velocities and suspended sediment concentrations (SSC – suspended solids concentration(s)) and the direction of sediment transport determined from these data, or inferred from the paths followed by tracer particles (see later).

2.1   Direct Measurement of SPM Fluxes

Before the advent of ADCPs (Chapter 4), the classic way of estimating fluxes of suspended particulate matter was through the use of winch-assisted or manual water-column profiling from an anchored vessel or moored platform (Chapter 7). SPM concentrations were measured with calibrated transmissometer instruments or OBS turbidity sensors (Chapter 7) at several depths in the column at the same time and the same depth-beneath-surface as measurements of water velocity, generally made using an impellor-type current meter (e.g., Uncles et al., 1985b). Repeated profiling over a tidal cycle, e.g., every 30 min, provided a time series that could be used to interpret sediment-flux behaviour and determine residual (tidally averaged) fluxes.

Kjerfve (1979) describes a methodology for analysing raw data from such classic profiling measurements that is straightforward to apply. The product of velocity and SSC at each depth yields the SPM flux at that time and depth, which can then be integrated through the water column to give the rate of transport per unit width of an estuary. For water, the volume flux at any depth is simply the water velocity, U (m s^{-1}), and the rate of water transport through the whole column, per unit width of column, is $H\overline{U}$ (m^2 s^{-1}), where $\overline{U}$ is the depth-averaged current velocity (assuming the currents are unidirectional) and H is the depth of water (m). Similarly, the suspended particulate matter flux at any depth is UP (kg m^{-2} s^{-1}), where P is the SPM concentration (kg m^{-3}), so that the rate of transport of SPM per unit width through the whole water column is $H\overline{UP}$ (kg s^{-1} m^{-1}).

If several anchor stations or moorings are utilised over an estuarine cross-section (e.g., Uncles et al., 1985a), then the measurements can be averaged over the section to give $A\overline{\overline{U}}$, where A is the cross-sectional area and $\overline{\overline{U}}$ the section-averaged velocity, so that the rate of water transport through the whole cross-section is $A\overline{\overline{U}}$ (m^3 s^{-1}), and the rate of transport of SPM through the whole section is $A\overline{\overline{UP}}$ (kg s^{-1}). A discussion of some of the errors associated with area averaging is given by Perillo et al. (1998). These fluxes and transports can be averaged over one or more tidal cycles to estimate the residual fluxes and transports, which can then be analysed in terms of specific estuarine processes to understand the mechanisms at work (e.g., Uncles, 1985a and b).

An inadequacy of deploying anchor or mooring stations over an estuarine cross-section is the frequent transverse complexity of longitudinal (i.e., along-axis or along-estuary) currents (Chapter 1 and Figure 9-1a and b), which necessitates the use of a large number of profiling stations to achieve an accurate estimate of along-estuary transport rates (e.g., $m^3 s^{-1}$ for water; $kg s^{-1}$ for SPM). However, modern acoustic instrumentation has transformed the measurement of these fluxes to provide data over multiple tidal cycles with great spatial resolution. Acoustic Doppler Current Profilers (ADCPs, Chapter 4), when used in conjunction with conventional water sampling for calibration, allow for the measurement of SPM flux continuously along transects. The rate of SPM transport ($kg s^{-1}$) through an estuarine section or through an inlet can then be accurately estimated from these data (an example of data for the Marsdiep Inlet, the Netherlands, is shown in Figure 9-1c; Nauw et al., 2014). Such a high spatial resolution for flux data was previously unobtainable with older instrumentation. Although strong spatial variations in fluxes may occur for each transect traversal, repeated sailings of the same transect serve to record these fluxes over the section or inlet and through time (e.g., Figure 9-1d illustrates data for the Thames Estuary, UK).

An increasing body of research has applied this acoustic methodology and investigated the assumptions and errors within the technique. Mariotti and Fagherazzi (2011) computed sediment fluxes in Willapa Bay, Washington State, US, using 2-MHz ADCPs that averaged data over 1 min and recorded every 15 min with a vertical cell size of 0.1 m. ADCP-measured velocities were combined with SPM concentrations (SSC) to yield sediment fluxes, UP; The SSC was estimated using the acoustic backscatter signal of the instruments according to well-established equations (Downing et al., 1995; Gartner, 2004; Hoitink and Hoekstra, 2005). Parameters in these equations were computed using SSC derived from calibration bottle samples from the sites. The same approach was used to measure sediment fluxes in the Caravelas estuarine system, Brazil, by Schettini et al. (2013).

The suspended sediment fluxes through the Marsdiep inlet between the mainland and the island of Texel, the Netherlands, were measured by Nauw et al. (2014) using a 1.0-MHz ADCP mounted 4.3 m below the sea surface on the hull of the ferry that repeatedly traversed the inlet. The acoustic backscatter intensity was used to estimate SSC profiles after calibration so that fluxes could be estimated from the product of the velocities and the SSC, i.e., UP (an example of their data is shown in Figure 9-1c); they found that the intensity of the acoustic backscatter was a function of the primary particle grain-size distribution.

Brand et al. (2010) used acoustic Doppler velocimeters (ADVs) to simultaneously measure current velocities, turbulence, SSC and then SPM fluxes within 0.7 m of the bed over shoals in South San Francisco Bay, US. They deployed 10-MHz and 6-MHz ADVs in burst mode and recorded data at 8 or 10 Hz.

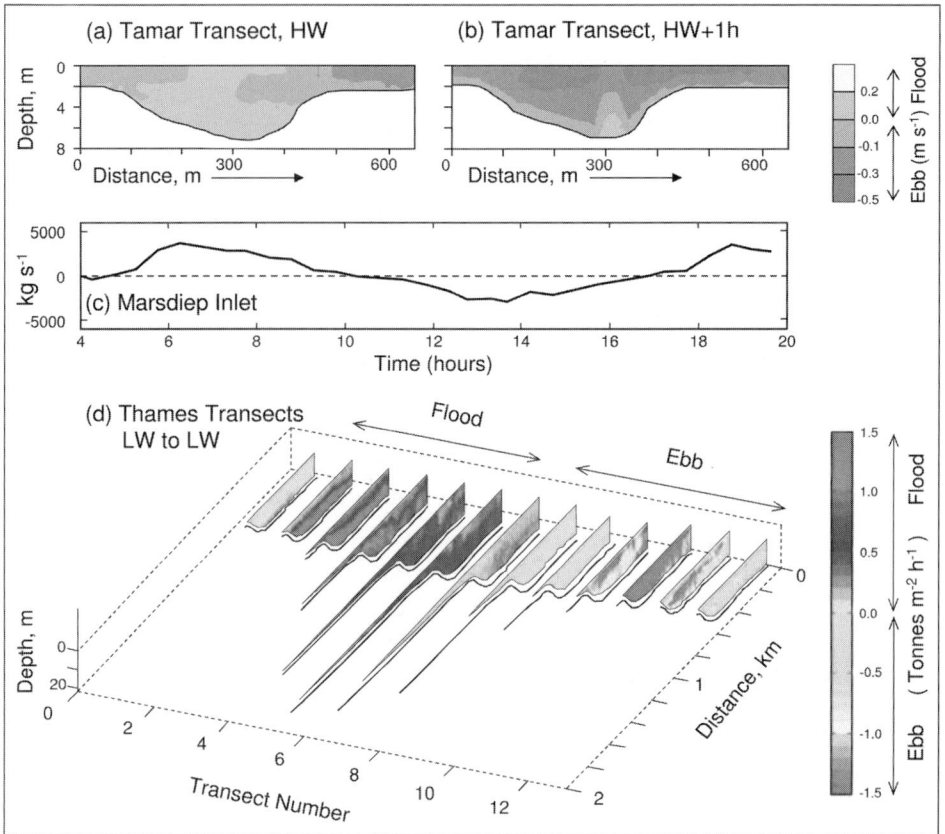

Figure 9-1 High-resolution, cross-estuary variations in along-estuary currents and rates of sediment transport measured using ADCPs over transects: (a) along-estuary currents over a cross-section in the Tamar Estuary (UK) at HW and (b) at HW+1h, showing ebb-directed currents over the shallow regions and flood-directed currents in the main channel at HW, and ebb-directed currents over most of the cross-section at HW+1h, with a core of near-bed, flood-directed current in the main channel (reprinted with modifications from figure 5a and c of Uncles, 2002, with permission from Elsevier); (c) ADCP-derived rate of SPM transport through the Marsdiep inlet during 15 September 2003, flood positive, extracted from a much longer time series (reprinted with minor modifications from figure 11c of Nauw et al., 2014, with permission from Elsevier); (d) a tidal time series, at approximately hourly intervals, of ADCP-derived SPM fluxes over a cross-section in the Thames Estuary, UK, modified for grey-scale presentation (with permission of DP World London Gateway Port). Note that the flood-tide grey scale is ambiguous for some flux values – a difficulty that is resolved by recognising that Transect 1 has an ebb flux everywhere except close to the bed, and that flood-directed fluxes generally increase from surface to bed at other times during the flood (Transects 2 to 7).

3   Bedload Sediment Transport

Bedload sediment transport is the result of sediment grains that move on or close to the bed, rather than in the main body of the water column, and are usually associated with propagating (migrating) bedforms (e.g., Figure 9-2a). The transport is usually associated

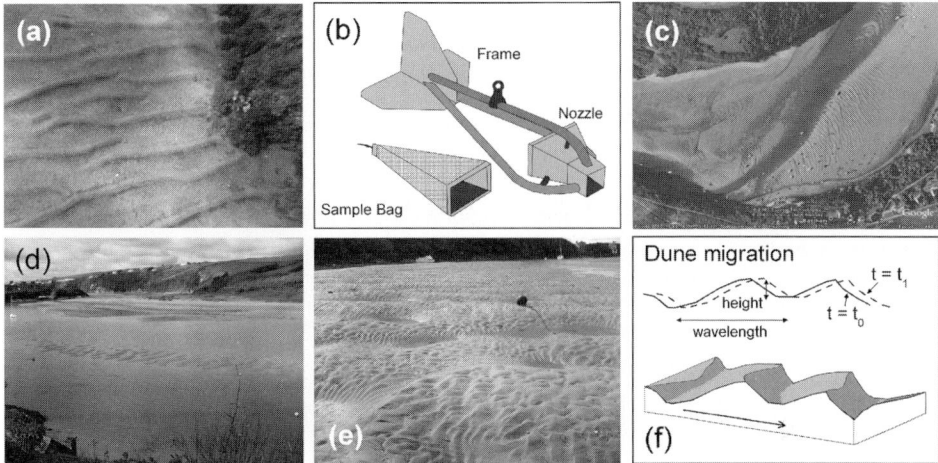

Figure 9-2 Bedload sediment transport; (a) coastal wave-induced sandy bedforms with approx. 0.1–0.2 m wavelength in approx. 3 m of water; (b) Helly Smith bedload sampler (modified from figure 1 of Emmett 1980, with permission); (c), Google Earth image of the lower Devonshire Avon Estuary (UK) showing a large exposed dune field at low water; (d) photograph of part of the same dune field, illustrating dune wavelengths that are evident due to inundation of troughs and exposure of crests; (e) the same dune field at low water viewed from a dune crest; (f) a schematic dune field that is propagating from left to right, illustrated at time t_0 and at a later time t_1 (dashed line), showing the dune wavelength and height. Photographs: R.J. Uncles, except (c).

with coarser grains (medium sand and greater) and can be driven by unidirectional (river) or oscillatory (tidal and wave-induced) water currents. The grains move by rolling on, or by saltating ('jumping') close to, the unmoving bed. The direct quantification of bedload has, over the years of active sediment research, been a notoriously difficult issue.

Two fundamentally different methods can be applied to the estimation of bedload sediment transport in aquatic systems. One is based on mechanical devices that have been designed to trap sediments that are moving in and close to the bed, and the second relies on knowledge of the propagation, i.e., migration rate, of bed features such as sand dunes (van Rijn, 1993).

3.1   Mechanical Traps

Bedload samplers that trap sediment grains as they are transported along the bed and store them for future retrieval and analysis are referred to as a direct-measuring samplers. They have the advantage of being straightforward to use. As an example, a commonly used sediment trap is the Helley Smith bedload sampler (Figure 9-2b). Typically, this is approx. 1 m in length from nozzle mouth to fin end, with a nozzle mouth that is approximately 0.1-m wide and approximately 0.1-m high, and a sample bag, approximately 0.5 m in length, made of 250 μm-mesh polyester (Emmett, 1980). The overall weight is approximately 30 kg. According to Emmett (1980) the sample bag can be 40 per cent filled with sediment of grain diameter >250 μm before the trapping efficiency of the sampler is reduced. There are other designs of bedload samplers that work on a similar principle (e.g., van Rijn, 1993).

The Fyke net, which is a bag-shaped net that is held open by hoops and usually used to catch eels, was utilised as a bedload sampler by Hemminga et al. (1996) to measure the near-bed and bedload transport of coarse detritus particles (>1 mm) in a tidal creek of the Westerschelde Estuary (The Netherlands). The net was retrieved and emptied every hour and turned around after high-water slack to sample the ebb current.

3.2   Bedform Movements

This method uses sequential measurements of bedform shape over a region to calculate the amount of sediment that has moved during successive time periods. Provided flow conditions and bedform shapes do not change during the interval between measurements, it is straightforward to calculate the sediment transport that results from bedform migration. A moving dune field (dune wavelength and height are shown schematically in Figure 9-2f) transports a volume of sediment in the direction of bedform migration (from left to right in Figure 9-2f). A simple approach to calculating the transport is (according to Engel and Lau, 1980, 1981; van Rijn, 1993):

$$s_b = \alpha_s(1 - p)\rho_s a\Delta \tag{1}$$

Where s_b is the bedload transport rate per unit width (kg s^{-1} m^{-1}), α_s is a 'shape' factor (0.5–0.6), p a porosity factor (approx. 0.4), ρ_s the sediment density (approximately 2650 kg m^{-3}), a the average migration speed (m s^{-1}) and Δ the average bedform height (m).

As an example of a surveyed estuary, the Devonshire Avon Estuary, UK, has an extensive field of subaqueous dunes in its lower reaches (Figure 9-2c, d and e); they are a common feature in strongly tidal, sandy estuaries (Ashley, 1990). Typically, in the Avon, these have heights of 0.05–0.2 m and lengths between 5 and 10 m (Masselink et al., 2009). In a three-month period they were observed to migrate 10–20 m, which corresponded to an annual, up-estuary volumetric transport of 2.6 m^3 per metre width of estuary, similar to the value modelled by Uncles et al. (2013). For the Avon, Masselink et al. (2009) surveyed the dunes with a laser Total Station (e.g., Leica, 2016) along four lines (240 m long and spaced by 16 m) every two weeks from July to October. The Total Station had a vertical and horizontal accuracy of the order of mm and cm, respectively, and measurements were made within 0.2 m of each survey line and approx. every metre or less when gradients were steep.

Similar studies were carried out by Goud and Aubrey (1985), who estimated bedload transport rates for a shallow, near-shore area of Nantucket Sound, US, using aerial photography to deduce the migration rates of shore-normal sand waves with decimetre-scale heights and wavelengths of the order of tens to hundreds of metres. They measured average migration rates of 10–20 m per annum over a 10-year period and used the migration distances and bedform dimensions to calculate an average bedload volume transport rate.

The subtidal dune morphology and dune migration rates in the Bahia Blanca Estuary, Argentina, were studied by Minor Salvatierra et al. (2015). They utilised a high-resolution multibeam (Chapter 3) bathymetric system (250-kHz GeoSwath Plus from Kongsberg GeoAcoustics Ltd., UK) with position data derived from a differential global positioning system (dGPS). Bottom samples of dune sediment were collected with a Shipek grab sampler (Chapter 6). They found that the bedforms, which had heights and wavelengths greater than 5 m and 130 m, respectively, migrated in the ebb direction with a mean velocity of 43 m year^{-1}. Similar studies for the Changjiang (Yangtze) Estuary, China, were reported by Wu et al. (2009).

4   Particle Tracking for Sediment Transport Studies

To obtain information on a sediment source and its associated transport pathways and eventual fate, a particle-tracking (i.e., sediment-tracing) approach can be employed. Sediment tracing involves the use of natural and/or artificial sediments labelled with an identifiable 'tag' or 'signature', through which they can be unequivocally identified in the environment (Drapeau et al., 1991). Sediment tracing is a flexible tool that can be used to identify point sources of sediment (Inman and Chamberlain, 1959; Cromey et al., 2002; Magal et al., 2008), elucidate transport pathways (Polyakov and Nearing, 2004; Kimoto et al., 2006; Carrasco et al., 2013), and assess zones or areas of accumulation or deposition (Kimoto et al., 2006). Questions that the tracking approach can usefully address in estuarine settings are numerous. For example, 'Does the material from this source accrete onto the marsh surface?'; 'Is the net sediment transport in runnels shoreward or seaward?'; or 'Is the mid-shore region stable under wave action?'

Sediment tracing techniques progressed significantly in the century following the first tracing study of Richardson (1902). The technique has recovered from significant setbacks; e.g., the environmental ban on the use of irradiated grains (Sigurbjornsson, 1994; Black et al., 2007) and fluctuations in popularity, primarily due to the resource-intensive nature of the method (White, 1998). Recent developments in tracer design and methodological approaches that have reinvigorated and revived the technique are outlined in Appendix 1. This resurgence of interest has led to novel applications and commercial enterprise within the geo-marine sector. A variety of tracing materials (tracers) are now available to perform tracing studies. Each tracer has its own benefits and limitations. Similarly, tracer-specific practical and analytical methodologies each have their own benefits and limitations.

An optimal tracer is one that can be deployed, tracked, recovered from environmental samples and enumerated successfully, and which has similar hydraulic properties to the native sediment. Key assumptions relating to tracers and their use that must be satisfied include

- The tracer's hydraulic and bio-organic properties closely match those of the native sediment and are such that the tracer is transported in the same manner as the native sediment.

- The tracer properties are stable and do not change through time.
- Introduction of the tracer does not disrupt the transporting system.

It is essential that these assumptions are acknowledged, tested and the results considered in the light of their potential effects on sediment-transport dynamics.

There are two principal types of active tracer that have been utilised in tracing studies: labelled (coated) natural particles and labelled synthetic tracer (Black et al., 2007). Historically, the most popular tracer signatures have been applied radioactivity, fluorescence and applied magnetism.

Irradiated sediment grains have been used extensively for sediment tracing (Davidson, 1958; Inman and Chamberlain, 1959; Sarma and Iya, 1960; Crickmore and Lean, 1962; Courtois and Monaco, 1968) due to the relative ease with which they can be monitored over large spatial areas using handheld Geiger counters. However, irradiated tracers are now not permitted in most environments on the basis of health and safety considerations, which significantly limit their use.

Applied fluorescent colour has proved a popular tracer signature and by far the majority of historic studies have used fluorescent colour (see Black et al., 2007, for a comprehensive review). Fluorescent materials provide a unique visual tool that enables noninvasive image analysis and visual inspection techniques to be utilised to identify tracer grains (Ciavola et al., 1998; Solan et al., 2004; Silva et al., 2007; Carrasco et al., 2013). A fluorescent signature can be applied to a natural material by the application of a fluorescent substance as a monolayer, or by direct staining of sediment with fluorescent dye. Coated natural minerals (through whichever tagging process) are a preferred tracer type because the hydraulic properties of the grains remain almost entirely unchanged for sands and gravels, and only marginally altered for silts.

Technological progress in the development and manufacturing of polymers and synthetic materials has led to the use of entirely synthetic tracers (Tauro et al., 2010). In an analogous fashion to coated natural material, synthetic tracers are labelled with an identifiable signature. Synthetic tracing materials that have been used in the past include fluorescent microspheres; e.g., Marsh et al., 1991; Angarita-Jaimes et al., 2008; and Pedocchi et al., 2008 (results from an early field example are shown in Figure 9-3a) and magnetic plastic beads (e.g., Harvey et al., 1989; Tanaka et al., 1998; Ventura et al., 2002). However, synthetic particles are often unable to replicate the particle-size distribution, particle density, shape, surface morphology, and surface reactivity of natural sediments (Zhang et al., 2003; Black et al., 2007). Also, concern has been raised over the deployment of entirely synthetic materials (e.g., plastic beads) to the environment, especially in ecologically sensitive areas (Thompson et al., 2004).

Regardless of environment or study context, when conducting a particle-tracking study, a series of methodological steps should be followed to ensure that best practice is adhered to. Seven key steps have been identified: (1) perform a background survey, (2) design or select a tracer, (3) match the tracer properties to those of the native sediment and test, (4) introduce the tracer to the environment, (5) conduct environmental sampling, (6) perform tracer enumeration and (7) analysis of results. The requirements of each step are discussed in subsections 4.1–4.7.

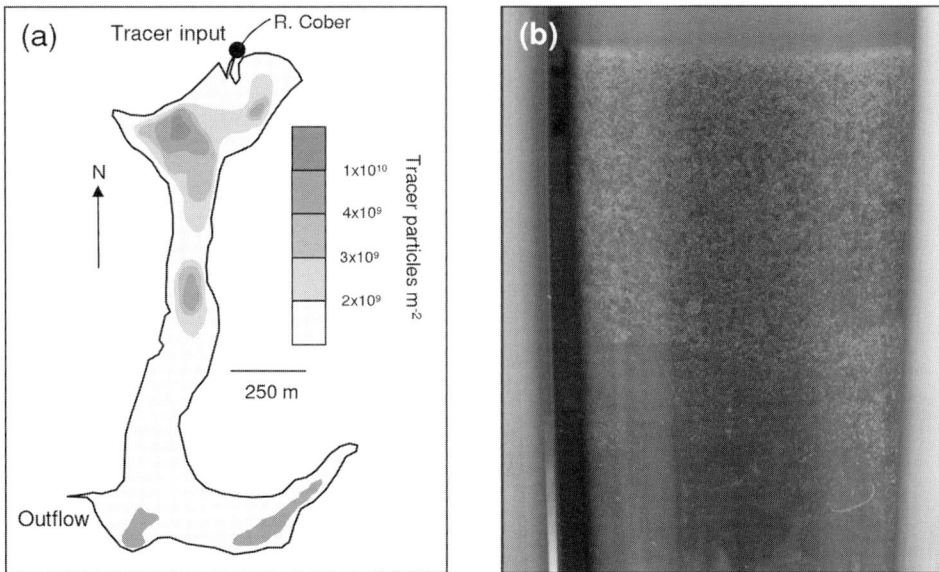

Figure 9-3 Particle tracking and the tagging of sediment; (a) an example, chosen for its graphical simplicity, of results from an early tracer experiment in a lake (Loe Pool, Cornwall, UK); artificial, fluorescent tracer particles were introduced into the River Cober and, after three days, core samples showed an accumulation of tracer in the northwest of the lake due to particle trapping by a wind-induced gyre (reproduced with minor modifications from Marsh et al., 1993; with permission); (b) an image captured under ultraviolet illumination that shows a core of cohesive peat sediment mixed with a fluorescent pink silt tracer (the tracer colour shows as a lighter grey here). The tracer has been hydraulically matched to the medium silt fraction of the cohesive peat sediment; it has then been thoroughly mixed with the peat and left to settle. The image demonstrates that the tracer grains have completely flocculated and become entangled with the peat floc aggregates. Thus, the peat flocs have been 'tagged' with an identifiable 'signature', enabling transport of the peat material to be assessed. The entrapment of the tracer grains does not materially affect the settling velocity of the peat flocs (Photograph: Partrac Ltd, with permission).

4.1   Step 1: Background Survey

The purpose of a background survey is threefold: (1) to determine the presence or absence of particulates within the study region that have the same or similar characteristics as the proposed tracer of use; (2) to provide a comprehensive assessment of the properties of the native sediment on which the tracer design can be based; and (3) to evaluate environmental baseline readings by testing any sampling techniques proposed, in order to ensure no false positives are recorded. It is crucial to know that the tracer to be used has a unique signature in the environment of use; if not, this compromises particulate-differentiation during tracer detection. Signature uniqueness can only be assessed through sampling and testing of the study-site sediments. The sediment to be tracked must be tested for its physical properties (e.g., grain-size distribution and density) because these influence its hydraulic behaviour (e.g., Dyer, 1986); it is essential that a tracer be designed to match these physical properties. Subsamples of collected sedimentary materials can be tested for their hydraulic parameters, but it is important to collect samples from across potential sediment source areas, transport pathways and deposition zones.

The background survey should also be used to collect relevant data regarding the forcing mechanisms at the site (e.g., current velocity). In addition, a qualitative expert geomorphological site assessment should be made because this may provide additional information regarding possible sediment transport rates and pathway(s) within the study area. Understanding the geomorphology and associated physical processes within the study area prior to introduction of the tracer enables an informed methodological strategy to be developed.

4.2   Step 2: Tracer Design or Selection

A tracer that matches the native sediment in terms of its hydraulic attributes should be selected or designed. The amount of tracer introduced into the environment is critical and needs to be considered at this stage; if too much material is deployed, then sediment transport processes may be unduly affected (Foster, 2000), whereas if too little is deployed the subsequent tracer recovery may be compromised (Courtois and Monaco, 1968; Ciavola et al., 1998), thereby limiting the conclusions that can be drawn from the data. However, for many studies the quantity of tracer that is utilised is dictated, pragmatically, by the financial budget of the project.

Many studies do not attempt to define a sediment budget because of the large uncertainties inherent in the quantification of sediment transport processes. Consideration is required of environmental and project restrictions, choice of tracer, the sampling and analysis methodology, and the desired precision. In general, the more dynamic the environment and the less controlled the study, the greater the quantity of tracer that is required, depending on the resolution of the measuring technique used (Liu et al., 2004). At this stage, the extent of the source area (e.g., where the tracer is introduced into the field), and a site boundary (sample limit) should be clearly defined.

4.3   Step 3: Matching the Tracer to the Native Sediment

Two hydraulically matched (equivalent) sediments will be cycled (eroded, transported, deposited) in the same way by a fluid flow (Dyer, 1986). On a practical level, matching the tracer hydraulic properties to the native sediment is straightforward; samples collected within the background survey and the tracer are tested for characteristics that influence transport, namely size distribution and density (and within silt tracing studies occasionally settling velocity), and the results statistically compared.

Commonly, the results of this similarity matching are expressed as a similarity ratio (e.g., $d_{50}(tracer)/d_{50}(native)$ for grain diameter). Tracing materials are designed to be unique within the environment; therefore, it is unlikely that a perfect hydraulic match will be achieved. Permissible differences between the hydraulic characteristics of the native sediment and tracer have been outlined in the literature. The median grain size of the tracer should be within ± 10 per cent of the native sediment (White and Inman,

1989), and the particle specific gravity should be within ±6 per cent (Black et al., 2007) to limit the effects of particulate-tracer differences on field observations.

Matching is particularly critical when tracking the finer sediment fractions (< 63 μm; Louisse et al., 1986). It is also more challenging given their cohesive nature (Brown et al., 1999), and the hydraulic matching process may be based on different rules to those of coarse sediments. Unlike sand and gravel particles, cohesive sediment is transported primarily as flocs (Droppo, 2001; Chapter 8). Thus, it is critical that any tracer that mimics cohesive sediment must be able to flocculate (on its own) and thereby resemble a natural floc aggregate (Spencer et al., 2011), which is an approach that can be called direct 'floc mimicking'. This approach is preferable because it accounts for the mineral sediments and the organic and inorganic floc constituents (Droppo, 2001). Unfortunately, however, tracers that are able to directly mimic a floc are not widely available, not least because within the direct floc-mimicking approach it is virtually impossible to establish that the tracer flocs behave in the same way as native flocs in terms of their temporal variations in aggregation and disaggregation (Louisse et al., 1986).

Significant theoretical and practical challenges remain in the development of very fine (< 4 μm) tracers that will work in practice. However, there is an alternative approach, which is termed 'floc tagging'. In this approach, the tracer particles must have similar physical and hydraulic characteristics to one or more of the constituent sediment size fractions found within the naturally flocculated material (i.e., size, density and settling rate). Tracking of the flocs is facilitated by directly labelling them. The key information is the particle size distribution of the native floc following acid digestion (e.g., Laubel et al., 2003). Because silts are by definition < 63 μm in size, a tracer of size, e.g., 25–55 μm would be suitable for the majority of projects. As an example, flocs of peat can be 'tagged' with an identifiable 'signature' that enables their transport to be determined (Figure 9-3b).

4.4   Step 4: Tracer Introduction

Tracer introduction methodologies are project and environment specific, but for all applications it is extremely important that the tracer particles are introduced into or onto the area of interest with minimal loss or redistribution. To ensure representative data are obtained, it is critical that the tracer is introduced over a depth extending to at least the base of the active transport layer (Inman and Chamberlain, 1959). There are two primary methods of tracer introduction; the foregoing introduction method (FIM), where tracer is introduced to the environment at one point in time, and the continuous introduction method (CIM), which involves continuously introducing tracer to a point at a steady rate and measuring tracer distributions downstream. Traditional beach-face studies (which all use the FIM method) require that the tracer is raked into the surface sediment layer, or introduced into a shallow trench (< 0.1-m deep). It is recommended that, where possible, tracer is introduced to the environment combined with native sediment in a 50:50 ratio to aid incorporation, particularly for cohesive sediment projects (Cronin et al., 2011; Spencer et al., 2011). The tracer should be mixed with a small amount of locally derived water prior to introduction onto the surface, or in a trench, to ensure that no redistribution of the tracer particles by aeolian transport occurs (e.g., Ciavola et al., 1998). Adding a small amount of detergent ($<$ 5 per cent) to the tracer–water admixture reduces the surface tension properties of the particles (Vila-Concejo et al., 2004), which generally is helpful to prevent tracer from being transported on the water surface as a result of surface tension.

Introducing tracer into deeper water is more challenging. Novel and ambitious methods have been utilised to deploy tracers to the bed. These include the use of water-soluble bags that dissolve in situ (Smith et al., 2007) and remotely operated chambers lowered from a ship (Courtois and Monaco, 1968; Ingle, 1966; White, 1987). Other specially designed or purpose-built devices have been utilised (Tudhorpe and Scoffin, 1987; Cheong et al., 1993; Collins et al., 1995; Van den Eynde, 2004); e.g., a weighted knife blade that rips open a bag of tracer on contact with the seabed, or a tube running from the ship to the sea floor (Joliffe, 1963; Crickmore, 1967). Ideally, in deep-water studies, the use of diver emplacement (SCUBA) is preferable due to the wealth of extra information provided by the divers. Where possible, in situ underwater videography or photography should be utilised to monitor and assess the success of the deployment (Cromey et al., 2002). It is recommended that coarse sand tracers should be utilised for shallow-water studies ($<$ 5–10 m), applied through a subsurface pipe, and dissolvable bags for deeper water studies ($>$10 m).

Silt tracers, applied to assess the entrainment and transport processes of deposited silts, present specific issues. It is essential to introduce the tracer onto the bed without loss of material to the water column. One way to introduce silt tracer to the seabed is via encapsulation of the tracer in ballasted ice because it provides a secure and robust delivery. This approach was pioneered by Krezoski (1989), who deployed a frozen pellet consisting of tracer and natural fine-grained sediment to the bed. Recent adaptations of this technique have included frozen plates, consisting of 50 per cent tracer mixed with 50 per cent natural sediment, deployed to the mouth of an estuary at low tide (Cronin et al., 2011) and premade, deep-frozen tracer blocks deployed from the side of a vessel (Black, 2012). Generally, the colder the blocks can be made the better they perform because this increases the time frames for encapsulation, enabling tracer introduction to greater depths.

Where studies are solely investigating suspended sediment transport there is often no requirement for the tracer to be deployed to the bed. The tracer can be used to simulate a floc or plume of suspended sediment directly by flushing tracer material in suspension down a subsurface tube. It is recommended that the tracer be premixed with saline or fresh water to create high-concentration tracer slurry that can be manoeuvred easily with trowels or shovels. High-flow water pumps, able to create a turbulent velocity field, are required to ensure complete disaggregation of the tracer particles within the slurry as it is discharged to the receiving water.

4.5   Step 5: Sampling

Sampling is an inherent and highly important part of the sediment tracing methodology. Recovering tracer for enumeration, or determining the presence of tracer in situ, requires spatial and temporal sampling. Due to the dynamic nature of sediment transport within all environments, dilution and dispersal of the tracer to beyond or below the detection limit can occur. Therefore, an adaptive sampling regime is desirable that considers sampling, relative to the tracer injection locality, in the entire near-field, mid-field and far-field, and incorporates a strategy for recurrent sampling through time. Poor sampling strategies can lead to flawed conclusions – the most common being that tracer is not present, whereas it might be that it has not been detected (Collins et al., 1995; Ferguson et al., 2002).

The two basic sampling methods are the spatial integration method (SIM), also referred to as Lagrangian sampling, where identified sample locations are sampled only once, and the temporal integration method (TIM), where identified sample locations are sampled repeatedly over time, via recurrent surveys. Predominately, the SIM is employed when sampling coarse sediments that have been transported as bedload (often associated with beach face, longshore transport types of studies), whereas TIM is most commonly used to sample sediments transported in suspension (Black et al., 2007).

A combination of random and systematic sampling techniques are recommended for tracing studies, which draw sampling units independently from each other with equal probability (Wang et al., 2012). Where no information is available regarding the direction of transport, practice dictates that a systematic sampling grid is used, which is usually arranged in two directions either side of the point of tracer injection. The samples are collected in a given order relative to the first collection point, requiring the use of systematic sampling zones. However, where transport direction is known a priori, the use of systematic grids has the potential to introduce bias (Webster, 1999), and a more complex approach, which uses both random and stratified sampling, may be more appropriate. In practice, the sampling grid layout and size are often dictated by landscape features, man-made structures, morphology, and budgetary and time constraints (Vila-Concejo et al., 2004). Because of this, and because the dispersion rates of tracer are unknown prior to deployment for the majority of studies, the collection of measurements at intermediate locations along the transport pathway, in addition to the target receptor, is also desirable (Ingle,1966; Inman et al., 1980; Ciavola et al., 1997). Increasing the sampling in terms of both spatial and temporal resolutions provides greater information regarding the distribution of tracer particles throughout the study area and improves any tracing study by providing a greater evidence base from which conclusions can be drawn. As a simple rule, as much information as possible should be sought within the resources of the study.

Tools commonly used to sample sediments in the marine, coastal and fluvial environments include sediment core and/or grab samplers (Chapter 6), sediment traps, water samplers and, recently, in situ magnetic sampling. In addition, sensor-based systems (e.g., field-deployable fluorometers) have found practical application to the in situ measurement of tracer material. By far the most popular tool within tracer studies is the sediment core collection itself. When collecting a core sample it is critical to sample through to the base of the active sediment layer to maximise the chance of tracer recovery following sediment mixing. These samples can then be used to determine the mass and volume of sediment transported, to investigate burial trends, and to calculate rates of sediment transport through the environment (Ciavola et al., 1998; Bertin et al., 2007).

If the thickness of the active transport layer is unknown, a visually identifiable tracer can be used as a horizon marker to ensure that the sample has been collected throughout it (King, 1951; Inman and Chamberlain, 1959; Komar, 1969; Inman et al., 1980); alternatively, where tracer content is known, Eq. 2 can be used (White and Inman, 1989):

$$K_0 = \frac{\Sigma N(z)\Delta_z(z)}{N_{\max}} \tag{2}$$

where K_0 is transport thickness; $\Delta_z(z)$ is the vertical thickness of a horizontal slice in the core; $N(z)$ is depth-dependent (z) tracer concentration, assumed in Eq. 2 to be horizontally uniform; and $N_{\max}$ is the maximum tracer concentration in the core. This transport thickness is only applicable if there is no increase in measured concentration with increasing depth (White, 1998).

Sampling strategies should strive to be flexible and adaptive to changes in environmental conditions that occur during the study. Consequently, a tracer that is able to be monitored effectively without using intrusive sampling techniques is advantageous. Semi-quantitative and qualitative sampling techniques, such as submersible fluorescence imaging (e.g., Solan et al., 2004) and night-time blue-lamp surveys (e.g., Ciavola et al., 1998; Vila-Concejo et al., 2004; Silva et al., 2007; Carrasco et al., 2013) are currently driving innovation within tracing studies. These techniques are best applied to monitor the spatial distribution of tracer particles in a nonintrusive manner, prior to using intrusive sampling techniques to quantify tracer content within the sampling grid.

4.6   Step 6: Tracer Enumeration

The tracer sampling programme is carried out to determine the tracer concentration (mass per unit volume) or dry mass of tracer for each sampling location and time step. The preferred enumeration method should be able to accurately analyse a large number of samples within a short period of time, which allows for many more samples to be collected from the field and analysed within a given resource budget – a facet that substantially improves many tracing studies (Guzman et al., 2013). Tracer enumeration techniques are wholly dependent on the type of tracer utilised. The benefits and limitations of the analysis techniques associated with each tracer illustrate the importance of taking a holistic view of the entire methodology at the study planning stage, in order to inform tracer material selection decisions. The methodologies used to enumerate the tracer mass or tracer content within environmental samples associated with two of the most popular tracing materials – fluorescent and magnetic tracers – are summarised in Appendix 2.

4.7   Step 7: Analysis

Tracer enumeration provides point-specific tracer mass or tracer concentration data at a single instant in time. The data should be generated in such a way as to provide an overview of the study at a series of time steps (e.g., at each sampling stage) and across the extent of the spatial area. Computer applications such as Geographical Information Systems (GIS) software (Esri, 2016), Surfer (Golden Software, 2016) and MATLAB (Mathworks, 2016) are useful tools with which to do this.

The direction of transport is determined via the quantity of tracer recovered from the sampling grid, which is best described as a percentage of the total tracer mass recovered, rather than the total tracer mass deployed (e.g., > 50 per cent of the tracer recovered was found to the south of a certain location). Because sediment transport is often multidirectional, the dominant transport pathway and receptor area are determined by the presence of the greatest quantity of tracer.

White (1998) gives an analysis of tracer dispersal that is based on the assumption that tracer properties measured at a sampled point (a core) are representative of those in a rectangular area surrounding the point, with the boundaries of the rectangle lying midway between sample points. Considering this simplest case of a rectangular sampling grid, such that each core, i, with cross-sectional area A_{ri}, lies within its representative rectangle, which has its centroid at (x_i, y_i), with sides Δx_i and Δy_i lying midway between adjacent cores, and an area $\Delta x_i \Delta y_i$, then the mass of tracer, M_i, within the representative rectangle can be estimated as (White, 1998):

$$M_i(t) = \frac{\Delta x_i \Delta y_i}{A_{ri}} \left(\sum_z N(x_i, y_i, z, t)/F \right) \tag{3}$$

where F is the number of tracer grains per unit mass of experimental tracer material and N, in Eq. 3, is the number of tracer grains in each vertical (z) slice of the ith sampled core, which is summed over all slices to give, once divided by F, the total mass of tracer in the core; multiplying this mass by the area of the representative grid area, and dividing by the core area, gives the estimated mass of tracer within the representative rectangular area, assuming spatial and temporal uniformity of tracer properties within it. Naturally, extrapolation away from the sampled point must be done with caution, and the distance between sample points must be appropriate.

The rate of tracer transport can be described using the rate of advance of the tracer front through the environment (Madsen, 1987) and by determination and tracking of the centre of mass of the tracer distribution (e.g., White and Inman, 1989; Vila-Concejo et al., 2004; Carrasco et al., 2013). In this latter approach the average transport velocity is calculated from the distance moved by the centre of mass of the tracer divided by the time between injection and sampling (White, 1998). Assuming that all collected samples have incorporated the whole active transport layer, and following enumeration of the tracer content from each sampling point, the location (X, Y) of the centre of mass of the tracer distribution can be determined from:

$$X(t) = \sum_i M_i(t)x_i / \sum_i M_i(t) \tag{4}$$

$$Y(t) = \sum_i M_i(t)y_i / \sum_i M_i(t) \tag{5}$$

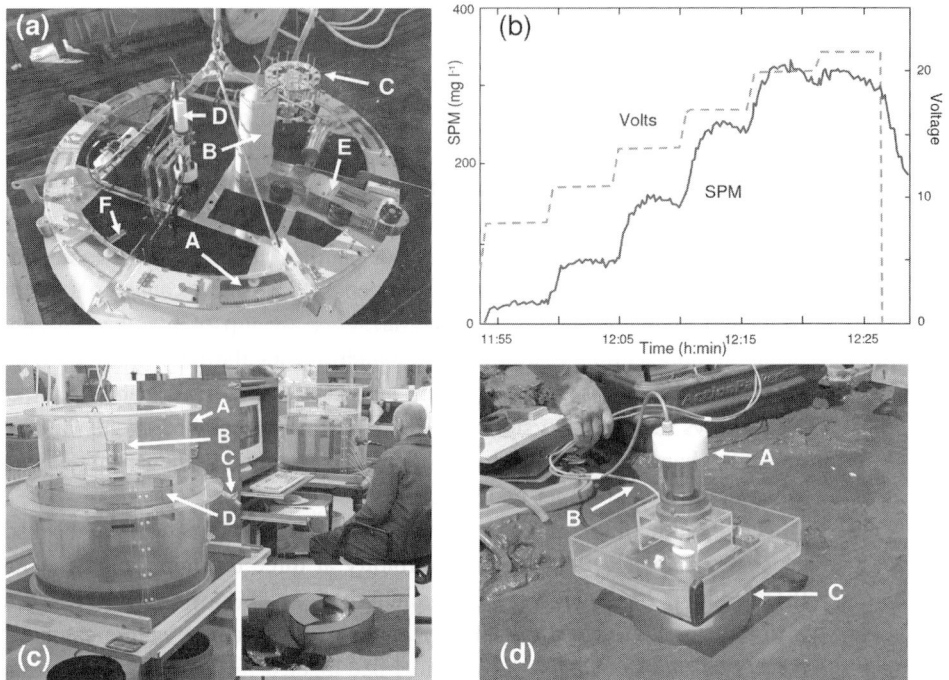

Figure 9-4 An illustration of benthic flumes for sediment erosion studies; (a) this shows a more recent (2016) version of the benthic flume operated by Thompson et al., (2011). The structure is 2.2 m in total diameter. Annotations 'A' to 'F' illustrate: 'A' – channel train drive (connected to 8 paddles); 'B' – submersible motor; 'C' – 12 port water sampler; 'D' – water quality data logger; 'E' – chain drive; 'F' – dissolved oxygen sensor. The flow sensor and three turbidity sensors are not visible; (b) an example of benthic flume erosion time series data, showing the applied motor voltage (dashed line – a proxy for lid rotation rate and thus, with calibration, flow velocity and bed stress) and total suspended particulate matter (SPM) concentration (mg l^{-1}, solid line). Successive periods of bed erosion are evident, as is sediment deposition following the switching off of the flume motor shortly after 12:25; (c) two 0.6-m-diameter PML benthic flumes, used for intertidal sediment work (described by Widdows et al., 1998, and Pope et al., 2006), during erosion experiments in the laboratory (the inset shows the sediment coring cases used for field-sample collection of intertidal sediment when the flume is not used in the field); annotations 'A' to 'D' illustrate: 'A' – rotating drive cylinder; 'B' – drive shaft and motor; 'C' – sampling ports; 'D' – paddle, one of four; ADV and OBS sensors are not shown; (d) a handheld 0.2-m-diameter PML mini-flume used for in situ muddy intertidal erosion studies (Bale et al., 2006); annotations 'A' to 'C' illustrate: 'A' – motor; 'B' – electrical lead to OBS sensor; 'C' – paddles, located at top of cylinder beneath 'tray'. Photographs: (a) G. Fones, Portsmouth University, with permission; (c) R. J. Uncles; (d) J. A. Stephens, Plymouth Marine Laboratory, with permission.

rate in a stepwise fashion via a series of acceleration ramps, and this is designed to sequentially suspend sediment and erode the bed (Figure 9-4b). Following completion of the experiment, the paddle rotation can either be stopped immediately, ramped down in a single step, or decreased in a stepwise fashion to zero (during this time period sediment redeposition will occur). The flume is then recovered.

The flume provides quantitative information on important bed sediment transport metrics. At some moment, as the paddle rotation rate is sequentially increased in response to an increased motor voltage, the flow-induced frictional drag (stress τ) is sufficient to induce sediment resuspension (Figure 9-4b), and this point is the critical erosion stress, denoted by τ_{crit} (units: Pa, i.e., N m^{-2}). Mechanistically, the vertical mass flux of sediment, E (kg m^{-2} s^{-1}), deduced from the time-evolution of suspended sediment concentration for normally consolidated estuary muds under an applied fluid bed stress, is eventually balanced by an increase in bed shear strength (τ_b) with depth (z) as erosion takes place, Eq. 6, and erosion then stops:

$$\tau = \tau_{crit} = \tau_b(z) \tag{6}$$

Further increases in flow velocity sequentially erode the bed vertically downward, and from flume deployments in which there are generally five or more discrete time steps it is possible to derive the relationship between flow stress, τ, and erosion rate, E.

Cohesive sediments are notoriously site specific, as well as being temporally variable (particularly on estuarine intertidal areas), and as a result, various expressions linking E to τ are found in the literature; e.g., Amos et al. (2004) provide a useful overview, including both linear and nonlinear relationships between E and τ, and Andersen (2001) measured erosion rates on a muddy, microtidal mudflat in Denmark and found a nonlinear relationship between excess bed shear stress ($\tau - \tau_{crit}$) and erosion rate. For example, Mehta and Parchure (2002) discuss relationships of the form:

$$E = E_M \left(\frac{\tau - \tau_{crit}}{\tau_{crit}} \right)^\delta \tag{7}$$

Eq. 7 is applicable when $\tau > \tau_{crit}$, where E_M is the erosion-rate 'constant'. Values of δ given by Mehta and Parchure (2002) range between 0.95 and 1.82.

Because the flume vertically erodes the sediment bed during erosion tests, several additional metrics are derivable from the erosion time series (e.g., Amos, 2004). The depth increment due to bed erosion (Δz) for each time-step can be computed from Eq. 8 using the dry bulk density of the bed (ρ_d), the incremental increase in the dry mass of material suspended in the water column (ΔM), and the flume bed area (A):

$$\Delta z = \Delta M / \rho_d A \tag{8}$$

In addition, the rate of change in bed strength (τ_{crit}) with depth, given by the best-fit regression line of τ_{crit} versus z from erosion-study datasets, offers a practical means of computing erosion depth for an applied bed stress and provides a more complete understanding of sediment stability than is provided from simply the critical erosion stress and erosion rate parameters.

Some understanding of the internal flow regime and associated sediment dynamics of in situ flumes is useful to understand their limitations and to aid data interpretation. Annular flumes offer constant channel geometry and an infinite flow length that result in a fully developed benthic boundary, which is an essential prerequisite for the application of flume-derived bed erosion studies to the natural environment (Amos et al.,

1992). Nonetheless, it is a recirculating, confined flow, and the internal flow patterns reflect this. Because the flow within the annulus is generated by the rotation of paddles, it is anticipated that these will impart additional turbulence to the flow that may affect bed sediment resuspension. Detailed analysis of high frequency (10.66 Hz) flow data by Amos et al. (1992) reveals a −5/3 power-law dependence in the energy spectra, indicating orderly structure in the macroturbulence of the annulus currents; there is evidence of a frequency-dependent, paddle-induced turbulence, but for mean flows less than approx. 0.6 m s^{-1} the energy of this turbulence is small compared with the energy of the spectrum as a whole. The flow becomes extremely chaotic if run at high paddle rotation rates, whereas at very high flow rates (> 0.8 m s^{-1}) cavitation can occur and the relationship between flow and bed erosion breaks down.

Another hydrodynamic issue is that of secondary circulations, which arise due to the shape of the channel (e.g., Burt and Game, 1985). In plan, the flow profile across the channel broadly resembles a flattened parabola, with wall effects limited to approximately 10 per cent of the channel section. Secondary circulations, which give rise to increasing bed stress values toward the outer wall, are known to occur, but the crucial observation made by Amos et al. (1992) is that they are mean-flow dependent, and only become of any significance when the mean flow is >0.32 m s^{-1}. The flume is rarely operated at flow velocities above this when eroding cohesive estuary mud, so that the impact of secondary flows on bed erosion is usually insignificant. Benthic flumes can, in principle, be used to assess bedload transport on coarser sediments by tracking ripple migration rates through the window, but the operator would need to be cognisant of the secondary flow manifestations and reductions in migration rate as a result of wall effects.

Finally, some consideration must be given to the fact that as erosion proceeds within the annulus, the water becomes increasingly loaded with suspended sediments, and this is known to cause a decrease in the stress at the bed ('drag reduction'; Dyer, 1986). Benthic flumes are initially calibrated to relate motor voltage or paddle rotation rate via flow velocity to bed stress. Eq. 9, for example, has been applied to the flume described by Thompson et al. (2011):

$$\bar{u}_* = 0.0167 + 0.097\bar{u} \tag{9}$$

The time-averaged shear velocity in Eq. 9 is $\bar{u}_*$, and the time-averaged velocity is $\bar{u}$ (both in units of m s^{-1}), which are derived from instantaneous flow-velocity measurements. The corresponding value for bed shear stress is derived using the expression:

$$\tau = \rho \cdot \bar{u}_*^2 \tag{10}$$

where ρ denotes fluid density.

It is necessary to correct Eq. 10 for the evolution of high suspended sediment concentrations during erosion runs because the shear velocity calibration is traditionally derived from clear water studies, which cannot take into account the 'drag reduction' effect. The change in bed stress is a function both of the sediment concentration and the bed stress imposed by paddle rotation itself (i.e., nonlinear behaviour). Eq. 11 has been used to revise the flume shear velocity (and subsequent bed stress estimates) in the presence of suspended sediment by Thompson et al., 2011 (concentration P in units of mg l^{-1}):

$$\bar{u}_{*P} = \bar{u}_* (1 - 0.0357\log_{10}P) \tag{11}$$

Numerous case studies use benthic flume technology to determine the stability of fine-grained sediments in estuarine environments; examples include those from the subarctic (Amos et al., 1996), the eastern seaboard of the USA (Maa and Lee, 1997) and the Venice Lagoon (Amos et al., 2004). The UK Land Ocean Interaction Study (LOIS) Littoral Investigation of Sediment Properties (LISP) study in the Humber Estuary during the years 1994–1997 extensively employed benthic flumes to document spatial and temporal patterns in the erodibility of north shore intertidal mud (see Black et al., 1998, and the papers therein); both large and small (of the order of 0.5-m diameter) benthic flume devices were deployed across the broad intertidal areas, including the PML (Plymouth Marine Laboratory) flume, which has been used for this and many other estuarine studies (e.g., Pope et al., 2006), both in situ and for erosion experiments in the laboratory (Figure 9-4c). The PML benthic mini-flume (Bale et al., 2006) has been used in situ to investigate mudbank stability in the upper Tamar Estuary, UK (Figure 9-4d).

These and other flume studies have illustrated that the erosion of estuarine muds conforms to previously found general patterns of bed erosion in laboratory studies, but revealed the field situation to be far more complex. The flume studies, in particular, revealed the relative importance of benthic microalgae and macrofauna as, respectively, agents of biostabilisation and biodestabilisation (Chapter 1), which were emerging topics in estuarine sediment transport research at that time.

Benthic flume technology has also been used in deeper settings and on the continental shelf. Black and Black (2014) report the use of a large and smaller flume system to measure the entrainment and deposition parameters of farmed fish faeces in water depths up to 40-m off the west coast of Scotland. Thompson et al. (2011) summarise results from the deployment of a large flume device in water depths up to 83 m in the central and southern North Sea, quite possibly the deepest successful deployment of flume technology to date.

5.2   Fine-Scale Deposition and Erosion

Seabed bathymetric elevations due to the processes of sediment resuspension and sediment deposition change on a daily basis in most coastal and estuarine settings (e.g., Lawler et al. 2001). More pronounced changes occur when wave energy can penetrate to the seabed during winter storms. Changes in seabed level greater than approx. 0.10–0.15 m can be detected using conventional single- and multibeam echo sounders (Chapter 3). From a navigational safety viewpoint, it is only changes at scales greater than this that are of importance. However, smaller-scale changes (mm–cm) are important to, e.g., benthic habitat stability, contaminated sediment impacts, dredging impact evaluation (e.g., siltation onto sensitive benthic communities) and benthic ecosystem function. Within these examples the transfer of sediments on the small scale can have considerable ecological and environmental impacts. There are also undoubted benefits from using the technology for sea-level rise studies, in which it is found to be particularly difficult to substantiate the accretion commonly associated with rising sea levels on prograding (growing seawards) coasts and infilling estuaries.

High-resolution recording altimeters are marine sonars capable of measuring small-scale (mm–cm) changes in sediment levels in coastal and estuarine environments, and whilst several manufacturers sell commercial varieties (e.g., the ALTUS system made by NKE, France – NKE, 2016; and the series of precision echosounders made by EchoLogger, 2016), these instruments have found only surprisingly limited use in estuarine research (but see Deloffre et al., 2007). The research of Bassoullet et al. (2000) on intertidal mudflat dynamics in the Baie de Marennes-Oléron illustrates well the utility of these sensors for understanding sediment dynamics; they used the NKE system, which measures flow depth, wave height and bed level. In tandem with flow and turbidity sensors, the variability of bed level was monitored directly at high resolution and continuously through time, enabling the relationships between the hydrodynamic drivers and the morphological response of the bed to be established. Of special note, the Altus system has a high (2 MHz) frequency, which also allows for fluid mud processes to be investigated.

Estuarine monitoring projects, where there is a need to monitor marginal accretion–erosion cycles and net morphological change, could usefully deploy a network of surveyed-in sensors to capture the spatial variability. No systems are available at present that can transmit the data in real time, but telemetry is no longer a technological impediment, and perhaps user demand could drive such a modification to present-day systems.

6   Concluding Remarks

This chapter has considered several advances in technology that have greatly increased the capability of scientists and engineers to measure sediment transport processes in estuaries and the coastal zone. Foremost amongst these is the development of acoustics-based instrumentation to determine and data-log velocities, including turbulent fluctuations, sediment concentrations and sediment fluxes with great temporal resolution and spatial resolution.

Another approach – sediment or particle tracking – provides a powerful tool for sediment transport studies, particularly when used in tandem with more traditional approaches, such as flow determination. Historically, fluorescent tracers have dominated these studies, although recently a dual signature tracer has been introduced that combines a fluorescent colour signature with a ferromagnetic signature. These tracers provide an advance on previously used 'mono' signature tracers; they have enabled unique monitoring, recovery and enumeration methodologies to be developed that reduce analytical timescales and associated costs.

Finally, the development of portable, in situ annular flumes and highly accurate altimeters has led to new and increasing insights into the interaction between flow and sediments and biology, and flow and sediments and morphology, as well as providing quantitative data with which to construct sediment-erosion relationships for computer models of sediment transport.

Acknowledgements

RJU thanks Professor Steve de Mora, Chief Executive of the Plymouth Marine Laboratory, for the award of a Senior Research Fellowship. We thank Dr A. Armstrong and Profs. J. N. Quinton and B. A. Maher of Lancaster University, UK, for their encouragement and support of this work.

References

Amos, C. L., Bergamasco, A., Umgiesser, G., Cappucci, S., Cloutier, D., Flindt, M., Denat, L., Cristante, S., 2004. The stability of tidal flats in Venice lagoon – the results of in situ measurements using two benthic flumes. *Journal of Marine Systems* 51, 211–242.

Amos, C. L., Grant, J., Daborn, G. R., Black, K., 1992. Sea Carousel – A benthic annular flume. *Estuarine, Coastal and Shelf Science* 34, 557–577.

Amos, C. L., Sutherland, T. F., Zevenhuizen, J., 1996. The stability of sublittoral, fine-grained sediments in a subarctic estuary. *Sedimentology* 43, 1–19.

Andersen, T. J., 2001. The role of fecal pellets in sediment settling at an intertidal mudflat, the Danish Wadden Sea. In: McAnally, W. H., Mehta, A. J. (eds.), *Coastal and Estuarine Fine Sediment Processes*. Amsterdam: Elsevier, 387–401.

Angarita-Jaimes, D., Ormsby, M., Chennaoui, M., Angarita-Jaimes, N., Towers, C., Jones, A., Towers, D., 2008. Optically efficient fluorescent tracers for multi-constituent PIV. *Experiments in Fluids* 45, 623–631.

Ashley, G. M., 1990. Classification of large-scale subaqueous bedforms: A new look at an old problem. *Journal of Sedimentary Petrology* 60, 160–172.

Bale, A. J., Widdows, J., Harris, C. B., Stephens, J. A., 2006. Measurements of the critical erosion threshold of surface sediments along the Tamar Estuary using a mini-annular flume. *Continental Shelf Research* 26, 1206–1216. doi:10.1016/j.csr.2006.04.003.

Bassoullet P., Le Hir, P., Gouleau, D., Robert, S., 2000. Sediment transport over an intertidal mudflat: Field investigations and estimation of fluxes within the 'Baie de Marennes-Oléron (France). *Continental Shelf Research* 20, 1635–1653.

Bertin, X., Deshouilieres, A., Allard, J., Chaumillon, E., 2007. A new fluorescent tracers experiment improves understanding of sediment dynamics along the Arcay Sandspit (France). *Geo-Marine Letters* 27, 63–69.

Black, K. S., 1989. *The In Situ Measurement of Sediment Erodibility: A review*. Unpubl. report submitted to ETSU, Department of Energy, Harwell, UK.

Black, K. S., 2012. *Using Sediment Tracers to Map Sediment Transport Pathways. A Primer Document*. Glasgow, UK: Partrac Ltd.

Black, K. S., Athey, S., Wilson, P., Evans, D., 2007. The use of particle tracking in sediment transport studies: A review. *Geological Society of London Special pubs.* 274, 73–91.

Black, K. S., Black, K. D., 2014 *Survey Report: Benthic Flume and Flow Velocity Measurements*. Internal report to the Scottish Government DEPOMOD project initiative.

Coastal Eng., ASCE, Sydney. Vol. 2, New York: American Society of Civil Engineers, 1215–1234.

Joliffe, I. P., 1963. A study of sand movement on the Lowestoft sandbank using fluorescent tracers. *Geographical Journal* 129, 480–493.

Kimoto, A., Nearing, M. A., Shipitalo, M. J., Polyakov, V. O., 2006. Multi-year tracking of sediment sources in a small agricultural watershed using rare earth elements. *Earth Surface Processes and Landforms* 31, 1763–1774.

King, C., 1951. Depth of disturbance of sand on sea beaches by waves. *Journal of Sedimentary Petrology* 21, 121–140.

Kjerfve, B., 1979. Measurement and analysis of water current, temperature, salinity, and density. In: Dyer, K. R. (ed.), *Estuarine Hydrography and Sedimentation*. Cambridge, UK: Cambridge University Press, 186–226.

Komar, P. D., 1969. *The Longshore Transport of Sand on Beaches*. Unpublished PhD thesis, University of California, San Diego, 142 pp.

Krezoski, J. R., 1989. Sediment reworking and transport in eastern Lake Superior: In situ rare earth element tracer studies. *Great Lakes Research* 15, 26–33.

Laubel, A., Kronvang, B., Fjorback, C., Larsen, S. E., 2003. Time-integrated sediment sampling from a small lowland stream. In: *Proceedings of the International Association of Theoretical and Applied Limnology* 28, 1420–1424.

Lawler, D. M., West, J. R., Couperthwaite, J. S., Mitchell, S. B., 2001. Application of a novel automatic erosion and deposition monitoring system at a channel bank site on the tidal River Trent, U.K. *Estuarine, Coastal and Shelf Science* 53, 237–247.

Leica, 2016. Leica Geosystems Ltd, Hexagon House, Michigan Drive, Tongwell, Milton Keynes MK15 8HT, UK. www.leica-geosystems.co.uk/en/Total-Stations-TPS_4207.htm [accessed June 2016].

Liu, P. L., Tian, J. L., Zhou, P. H., Yang, M. Y., Shi, H., 2004. Stable rare earth element tracers to evaluate soil erosion. *Soil and Tillage Research* 76, 147–155.

Louisse, C. J., Akkerman, R. J., Suylen, J. M., 1986. A fluorescent tracer for cohesive sediment. In: *International Conference on Measuring Techniques of Hydraulics Phenomena in Offshore, Coastal and Inland Waters*, London, 9–11 April 1986, 367–391.

Maa, J.P.-Y., Lee, C.-H., 1997. Variation of the resuspension coefficients in the lower Chesapeake Bay. *Journal of Coastal Research* 25, 63–74.

Madsen, O. S., 1987. Use of tracers in sediment transport studies. *Coastal Sediments* 87, 424–435.

Magal, E., Weisbrod, N., Yakirevich, A., Yechieli, Y., 2008. The use of fluorescent dyes as tracers in highly saline groundwater. *Journal of Hydrology* 358, 124–133.

Mariotti, G., Fagherazzi, S., 2011. Asymmetric fluxes of water and sediments in a mesotidal mudflat channel. *Continental Shelf Research* 31, 23–36. doi:10.1016/j.csr.2010.10.014.

Marsh, J. K., Bale, A. J., Uncles, R. J., Dyer, K. R., 1991. A novel technique for the study of suspended particle behaviour in aquatic environments. In: *Proc. Int. Conf. on Transport of Suspended Sediments and Its Mathematical Modelling*, Florence, Italy: Publ. Int. Assoc. for Hydraulic Research, 665–681.

Marsh, J. K., Bale, A. J., Uncles, R. J., Dyer, K. R., 1993. A particle tracing experiment in a small, shallow lake: Loe Pool, UK. In: McManus, J., Duck, R. W. (eds.), *Geomorphology and Sedimentology of Lakes and Reservoirs*. Chichester: John Wiley and Sons Ltd., 139–153.

Masselink, G., Cointre, L., Williams, J. J., Blake, W., Gehrels, R. W., 2009. Tide-induced dune migration and sediment transport on an intertidal shoal in a shallow estuary in Devon, UK. *Marine Geology* 262, 82–95.

Mathworks, 2016. The Mathworks. 1 Apple Hill Drive, Natick, MA 01760–2098, USA. www.mathworks.com [accessed July 2016].

McLaren, P., Hill, S. H., Bowles, D., 2007. Deriving transport pathways in a sediment trend analysis (STA). *Sedimentary Geology* 202, 482–498.

Mehta, A. J., Parchure, T. M., 2002. Surface erosion of fine-grained sediment revisited. In: Winterwerp, J. C., Kranenburg, C. (eds.), *Fine Sediment Dynamics in the Marine Environment. Proceedings in Marine Science* 5, 55–74.

Mehta, A. J., Partheniades, E., 1982. Resuspension of deposited cohesive sediment beds. In: Flemming, B. W., Delafontaine, M. T., Liebezeit, G. (eds.), *Muddy Coast Dynamics and Resource Management, 18th Conference on Coastal Engineering*, ASCE, 1569–1588.

Minor Salvatierra, M., Aliotta, S., Ginsberg, S. S., 2015. Morphology and dynamics of large subtidal dunes in Bahia Blanca Estuary, Argentina. *Geomorphology* 246, 168–177. http://dx .doi.org/10.1016/j.geomorph.2015.05.037.

Nauw, J. J., Merckelbach, L. M., Ridderinkhof, H., van Aken, H. M., 2014. Long-term ferry-based observations of the suspended sediment fluxes through the Marsdiep inlet using acoustic Doppler current profilers. *Journal of Sea Research* 87, 17–29. doi:10.1016/j .seares.2013.11.013.

NKE, 2016. 6 rue Gutenberg, ZI Kerandre, 56 700 Hennebont, France, www.nke-instrumentation .com [accessed July 2016].

Nowell, A. R. M., Jumars, P. A., Eckman, J. E., 1981. Effects of biological activity on the entrainment of marine sediments. *Marine Geology* 42, 133–153.

Pedocchi, F., Martin, J., Garcia, M. H., 2008. Inexpensive fluorescent particles for large-scale experiments using particle image velocimetry. *Experiments in Fluids* 45, 183–186.

Peirce, Y. J., Jarman, R. T., de Turville, C. M., 1970. An experimental study of silt scouring. *Proceedings of the Institution of Civil Engineers* 45, 231–243. http://dx.doi.org/10.1680/ iicep.1970.7155.

Perillo, G. M. E., Piccolo, M. C., 1998. Importance of grid-cell area in the estimation of estuarine residual fluxes. *Estuaries* 21, 14–28. doi:10.2307/1352544.

Polyakov, V. O., Nearing, M. A., 2004. Rare earth element oxides for tracing sediment movement. *Catena* 55, 255–276.

Pope, N. D., Widdows, J., Brinsley, M. D., 2006. Estimation of bed shear stress using the turbulent kinetic energy approach – A comparison of annular flume and field data. *Continental Shelf Research* 26, 959–970. doi:10.1016/j.csr.2006.02.010.

Richardson, N. M., 1902. An experiment on the movements of a load of brickbats deposited on Chesil Beach. *Proceedings of the Dorset Natural History Field Club* 23, 123–133.

Sarma, T. P., Iya, K. K., 1960. Preparation of artificial silt for tracer studies near Bombay Harbour. *Journal Scientific Ind. Research India* 19, 99–101.

Schettini, C. A. F., Pereira, M. D., Siegle, E., Bruner de Miranda, L., Silva, M. P., 2013. Residual fluxes of suspended sediment in a tidally dominated tropical estuary. *Continental Shelf Research* 70, 27–35. doi:10.1016/j.csr.2013.03.00.

Scoffin, T. P., 1968. An underwater flume. *Journal of Sedimentary Petrology* 38, 244–246.

Sigurbjornsson, B., 1994. Use of Nuclear techniques in Food, Agriculture and Pest Control. Lecture 8 of the National Seminar on Nuclear Energy in Everyday Life, 28–29

June 1994, Cairo, Egypt. www.iaea.org/inis/collection/NCLCollectionStore/_Public/26/066/26066429.pdf?r=1 [accessed July 2016].

Silva, A., Taborda, R., Rodrigues, A., Duarte, J., Cascalho, J., 2007. Longshore drift estimation using fluorescent tracers: New insights from an experiment at Comporta Beach, Portugal. *Marine Geology* 240, 137–150.

Smith, S. J., Marsh, J., Puckette, T., 2007. Analysis of fluorescent sediment tracer for evaluating nearshore placement of dredged material. In: *Proceedings of the XVIII World Dredging Congress* (von Newman Printing, Bryan), 2, 1345–1358.

Solan, M., Wigham, B. D., Hudson, I. R., Coulon, C. H., Norling, K., Nilsson, H. C., Rosenberg, R., 2004. In situ quantification of bioturbation using time-lapse fluorescent sediment profile imaging (f-SPI), luminophore tracers and model simulation. *Marine Ecology Progress Series* 271, 1–12.

Spencer, K. L., Suzuki, K., Hillier, S., 2011. The development of rare earth element-labelled potassium-depleted clays for use as cohesive sediment tracers in aquatic environments. *Journal of Soils and Sediments* 11, 1052–1061.

Tanaka, S., Yamamoto, K., Ito, H., Arisawa, T., Tagaki, T., 1998. Field investigation on sediment transport into the submarine canyon in the Fuji coast with new type tracers. *Coastal Engineering* 20, 3151–3164.

Tauro, F., Aureli, M., Porfiri, M., Grimaldi, S., 2010. Characterization of buoyant fluorescent particles for field observations of water flows. *Sensors* 10, 11512–11529. doi:10.3390/s101211512.

Thompson, C., Couceiro, F., Fones, G. R., Helsby, R., Amos, C. L., Black, K., Parker, E. R., Greenwood, N., Statham, P. J., Kelly-Gerreyn, B., 2011. In situ flume measurements of resuspension in the North Sea. *Estuarine, Coastal and Shelf Science* 94, 77–88.

Thompson, R. C., Olsen, Y., Mitchell, R. P., Davis, A., Rowland, S. J., John, A. W. G., McGonigle, D., Russel, A. E., 2004. Lost at sea: where is all the plastic? *Science* 304, 838. doi:10.1126/science.1094559.

Tudhorpe, A. W., Scoffin, T. P., 1987. A device to deposit tracer sediment evenly on the deep sea bed. *Journal of Sedimentary Petrology* 57, 761–762.

Uncles, R. J., Elliott, R. C. A., Weston, S. A., 1985a. Dispersion of salt and suspended sediment in a partly mixed estuary. *Estuaries* 8, 256–269. doi:10.2307/1351486.

Uncles, R. J., Elliott, R. C. A., Weston, S. A., 1985b. Observed fluxes of water, salt and suspended sediment in a partly mixed estuary. *Estuarine, Coastal and Shelf Science* 20, 147–167. doi:10.1016/0272-7714(85)90035-6.

Uncles, R. J., Stephens, J. A., Harris, C., 2013. Towards predicting the influence of freshwater abstractions on the hydrodynamics and sediment transport of a small, strongly tidal estuary: The Devonshire Avon. *Ocean & Coastal Management* 79, 83–96.

Van den Eynde, D., 2004. Interpretation of tracer experiments with fine-grained dredging material at the Belgian Continental Shelf by the use of numerical models. *Journal of Marine Systems* 48, 171–189.

Van Rijn, L. C., 1993. *Principles of Sediment Transport in Rivers, Estuaries and Coastal Seas.* Amsterdam: Aqua Publishers.

Ventura, E., Nearing, M. A., Amore, E., Norton, L. D., 2002. The study of detachment and deposition on a hillslope using a magnetic tracer. *Catena* 48, 149–161.

Vernon, J., 1963. *Fluorescent Sand Tracer Tests, Zuniga Shoal, San Diego, California.* Los Angeles: University of Southern California.

Vila-Concejo, A., Ferreira, O., Ciavola, P., Matias, A., Dias, J., 2004. Tracer studies on the updrift margin of a complex inlet system. *Marine Geology* 208, 43–72.

Wang, J. F., Stein, A., Gao, B. B., Ge, Y., 2012. A review of spatial sampling. *Spatial Statistics* 2, 1–14.

Webster, R., 1999. Sampling, estimating and understanding soil pollution. In: Gomez-Hernández, J., Soares, A., Froidevaux, R. (eds.), *Geostatistics for Environmental Applications. Quantitative Geology and Geostatistics*. Dordrecht: Kluwer Academic Publishers, 155–166.

White, T. E., 1987. *Nearshore Sand Transport*. Unpublished PhD Thesis. University of California, San Diego.

White, T. E., 1998. Status of measurement techniques for coastal sediment transport. *Coastal Engineering* 35, 17–45.

White, T. E., Inman, D. L., 1989. Measuring longshore transport with tracers. In: Seymour, R. J. (ed.), *Nearshore Sediment Transport*. New York: Plenum Publishers, 287–312.

Widdows, J., Brinsley, M. D., Bowley, N., Barrett, C., 1998. A benthic annular flume for *in situ* measurement of suspension feeding/biodeposition rates and erosion potential of intertidal cohesive sediments. *Estuarine, Coastal and Shelf Science* 46, 27–38. doi:10.1006/ecss.1997.0259.

Wu, J., Wang, Y., Cheng, H., 2009. Bedforms and bed material transport pathways in the Changjiang (Yangtze) Estuary. *Geomorphology* 104, 175–184. doi:10.1016/j.geomorph.2008.08.011.

Young, R. A., 1977. Seaflume: A device for in situ studies of threshold erosion velocity and erosional behaviour of undisturbed marine muds. *Marine Geology* 23, M11–M18.

Zhang, X. C., Nearing, M. A., Polyakov, V. O., Friedrich, J. M. 2003. Using rare earth oxide tracers for studying soil erosion dynamics. *Soil Science Society of America Journal* 67, 279–288.

Appendix 1

Some of the recent technological and methodological developments that have prompted a resurgence of interest in the sediment tracing technique are briefly described here.

1 Technological Developments and Their Impact

1.1 Tracer Design

1.1.1 Synthetic Tracers

The development of commercially available synthetic tracers has increased the application of the sediment tracing technique. Synthetic tracers provide a low-cost sediment tracer.

1.1.2 Improved Labelling Techniques

Labelling or tagging natural sediment with an identifiable signature has led to the development of tracers that better replicate the hydraulic properties of the native sediment load, remain stable through time and do not interfere with the transporting system. This has increased the validity of the sediment tracing technique.

1.2 Sampling Advancements

1.2.1 Passive Sampling Techniques

Passive sampling techniques are driving innovation within sediment tracing studies; e.g., the use of in situ fluorometers, imaging techniques and magnetic susceptibility. These techniques provide information regarding the spatial and temporal distributions of tracer particles without direct sampling of the sediment or soil, which reduces the interference with the system and the requirement for resource-intensive, direct sampling and sample analysis.

1.3 Analytical Advancements

1.3.1 Novel Tracer Enumeration Techniques

Tracing studies have regularly used resource-intensive, sample-analysis techniques. These techniques restrict sampling frequency in the field, due to the analytical time scales and/or the related cost of sample analysis. Recent techniques have reduced the resource-intensive nature of sample-analysis techniques.

Appendix 2

Some of the tracer-enumeration methodologies available for the most popular tracing materials, e.g., fluorescent tracers, magnetic tracers and rare earth element tracers, are described here.

2 Tracer Material and Enumeration Techniques

2.1 Enumeration of Fluorescent Tracers

2.1.1 Counting by Eye

The spectral characteristics of fluorescent tracers enable unequivocal identification of the tracer particles within environmental samples using a blue light source or ultraviolet lamps (UV, with an emission wavelength of 395 nm). The sample is thinly spread and the tracer particles counted. This technique is not viable when using silt tracers.

2.1.2 Filtration/Sedimentation Followed by Digital Image Analysis

A similar methodology to that described earlier, except an image of the sample is captured under UV illumination and a digital-image analysis system is then used to count the fluorescent particles within the image.

2.1.3 Fluorescent Coating Dissolution, Centrifugation and Spectrofluorometry

Measuring the fluorescence intensity of an environmental sample via spectrofluorometric analysis allows the tracer content within an environmental sample to be determined and enables the determination of tracer content when using both silt- and sand-sized tracers.

2.1.4 Flow Cytometry

The fluorescence intensity is measured at the interrogation point of a flow cytometer. This technique enables silt-sized particles to be analysed yet is unable to analyse sand-sized particles. In addition, the sample size that can be analysed is restricted, which requires subsampling.

2.2   Enumeration of Magnetic Tracers

2.2.1   Magnetic Separation

Magnetic tracers have the advantage that an entire sample can be flushed through a magnetic separator (Frantz Isodynamic Separator; Franz, 2016) to recover tracer by separating the nonmagnetic and magnetic fractions from the sample. Use of all sample material obviates any necessity for subsampling, which avoids subsampling errors and greatly increases the probability of finding tracer.

10 The Use of Autonomous Sampling Platforms with Particular Reference to Moored Data Buoys

J. R. Fishwick, J. D. Turton

1 Introduction

In 1971 the UK's Natural Environmental Research Council (NERC) identified the likely need for data buoy technology in the UK; to this end they commissioned a feasibility and design study for a prototype buoy (Rusby, 1976). This effort culminated in the deployment of a trial system off the Suffolk coast in November 1975. During October 1987 the UK was hit by a major storm, with wind speeds in excess of 100 knots ($>$ 50 m s^{-1}) being recorded along the south coast (Burt and Mansfield, 1988). It was estimated that the damage caused by the storm was in the region of £2 billion (GBP), and questions were asked regarding the ability of the UK to predict such events. In a response to the need for observations from the northeast Atlantic to enable better storm predictions, the UK Meteorological Office (Met Office) embarked on their Marine Automatic Weather Station (MAWS) network on Ocean Data Acquisition System (ODAS) buoys, lightships and several remote islands. Although the network existed prior to the Great Storm of 1987, it was primarily developed and extended after this time. The MAWS network monitors meteorological conditions as well as sea state information along the UK continental shelf of the north-eastern Atlantic, extending south into the Bay of Biscay. The information acquired from this network enables the Met Office to better predict the weather and sea conditions affecting the UK.

Further to the ODAS network and in response to a report in 1996 by the Fishing Vessels Safety Authority, in which it was highlighted that better weather forecasts would decrease the number of lives lost at sea, the Marine Institute in Ireland invested in a buoy network. The first buoy was deployed in 2000, and the Irish Marine Weather Buoy Network now consists of four buoys around the coast and one out to the west of Ireland. In addition to providing real-time data for forecasting purposes, these networks have now accumulated many years of data for marine climate studies.

Over the past decade, major advances in technology have enabled buoy platforms to measure many more variables at high temporal resolution. Buoys, together with other autonomous systems, are becoming the mainstay of marine scientific exploration, partly because of the ever increasing running costs of vessels. These systems include buoy networks, such as the Centre for Environment, Fisheries and Aquaculture Science (Cefas) SmartBuoys, as well as the platforms operated by Plymouth Marine Laboratory (PML) within the context of the Western Channel Observatory (WCO; Fishwick et al., 2011; Smyth et al., 2015). These and other scientific data systems provide new insights into ecosystem function and marine biodiversity, as well as monitoring anthropogenic effects. The data sets generated over long time periods (SmartBuoys from 2000) create the potential to investigate climate change indicators as well as ecosystem model development and model validation. Commercially, buoys are used to monitor meteorological conditions and sea states at offshore installations within the oil, gas and renewable energy industries. In addition, sediment transport is monitored by the dredging industry to gain information relevant to seabed dredging activities.

This chapter addresses many aspects of autonomous sampling platforms and gives consideration to the major elements necessary for a successful monitoring programme within the estuarine and coastal environment.

2   Platform Design and Legal Obligations

At the start of any project, before any specifications of size or design of a buoy platform are considered, the appropriate applications for permission must be sought. It is important to identify who owns the seabed at the proposed deployment location, and it is also advisable to select a location away from any vessel movements. In the first instance, within the UK, contact with the Marine Management Organisation (MMO) is necessary to establish if a licence is required. Discussions with local harbour masters and port authorities may also be required, and as far as is possible consultation with other marine users to alleviate any possible inconveniences caused by the buoy deployment. This will prove prudent in the long term as it will reduce the risk of collision or vandalism to the buoy. Advice should be obtained from the MMO and port authority/harbour master regarding the final location and any particular light flash sequences they require. This information needs to be then given to the UK Hydrographic Office along with deployment dates for the buoy to be marked on the navigation charts.

When it comes to the specific design of the buoy platform, consideration must be given to matching the size and design of the buoy to the particular application. Platform manufacturers usually offer a range of buoy platforms and will be able to give advice on the most appropriate design for any specific application. Whilst many buoy platforms on offer appear to have some standard design features, there are differences that should be considered when selecting the most suitable design.

Traditionally, buoy platforms were constructed completely out of metal (usually steel). Most modern buoys, however, are a composite of metals and plastics. This has proven advantageous not only in cost but also in weight and handling of the buoy platform. Whilst the centre of the buoy hull is usually metal the external flotation is either plastic or foam. Plastic floatation is usually provided in the form of several independent sections bolted together, except possibly on the smaller buoys. There are several reasons for this, including ease of transportation and manufacture, but primarily this offers some form of protection against collisions. In the event of the buoy being hit by a vessel and the floatation becoming compromised, then only the damaged section will fill with water. This will help keep the structure afloat, albeit not necessarily vertical, until recovery operations can be mobilised. Some manufacturers offer a foam-filled plastic float so that if the floatation is compromised then the sections will not fill with water; however, depending on the amount of damage, these sections may still require replacing. There are a few buoy platforms that offer a polyelastomer foam float, and this will act very much like a fender in the sense that it can be compressed to less than 50 per cent of its original size and then regain its original form.

When the buoy platform as a whole is considered, there are many advantages and disadvantages to any one provider, and indeed some providers may not be able to accommodate all specific design requirements. It is always advisable to discuss the requirements with as many providers as can be identified, and the use of a tender process for final selection may be necessary. The design of the superstructure is an important part of the overall process, and it is worth noting that commercially available structures come in a large range of shapes, sizes and materials. Whilst most providers will offer standard 'off-the-shelf' products, some will also offer the capability of bespoke design and build for the specific requirements of a project. If a standard product can be used this will usually decrease the complexity, cost and delivery times.

An important point to consider is that the bigger and heavier the superstructure, the more unstable the buoy platform will become. Any good manufacturer/provider will be able to advise on the location of equipment to ensure the platform is as stable as possible, with a guideline that the heavier items, for example the batteries, need to be mounted as low on the structure as possible. If using wind turbines for power generation, then for health and safety reasons the superstructure will need to be tall enough to accommodate them above head height. If the structure needs to be tall then it is possible to ballast the buoy at the bottom, although the more weight that is required, the more buoyancy the floatation will need to possess. Most manufacturers will factor all these considerations into the final design and have the ability to model the platform characteristics prior to final build. One final but important consideration when designing a platform is that sufficient resources are available to deploy and maintain it. Vessel time is highly expensive, and the larger the vessel required the higher the cost.

It is imperative that the buoy platform fulfils the requirements laid down by the International Association of Lighthouse Authorities (IALA); the UK is within the Region 'A' buoyage system. As any scientific buoy will be identified as a special mark, it should therefore be completely yellow in colour, display a St. Andrews cross top mark and have a yellow flashing light. The light characteristics can be any rhythm except that it must not match any of the white flashing lights of cardinal buoys. The light range will usually be a minimum of 3 nautical miles. Other navigation warning systems should be considered, such as radar reflectors and Automatic Identification Systems (AIS). It may be that the buoy structure offers a sufficient radar target; however, the greater the size of the radar signature the more likely an approaching vessel will be to see the buoy. If platform size is an issue then active radar reflectors are available, although these will require power from the buoy system.

3   Power Generation

Central to any autonomous platform is the ability to generate and store the necessary energy to power the onboard payload, as well as being able to buffer periods of low power generation. When considering what power generation system is required, it is important to pay careful attention to the power budget of the complete system, as well as the environmental conditions expected at the mooring site; for example, solar charging will be dependent on day length and likely light intensities.

3.1   Battery Considerations

There are several important considerations when it comes to installing the most appropriate battery configuration; these include power budget, weight, cost and charging method. Historically, the battery of choice has been a sealed deep discharge type, lead acid battery. Other newer technology is starting to be utilised in the form of lithium ion batteries. These batteries are lighter, provide more power per unit volume and can handle deeper discharges than their lead acid counterparts.

3.2   Solar Charging

This technology is referred to as photovoltaic (PV), 'photo' meaning light and 'voltaic' meaning chemical production of electricity, which uses solar panels to harvest the sun's energy and convert it into electricity. Maximum Earth-normal surface solar irradiance is approximately 1000 watts of power per square metre at sea level on a clear day; however, solar panels only have an efficiency rating of between 20 and 30 per cent. This is largely due to the panels only being able to harvest a small section of the whole solar spectrum. Solar panels work using the photoelectric effect, which is the ability to emit electrons when light is shone on them. The material most commonly used is silicon, which is a semiconductor, and to create the electrical imbalance required, two nonpure forms are used. Silicon is a reflective material and therefore would reflect a considerable portion of the Sun's energy. To prevent this, a layer of absorbent material is placed over the top of the silicon. The final stage of construction of a solar panel is the encapsulation process, and it is important when using solar panels in the harsh marine environment that the finishing materials are suitable. It is recommended that solar panels designed for the marine industry are used and the advice of manufacturers should be sought.

When designing the solar power system for an autonomous buoy, consideration should be given to the amount of power required and the anticipated energy available from the sun in low light periods (winter months). Other considerations include the orientation of the buoy; if the mooring will not keep the buoy in a fixed orientation then solar panels are best placed around the circumference of the superstructure to maximise light harvesting. Also, in more benign and calmer areas, the angling of the solar panels upward from the vertical will enhance the light energy collected. On some buoy hulls

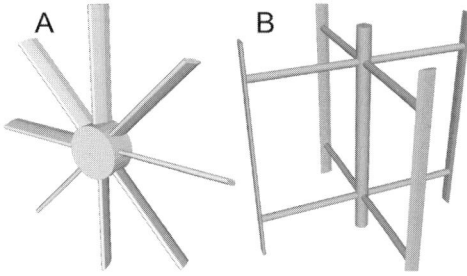

Figure 10-1 An illustration of the two main types of wind turbines: the traditional style turbine with blades mounted around a horizontal axis is shown as 'A'. A system in which the blades are mounted around a vertical axis is shown as 'B'.

the solar panels are nearly horizontal, and although sufficient during summer months, they may give little or no power generation during the winter. A final consideration is the stability of the platform and the amount of both weight and windage added by the solar panels; therefore, advice must be sought from the platform manufacturer regarding the size and quantity of solar panels.

3.3 Wind Turbines

Depending on the area of operation, but particularly in more temperate and high latitude zones, such as the UK, solar charging alone may not offer the best solution for power generation. Wind turbines are now commonplace and can be found on boats in marinas all around the UK. As with the solar panels, it is advisable to install a wind turbine that has been specifically manufactured for the marine environment. This will have protected bearings and be made from resilient materials.

Although there are many wind turbines available, the fundamental designs fall into two categories (shown diagrammatically in Figure 10-1). Whilst many variants of these two design types exist, the main difference is fundamental. The most common design is that of a traditional windmill style, where the blades are mounted around a central horizontal axis ('A' in Figure 10-1). The second design, which is much less common, comprises vertically mounted blades ('B' in Figure 10-1). There are several key advantages and disadvantages to each type, and these should be fully considered before opting for any particular turbine. First and foremost it is critical to understand that a buoy system is autonomous and can be subject to very high wind conditions; in such conditions a vessel operator would usually mechanically disable the wind turbine. Attention should therefore be paid to the specifications of a turbine. Some models include high wind speed decoupling from the generator. It is also advisable to fit an inline fuse between the wind turbine and the charge regulator to protect the rest of the power system in high wind periods.

Whilst horizontal wind turbines are much more efficient and produce more power than vertical turbines they are usually much larger. A significant health and safety risk is present with any wind turbine on a buoy system; therefore, they are normally only fitted to larger buoy platforms where they can be safely shielded, thereby preventing personnel from coming into contact with the blades. Vertical bladed turbines, whilst much less efficient, occupy a much smaller footprint, therefore making them more easily shielded. Although they do not offer the same charge characteristics, they are stated to be more reliable in higher wind conditions, and this, coupled with the smaller size, makes them an attractive form of wind turbine for buoy platforms. The vertical systems are, however, more expensive than the horizontal models and so budgetary considerations may need to be taken into account.

3.4   Power Generation from Buoy Motion

At the time of writing a new technology is under development, which will provide another source of renewable energy for many platforms, including buoys. This is a device that captures the chaotic energy from the motion of the buoy platform and converts it into rotational energy to drive a generator. These devices are in the early stages of development and as yet offer no definitive charge characteristics; however, they appear to offer huge potential.

3.5   Final Considerations and a System Example

As well as the method of charging and the batteries to be used, other important factors to consider are charge regulators, fuses, and circuit and battery protection. If the buoy is to rely only on solar charging, then the regulators' only requirement is that they are rated for the total power achieved by the solar panels. If the buoy is to have a combination of solar and wind power generation, then the best regulators to use will usually be the ones provided by the wind turbine manufacturer. This is because the regulator will be optimised for the wind turbine. However, each regulator will have different maximum power capability for solar charge, and this should be considered. Charge regulators are available that output the charge characteristics of all charging systems and the battery voltages, usually in the form of an ASCII text string. This information may be acquired by the buoy control system and transmitted back to shore, allowing the operator to assess power budgets and diagnose charge problems and failures. It is generally thought that the solar charging systems are the main systems used, with any wind turbines acting as a secondary system. Solar charging is a more proven technology and usually proves to be more reliable. Due to this, the regulator should be protected from the wind turbine with an inline fuse. Solar panels should also have inline surge protection circuits to protect the regulators. In addition, protection of the batteries is important, which should include protection against overcharging and overdischarge.

Depending on the level of redundancy required, it is advisable, particularly on larger buoy platforms, to have two independent power systems. A working example of the power systems currently in use on the scientific data buoys operated by the Plymouth Marine Laboratory in the western English Channel is given here. The power system installed on each of the data buoys comprises two identical systems (Figure 10-2). Each system, per buoy, includes 140-W solar panels (two panels at 70 W) and a wind turbine

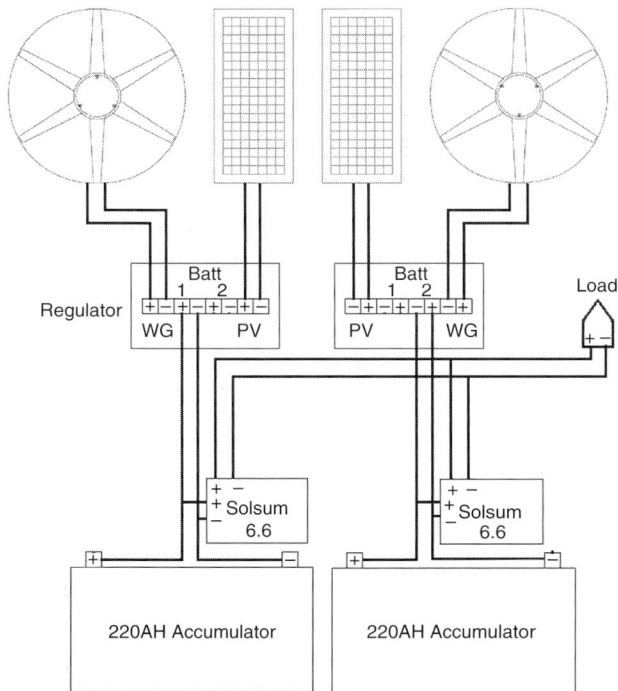

Figure 10-2 An illustration of the power system configuration that is typically deployed by Plymouth Marine Laboratory on its data buoys in the Western Channel Observatory.

with a maximum output of 300 W. The solar and wind generators interface with a charge regulator that also offers an output of charge characteristics. In addition, the regulator maintains the level of charge to the batteries at suitable rates and prevents overcharging. Because both batteries are connected in parallel on the load side, a second regulator is mounted on each battery to act as a diode, permitting current to pass only from the battery and preventing cross charging. The second regulator also protects the batteries from over discharge by cutting the power at around 11.4 V and only restoring it once the batteries reach around 12.5 V (Smyth et al., 2010).

4   Data Handling and Control Considerations

There are many data logging and control systems on the market with varying degrees of complexity and user configuration options. The unit of choice is largely dictated by the individual application and the operator's preference. There are a number of important points to consider when selecting the most appropriate unit, not least of which is the budgetary requirement. From the outset, when planning and designing a data buoy project, the objectives should be clearly identified, as these will dictate the whole design process. The purpose of the buoy and the parameters to be measured should be decided at an early stage; it is also advisable to build into a system the scope for additional sensors and expandability.

If a particular sensor manufacturer is selected it is reasonable to assume that they will also be able to provide the logging and control system; however, this can result in the manufacturer's sensors being the only ones able to interface with it. The ability to interface with any number of different sensors is highly complex because sensors interact in different ways with the control unit. Sensors can be either analogue or digital, with different voltage ranges and communication requirements. They may also need to be sent various commands to take measurements, operate antifouling systems and shut down appropriately. This usually requires each sensor to be controlled independently from the other sensors on the platform. Manufacturers of control units usually enable interfacing with a wide variety of sensors by developing what is termed device handlers. These are small software files that tell the control unit how to interact with any given sensor. Over time a manufacturer may create a large library of these device handlers, but it is worth checking that the handlers exist for the intended sensors, or extra development costs will usually be incurred. It is also important to check that the control unit has the capability of interfacing with the communication system that is intended for use on the buoy.

Depending on the sampling schedule and the sensor measurement frequencies, large datasets can be quickly generated, and this can cause problems when transmitting the data to shore, especially if the communications incur cost. Some of the higher-end logging and control systems allow the user to perform basic statistics on the data prior to transmission. Care should be taken to clearly identify what data the user requires to be stored on the buoy because some units will only store the transmitted messages, and the raw data can be lost.

To allow full control of the numerous sensors and transmission systems deployed, Plymouth Marine Laboratory has opted to run an embedded PC on the scientific buoys of the Western Channel Observatory. The first system was a PC104 Linux-based system that allowed total control of the buoy – from the battery monitoring to the webcams. This system did, however, come at a cost, in that it was extremely power hungry, using over 25 ampere hours (Ah) per day. As technology has moved on, the latest control PC is a BeagleBone credit-card sized Linux computer. This upgrade has offered more computing power and the added ability to put the system to 'sleep', thereby halving the electrical power consumption. An advantage of building a bespoke PC control unit is that the user will have full control of the system because the control software will also need to be provided by the user. This allows, for example, the raw data to be stored on memory cards whilst averaged data sets can be transmitted. Any changes to setup or sampling frequency can also be easily communicated to the buoy.

5   Communications

5.1   Radio

The use of radio modems operating in the VHF and UHF has traditionally been the most common form of communication. This is largely due to the fact that once the systems have been purchased the communications are free, with the exception of an annual licence payment. With the correct combination of modems and antennas, 'line of sight' communications may reach in excess of 20 nautical miles. The usual method utilises a high gain 'Yagi' aerial on the shore-based node because this type of aerial is highly directional. It therefore needs to be on a stationary platform that is aligned to the buoy. Because the buoy is not a stationary platform a less-directional aerial is required. A dipole aerial is often used, and this will transmit and receive in a 360-degree circle with a transmission lobe in the shape of a toroid, due to the fact that this type of aerial concentrates its transmission power in the horizontal. It is important to note that the dipole aerial should be mounted vertically, and the corresponding Yagi aerial should have the cross members in vertical alignment. Due attention should be given to the location of each aerial because the greater the height the better the performance from the equipment.

Radio modems may operate in both the VHF and the UHF frequency ranges, and the selection should be based on the user's requirements. The VHF range offers a greater transmission range; however, the UHF range is less susceptible to interference, so a more reliable transmission is achieved. Also, it is worth noting that the UHF range requires shorter aerials that, depending on the platform, may be more desirable. Some frequencies are unlicensed, and anybody can freely transmit on these bands within certain power limits (usually < 0.5 W). There is therefore a risk that someone else will be transmitting on the same frequency at the same time, thereby blocking the transmissions. It is strongly advisable to licence your apparatus on a dedicated frequency through OFCOM (UK). The VHF band is heavily used in the marine environment, and therefore the frequencies are cluttered and so present a higher risk of interference from other users. The UHF range is less utilised because the transmission distances are shorter and so may offer a more reliable communication system.

5.2   GSM – Mobile Network

The GSM modem works on the mobile phone network system, and therefore it is important to first make sure that there is data connectivity at the planned deployment location. Within the UK there are several network providers, so it is prudent to check online for the one with the best coverage in the deployment area before taking out a data contract with any one provider. This technology can provide an internet connection to the buoy and data are usually uploaded to a server address, giving good data rates and reducing the necessity to compress data transmissions. Problems associated with this form of communication are that networks can drop out, and unlike other countries, the UK system does not have inbuilt capabilities to utilise another network (although this technology can be built into a system). It is not uncommon for networks to become oversubscribed in periods of heavy use. An example of this is a situation where no data is received from a buoy when a cruise ship comes into port because many passengers are making use of their mobile phone reception.

5.3   Satellite

Satellite communications have always had a reputation for being expensive, and even though the costs are decreasing compared with other communication methods, it still remains an expensive option. The most commonly used satellite network is the Iridium network, but there are others; for example: Inmarsat, ORBCOMM and Globalstar. Once a network has been selected, a contract with an airtime provider needs to be set up. With the Iridium system the buoy sends a SBD (Short Burst Data) message that takes a similar format to a small text message; the data are usually averaged and compressed because whole data set transmissions are unrealistically large. This communication system does provide two-way communication, although a logging/control unit needs to be on the buoy to be able to make system changes as instructed by small text message commands. The method would normally only be used in coastal and offshore areas that are beyond the range of VHF/UHF or mobile networks.

5.4   Wireless Ethernet Communications

With recent technological advances, wireless Ethernet, which is similar to a domestic WiFi connection, has started to emerge. This allows the control system on the buoy platform to form a part of the existing computer network with very high data rates. Because these systems are being utilised in many commercial land-based applications, the drive and demand exists for continual development. Systems are now readily available that can establish a link over 20 km of 'line of sight', and this is providing a new and much more powerful way to communicate with autonomous systems. Sensors are now becoming available that connect straight to the wireless Ethernet unit and essentially become part of the network. This is possibly the starting point for developing an autonomous buoy with the ability to just plug additional sensors into the existing network infrastructure. Obviously, a line-of-sight, land-based station needs to be identified, and the radio signal could potentially be blocked and interfered with by shipping and other objects. This is a promising technology with the potential to revolutionise autonomous systems.

6   Meteorological Measurements

Marine meteorological observations are regularly made in support of weather forecasting and climate applications. They include in situ measurements from ships, moored and drifting buoys, fixed platforms and light vessels as well as from satellites. Traditionally, in situ observations have been made manually, but in recent years the use of Automatic Weather Stations (AWS) on buoys, light vessels, ships and offshore platforms has become the primary means of making in situ observations over both the open ocean and in coastal/shelf waters. Further information can be found in the World Meteorological Organisation (WMO) Guide to Meteorological Instruments and Methods of Observation (WMO-No. 8).

Instrumented moored buoys provide a key source of observations for a wide range of meteorological and oceanographic variables and are operated in both deep water and coastal regions around the world. Typically, meteorological data buoys are designed to measure the following variables: wind speed and direction, atmospheric pressure, air temperature, relative humidity, sea temperature and sea state (waves), although other

Table 10-1 E-Surfmar meteorological measurement requirements from moored buoys.

Variable	Requirements			
	Range	Resolution	Mean Error	Typical sample period
Wind Speed (including gust)	0 to 75 m s^{-1}	0.5 m s^{-1}	±1 m s^{-1} $<$ 20 m s^{-1} $\pm5\%$ $>$ 20 m s^{-1}	10 min
Wind Direction	0 to 360°	1°	$\pm10°$	10 min
Maximum Gust	75 m s^{-1}	0.5 m s^{-1}	$\pm10\%$	highest 3s average over 10 min
Air Pressure	900 to 1050 hPa	0.2 hPa	±0.5 hPa	1 min
Air Temperature	−20 to +40°C	0.1°C	$\pm0.2°$C	1 min
Sea Surface Temperature	−5 to +25°C	0.1°C	$\pm0.2°$C	1 min
Relative Humidity or Dew-Point	38–100%	1%	$\pm5\%$ $<$ 85% $\pm3\%$ $>$ 85%	1 min
	−20 to +40°C	0.1°C	$\pm0.4°$C	1 min
Wave Height	±10 m rel. MSL	0.1 m	$\pm10\%$ (or ±0.2 m if greater)	20 min
Wave Period	0 to 20 s	0.1 s	$\pm5\%$ (or ±0.5 s if greater)	20 min

variables are sometimes measured (e.g., rainfall, downward shortwave and/or longwave radiation). Such buoys are usually fitted with GPS or Argos transmitters to determine their positions and to track them in the event of mooring failures.

With regard to the accuracy and reliability of the measurements, meteorological observations from moored buoys are now considered to be of better quality than those from ships (Wilkerson and Earle, 1990; Ingleby, 2010) and are generally regarded as providing the highest quality observations for a wide range of marine meteorological variables. In addition to their use by forecasters and their assimilation into numerical weather prediction models, the data are also used to provide information on the climatology of oceanic areas, 'ground truth' reference data for satellite calibration/validation, and supply estimates of surface fluxes. At the European level, marine meteorological observations are coordinated under the EUMETNET Surface Marine Programme, E-Surfmar (EUMETNET, 2016), where the accuracy, range and resolution requirements for moored buoy meteorological measurements are specified (given in Table 10-1). The errors shown in Table 10-1 refer to the mean errors of the measurement system (sensors, interface and data processor) and do not include those due to poor or unrepresentative exposure. For example, wind sensors should be sited as far away from any structures or obstructions as possible, air temperature and humidity sensors should be mounted within a naturally aspirated radiation shield. The measurements are normally made hourly and reported on the hour.

For wave height and period there are a number of different parameters that may be measured; e.g., significant wave height, maximum wave height, average wave period, peak wave period etc., as well as wave direction (discussed in greater detail in Chapter 5). For wave measurements it is important that the mooring is designed such that it does not unduly influence the buoy's motion and allows the buoy to follow the waves; hence, moored buoys with taut moorings are unsuitable for making wave measurements. On Met Office moored buoys in water depths of 30 to 100 m an all-chain mooring is used with a subsurface float where appropriate; in deeper waters an inverse catenary mooring with a subsurface float is used.

The data are usually transmitted hourly through various satellite communications systems (e.g., Argos, Iridium, Meteosat/GOES) and after processing are exchanged internationally via the WMO Global Telecommunications System (GTS). In some regions of the globe (e.g., the Indian Ocean) vandalism of moored buoys is a major problem for buoy operators.

Barometric pressure can be measured using digital pressure transducers. Wind-induced effects are one of the main sources of error in measuring air pressure. Variations due to strong and gusty winds can be overcome by using a static pressure head to 'filter out' the effect of dynamic pressure or by mounting the pressure transducer inside a box or compartment, provided the box/compartment is not pressure tight. For operational meteorology it is important to know the height of the pressure transducer, in order that the pressure can be corrected to sea level (typically, the pressure decreases by approx. 0.12 hPa for every metre of elevation).

Traditionally, winds have been measured from buoys using marine-grade cup and vane or propeller anemometers. However, these can be subject to saltwater ingress and bearing failure, so acoustic anemometers are increasingly being used – having no moving parts they have a longer operating lifetime (Turton and Pethica, 2010). Because moored buoys are not fixed and can rotate in the water, sonic anemometers can only measure wind direction relative to their orientation; therefore, there is a need to interface the anemometer to a compass system to measure the orientation of the buoy and derive the true wind direction. Similarly, it is important to know the height of the wind sensor because the wind speed will normally increase logarithmically with height.

7   Sea State Measurements

One of the key measurements from moored buoys is waves, which can be measured from both meteorological and 'waverider' buoys (the latter being designed specifically for wave measurements; Chapter 5). In these systems, the vertical acceleration is measured by means of an accelerometer mounted on a gyroscope (or a partially stabilised platform). The simplest processing technique is a counting analysis in which the time series of sea surface elevations are processed to give the heights and periods of individual waves. From this it is possible to determine such parameters as the significant wave height (H_s), mean wave height (H_m), maximum wave height (H_{max}) and mean wave period (T_m). Typically, the sampling periods for such analyses are between 15 and 30 min, with a 4 Hz sampling frequency. Wherever possible, it is advantageous to make colocated wind measurements from the same platform that is used for the wave measurements. The high-frequency response of the buoy to waves is limited by the dimensions of the buoy; for wavelengths smaller than the buoy's diameter the wave motion cannot be followed. At the other end of the frequency spectrum, the low-frequency response is determined by the combined buoy and mooring system.

Waverider buoys have been widely used by commercial companies and local authorities for many years, for which the Datawell Mk II waverider buoy is regarded as the 'industry standard', although other models, such as the Axys TRIAXYS^{TM}, are also widely used (Chapter 5). These are generally smaller, near-spherical buoys (up to approximately 1 m in diameter) that use both vertical and horizontal accelerometers to determine the pitch and roll of the buoy, which enables directional wave information to be derived. Such buoys are usually deployed relatively close to shore and are therefore able to use HF/VHF radio or GSM communications. Being significantly smaller (and lighter) than meteorological moored buoys, and operating in relatively shallow coastal waters, a rubber bungee usually forms the upper part of the mooring, which allows the buoy to follow the waves.

The distribution of wave energy with frequency is calculated by transforming the digital time series to the frequency domain using Fast Fourier Transform techniques. It is recommended (Swail et al., 2010) that directional spectral wave measuring systems should reliably estimate the 'first 5' parameters for each frequency band. The first variable is the wave energy, which is related to the wave height, and the others are the first four coefficients of the Fourier series that define the directional distribution (wave direction, wave spread, skewness and kurtosis). Spectral wave measurements can also be made from larger buoys but these are likely to be poorer at 'following' the waves. Although this problem can, in principle, be remedied by using appropriate transfer functions and careful data processing, there will still remain important parts of the wave spectrum that are inadequately measured.

The Defra strategic wave monitoring network (WaveNet), managed by Cefas, provides a single source of real-time (and archived) wave data from a network of wave buoys around the British Isles; it is operated by various organisations in support of coastal flood defence and forecasting. In addition, the Channel Coastal Observatory provides access to data (real-time and archived) from waverider buoys operated by various bodies around the English coast.

8   Temperature, Salinity, Density, Turbidity, Optical (AOP and IOP) and ADCP

This section identifies the measurement of important physical variables with specific reference to autonomous systems. The methodology for each measurement is covered elsewhere in this book (Chapters 4, 7 and 8).

8.1   Temperature, Salinity and Density

One of the most fundamental instruments in oceanography is the CTD or Conductivity, Temperature, Depth probe (Chapter 4). This sensor has been used for many years, and as the technology has improved the resolution and accuracy of the measurements has also increased. There are a multitude of these sensors available from numerous manufacturers, and it is important to look carefully at the specifications before purchase. They work by measuring the water temperature and conductivity relative to depth, and from this information the salinity and density are calculated. It is worth downloading the manuals for any specific sensor, or speaking to the manufacturer first, to identify how the temperature and conductivity are measured and what formulae are used for the calculation of salinity and density. A general rule of thumb is that you get what you pay for.

Measuring these variables through the water column has traditionally been achieved by means of a thermistor/CT chain. This consists of multiple sensors suspended at set depths below the data buoy; the higher the depth resolution required then the more sensors that are required. Because these sensors have a low power requirement they can be left unattended for long periods of time and then retrieved for data downloading. If the data are required in real time then the use of either underwater modems or electrical connections between sensors will be required and a significant level of complexity added; however, several commercial manufacturers are able to set up such a system.

8.2   Turbidity and Optical Measurements

A commonly used instrument for the determination of turbidity on an autonomous platform is the Nephelometer (scattering measurement, Chapter 7). This sensor works by measuring the amount of scattered light at a fixed angle from a known light source. The data produced are measured in Nephelometric Turbidity Units (NTU); this measure is highly dependent on several factors, including particle size, shape, colour and reflectance. For these reasons any relationship to a total suspended material measurement will, at best, be local to the environment in which the sensor is deployed. When identifying a suitable sensor it is important to check the wavelength of the light source to ensure that it is not affected by any other constituents in the water; for example, a source in the blue will be absorbed by the Coloured Dissolved Organic Material (CDOM) component in the water, which in estuaries and coastal environments can be considerable.

As satellite resolution increases (i.e., pixel sizes are reduced) and with the use of airborne sensors, remote sensing in estuaries and coastal areas is becoming more effective (see Chapter 11). However, with any remote sensing system the need to provide ground-truth is critical to the interpretation of the measurements. Autonomous buoys can be used to deploy an array of sensors that measure both Apparent Optical Properties (AOPs) and Inherent Optical Properties (IOPs) for ground-truth data acquisition. The AOP sensors can be either wavelength specific (multispectral), matching the wavelengths of the remote sensor, or hyperspectral, which means that they scan the complete spectrum at each sample. It is important to measure the above-water downwelling irradiance (E_d), the below-water downwelling irradiance (E_s) and the below-water upwelling radiance (L_u). These measurements can be used to calculate the water-leaving radiance, which is what the remote sensor will be measuring (Chapter 11). Because the angle, relative to the vertical, at which these measurements are made is important (5 degrees or less), consideration should be given to the sampling frequency of the sensor being used. An advantage of the multispectral sensor is that it measures all the specified wavelengths at the same time and at several Hz. When the data are then corrected for the angle at which they were collected, more simultaneous data will be available than with a hyperspectral sensor. The IOP measurements are made to gain as much understanding as possible about the scattering and absorption properties of the water mass. An IOP sensor is one that uses a light source to quantify absorption and scattering properties. Several sensors are now available that can yield an understanding of the wavelength-specific scattering and absorption properties, as well as the volume scattering function.

As with all optical sensors, the measurements are only as good as the clarity of the lens; once any dirt or biological fouling forms on the sensor lenses the calibration and the data are no longer reliable (Chapter 7). A later section of this chapter will deal with the effects of this problem and its solutions in more detail, but primarily the sensors deployed should have some kind of shutter and wiper system installed.

8.3   Acoustic Doppler Current Profiler (ADCP)

An ADCP uses the Doppler effect of back scattered sound waves to measure current velocities within defined depth boxes throughout the water column (see Chapter 4). This sensor can be deployed on the buoy itself, looking down through the water column, or on the mooring. However, these deployment locations can result in data that are difficult to interpret because the motions of the buoy and mooring will affect the velocity measurements. The preferred method of deployment is on the sea bed, facing up, although this too is not without disadvantages. For the sensor to accurately measure the water currents it needs to have an unobstructed view of the water column. Therefore, the sensor must be placed a sufficient distance from the surface buoy to not have the buoy or its moorings in the sensor's field of view. This creates a situation in which the sensor no longer has the protection of the surface buoy and is exposed to trawls, anchors and other damaging interference. To offer as much protection as is possible, the ADCP can be mounted in a trawl-resistant cage, which is basically a square-based pyramid whose sloping sides will cause the trawl to bounce over the sensor. In an ideal situation, the ADCP will be tethered both physically and electrically to the buoy, which will facilitate the powering and communications with the sensor and prolong deployment times. In addition, because the sensor is on the sea bed, it will be better protected from the fast-growing biological fouling that is often associated with sunlit surface waters.

9   Biofouling and Preventative Methods

One of the major problems with long-term deployments in an aquatic environment is biological growth on the platform and sensor arrays. This growth will significantly add to the weight of the buoy platform and therefore reduce the residual buoyancy; moreover, once attached to the underwater sensors, it can seriously impair the scientific measurements being taken. The submerged sections of the buoy platform (hull) can be antifouled with the same systems that are utilised on boat hulls, and these are readily available from local chandlers. There is a great deal of effort being made to develop better antifouling systems, particularly for vessel hulls, because even a small amount of biological growth can have considerable impacts on fuel efficiency. A vessel that has been antifouled and then moored for any length of time will have considerable growth on its hull. The majority of paints are designed to inhibit biological growth and rely on the motion of the vessel through the water to remove any potential growth. Depending on the type of vessel and its speed through the water, this growth will usually detach from the hull once the vessel is underway. Obviously, a moored buoy will not have the same degree of ability to shed the biological fouling on its hull; therefore, it is a good idea to power-wash the hull annually and to reapply the antifouling system (if required). A more recent product that is now being used on many leisure boat hulls is coppercoat, which is a resin that is mixed with copper powder to form a hard copper surface on the hull. This surface discourages the growth of biological material but, as with other products, does not completely eradicate it, so an annual power wash is still required. The manufacturer claims that once this product is applied, no further applications are required for at least ten years.

As scientists require an ever increasing amount of long-term monitoring solutions, attention has been given to developing sensors that can be deployed for long periods of time. Factors that have to be considered include power requirements, durability, calibration, measurement accuracy and stability, as well as mechanisms to deal with biological growth. In recent years several antifouling solutions have become commercially available, and these are usually in the form of bespoke systems developed by instrument manufacturers to enhance the long-term usability of their sensors. Such systems have several common features that have been field tested and proven to increase sensor deployment times (Delauney et al., 2010). Certain types of antifouling are more appropriate to specific probes and sensors than others, and this is why on any particular instrument one may find several solutions all working together. The use of copper is a common antifouling technique, and sensor housings are often clad with copper. This can be a factory solution; however, with the use of copper tape/sheeting any sensor housing can be covered in a copper layer with relative ease (care must be taken when applying the tape as the edges are usually sharp). The major disadvantage of tape when compared with a solid copper sheathing is that over long time scales (> 3 months) the tape can start to detach from the sensor; if the sensor is then raised for cleaning the process will usually remove the copper tape.

Some of the more difficult sensor types to keep clean are those that utilise optical windows for making measurements; e.g., transmissometers, fluorometers, optical scattering sensors and light intensity sensors (Manov et al., 2003). Obviously, any material that covers the optical window must not affect its optical characteristics because this would invalidate its calibration parameters. The most common form of antifouling for optical sensors is a shuttering system. These shutters are usually, but not always, made of copper and include some kind of wiper. When the sensor is inactive the shutter will be in a closed position, covering the window and approximately 1 mm away from it; however, when the sensor is powered on the shutter will open and in doing so a rubber

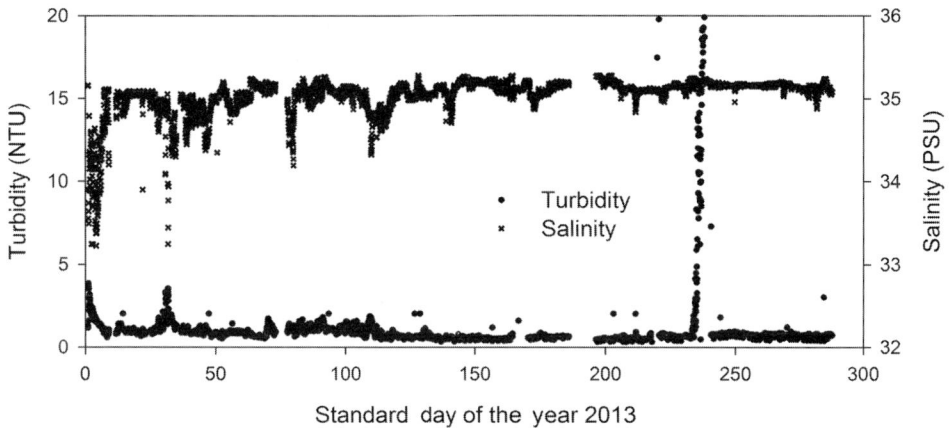

Figure 10-3 An example of data collected from the L4 buoy in the Western English Channel. The graph plots salinity (dimensionless but expressed here as psu) and turbidity (NTU) against days of the year during 2013.

wiper will pass across the lens to remove any biological films. Optical sensors that are not properly protected against biological fouling will exhibit a drift in their calibration as the optical characteristics change through time.

At the Plymouth Marine Laboratory (PML), data from the scientific buoys in the western English Channel are transmitted back to PML on an hourly basis, and despite having copper shuttering with rubber wipers the sensors still eventually fail. Evidence of this can be seen in near-real time from the graphical displays of the data – the failure is usually simultaneous across all the optical sensors. The effect of biofouling on the optical window of the turbidity sensor from the WCO L4 buoy is evident from data for 2013 (Figure 10-3). The sensor wiper failed to adequately clear the window on day 234, and within two days the instrument had saturated, yielding maximum turbidity readings of 25 NTU. This instrument saturation was also replicated in the chlorophyll *a* and CDOM fluorometers (not shown). On day 240 the sensor was manually cleaned and the wiper restored to its full efficiency (Figure 10-3). Biological fouling is dependent on many factors (e.g., sunlight, temperature and nutrients), so that sensors require maintenance at different frequencies, depending on these factors. Data from the WCO L4 buoy for 2013 also show that the early part of the year exhibited several sediment-laden freshwater intrusions, indicated by reduced salinity and corresponding increases in turbidity (Figure 10-3).

Chemical antifouling systems are becoming increasingly commercially available. These are primarily in the form of an injection of antifouling agent, such as bleach or acid, and work best on enclosed flow cells, where the substance can be used in small quantities and its dispersal in the water will not cause it to become ineffectively weak. This type of system is utilised on membrane sensors, such as those used for dissolved oxygen, to prevent the membrane from fouling, and is effective for any measurement probes that can be housed within a flow cell. The technology can even be used at depth, with some commercially available sensors claiming a 200-m depth rating. The major

Figure 10-4 Oceanographic sensors deployed by the Plymouth Marine Laboratory on its data buoys in the Western Channel Observatory. The sensors are completely covered with copper to deter biofouling. In the image, mechanical antifouling devices are annotated by: 'A' – a mechanical brush arm; and 'B' – a shutter system with a rubber wiper. The bleach reservoir used to inject bleach into the flow cells between data acquisitions is annotated by 'C'.

drawback is that the bleach/acid reservoirs need to be regularly replenished – the higher the volumes injected the more frequent the maintenance fieldwork; clearly, it is highly advantageous to be able to determine the injected volume and then adjust it remotely, depending on the biological fouling conditions (e.g., summer or winter).

A more recent technology to be developed within the field of antifouling utilises electrolysis of seawater to produce chlorine. This process can be implemented in small-scale applications within flow cells and across optical windows and also on larger scales using a chlorinating chamber that contains the underwater sensor suite. Whilst this larger scale method is extremely effective, it does rely on the sensors being housed within an enclosed underwater container during the chlorination period. Chlorine systems that utilise a winch system to bring the sensors into an enclosed environment are now becoming commercially available. With this system and the bleach/acid systems it is essential that the amount of exposure a sensor receives to these corrosive substances is minimal to prevent the sensor itself from becoming damaged. For the chlorination system it has been shown that as little as 1 hour a day spent in the chlorination chamber can eradicate any form of biological fouling on sensors that have no other mechanisms to combat it.

The oceanographic sensors deployed within the WCO are completely covered in copper (Figure 10-4). Two types of mechanical cleaning are utilised: in the first, a mechanical arm is deployed that passes through the flow cell and cleans the lenses on both faces (shown as 'A' in Figure 10-4); in the second, a copper shutter with a rubber wiper fitted is deployed to clean the optical window of the instrument ('B' in Figure 10-4). The system also comprises a bleach reservoir for the bleach injection system (the area labelled 'C' in Figure 10-4).

Probably the most obvious form of antifouling is to not have the sensors submerged in the water. This is a method used for sampling surface waters in which a pump and hose supply water through a flow cell arrangement, and after sampling is complete, the flow cell is flushed. This methodology has some significant drawbacks; in particular, the temperature difference between the sensor electronics and the water sample can cause instability, especially if the sensor is exposed to extremes of temperature. The electronics will remain at a more stable temperature whilst submerged in the water. Some types of membrane and sampling probes can also be damaged if they dry out.

Finally, antifouling systems are a key part of any autonomous platform, and sufficient time should be given during the design and specification phase of a project to select the most appropriate sensor suite. Factors that may determine these decisions are firstly cost, variables to be measured, buoy location, potential for biological fouling, and also the practicalities of regular maintenance fieldwork to possibly refill reservoirs of bleach or acid.

10   Mooring Design

The mooring of a buoy platform is a highly complex issue, and it is not the aim of this section to provide any specific designs to be implemented but merely to discuss and present some mooring considerations. The moorings are a significant part of a buoy deployment project, and it is advised to consult professional design services; the buoy provider should be able to assist with either the mooring design or a third-party contact to help with this. It is also important not to underestimate the potential cost of the mooring system because this depends on many factors and can form a considerable part of the overall project budget.

The simplest mooring to both design and deploy is a single-point mooring, which usually consists of a suitable weight on the seabed with a length of chain to the buoy. The chain size (gauge) and length is determined by the water depth and the buoy drag/displacement. Mooring line lengths are important; if these lines are too short the line between the weight and the buoy can become taut and then expose the buoy to much greater forces, potentially causing serious damage to the structure. Chain can be very expensive and also too heavy for the surface buoy to support; if this proves to be the case, then a chain/rope combination can be used. It is important that the rope does not come into contact with the seabed because particles of sediment will penetrate between the fibres and cause abrasion.

A single-point mooring will allow the buoy to move both rotationally and in absolute spatial location. If a more specific orientation and location are required then a multipoint mooring can be employed. In the confines of a shallow estuary a multipoint mooring would be difficult to successfully implement, if not impossible, but in a more coastal environment it becomes more achievable. A multipoint mooring can take the form of a two-, three- or even four-point system, and these are much more complicated to design and lay. Weights, distances and mooring line lengths become more critical to the design because of the need to prevent weights being pulled together in storm events and to prevent entanglement of the mooring lines. Obviously, the greater the number of mooring lines attached to the buoy then the greater the weight, so that chain/rope combinations become more desirable in these situations. Another point to consider is the use of anchors, either stand-alone or in conjunction with weights. Because the direction of pull should always be the same, an anchor can offer a more substantial hold, depending on seabed composition. To provide slack in a mooring system, the lines need to be sufficiently long to deal with large tides and high waves; however, this in turn can result in too much slack at low tide water levels. A method to combat this is to add a subsurface float to the mooring line, thereby allowing slack in the system to be taken up and the lines to be pulled away from the buoy.

A large number of factors influence the type of mooring used for any specific deployment, including water depth, current speeds and direction, anticipated maximum wave heights, buoy size and shape, windage, vessel capability, buoy orientation requirements, and material specifications. All these factors, and more, create a very complex design brief, and this can only be successfully achieved with a great deal of experience. Never underestimate the power of the sea, even down to its influence on the shackle links between components; these must be properly seized, ideally by welding them shut. The most important point to be taken from this section is the requirement to seek professional advice on mooring design and specification.

11   Recent Autonomous Technologies

With increasing fuel costs the operation of scientific research vessels is becoming more and more expensive, and as a result, considerable resources are being directed into the development of autonomous vehicles (see also Chapter 4). This added pressure comes at a time when our requirement for data is at an all-time high, with highly sophisticated models requiring training and validating, and with our quest to unravel the uncertainties regarding global issues such as climate change and ocean acidification. As our understanding of the role that the world's oceans and seas play in global climate change evolves, these data and their measurement technologies become more important. Real-time data are required for assimilation into forecasting models, not only for sea state and meteorological conditions but also for the prediction of extreme events that threaten lives.

Although beyond the scope of this handbook, a major initiative that has developed over the last decade is the Argo programme, which presently maintains over 3,500 autonomous profiling floats; these report CTD data throughout the world's ice-free deep oceans, with increasing numbers of floats also reporting biogeochemical variables such as dissolved oxygen and chlorophyll fluorescence (Roemmich et al., 2009; Freeland et al., 2010).

Autonomous vehicles are now commercially available, and these have become a useful tool in the observing system, mainly due to the latest battery technologies and the miniaturisation of sensors. Autonomous underwater vehicles (AUVs) are usually powered through the water by an onboard propulsion system and can vary significantly in size from small torpedo-style systems to small submarine size. These systems offer different payload capacities and are usually limited to several-day missions, mainly due to battery life. Buoyancy gliders are becoming increasingly used within observing systems, and unlike AUVs, they do not rely on mechanical propulsion. A buoyancy glider works by altering its density to rise and fall through the water column, and by angling its lateral fins, the glider can make forward motion as it ascends and descends (see also Chapter 4). Because an object's density is dependent on mass and volume, the glider can change its density by altering its volume, which can be increased and decreased by pumping oil into and out of a bladder. Because this requires relatively little power compared with turning a propeller, mission duration can be of the order of a few months. A drawback of this kind of propulsion for coastal work is water depth – most buoyancy gliders are developed for depths greater than 200 m. However, it is possible to use certain models in coastal waters, although the shallower the waters, the less headway they will make. When deployed in shallow waters the mission duration will also decrease because the oil bladder will be filled and emptied at a significantly higher rate, requiring greater power consumption. Payloads are also much smaller, and they rely on miniaturised sensors, with batteries costing over £1000. It can be a costly exercise to operate this type of vehicle over long time periods.

The final platform to consider is the autonomous surface vehicle (ASV), of which the 'wave glider' is an example. These are platforms that, through the use of mechanical systems and hydrodynamics, convert wave energy into forward motion. They have a surface-float/small-boat expression, and this allows for the use of solar and wind energy to provide power for the sensors and communications. Utilising a surface vehicle permits continuous communications, which is usually through satellite-based systems (e.g., Iridium). These platforms have the capacity to either mount sensors below the surface float or tow a small array of sensors. Small winch systems are currently under development for these platforms to facilitate the taking of measurements through the water column. A disadvantage with them is that they are at risk of collision with vessels, especially in the coastal zone. Some ASVs carry AIS systems to warn ships of their presence, but this takes vital power from sensors, and with limited antenna height, the range at which such information can be received is much reduced.

The autonomous vehicle systems briefly discussed here have significant potential to provide much needed data from coastal seas. However, the application and suitability of each platform should be carefully considered before purchase. In addition, the use of mobile autonomous systems will always raise safety concerns because the risk of collision with other vessels is significantly elevated in coastal areas. The legislation surrounding autonomous vessels is still unclear, and as such it would be prudent, prior to purchase or deployment, to discuss potential use with the relevant port authorities; in the UK these are the Maritime and Coastguard Agency (MCA) and the MMO.

12   Final Remarks

Although scientific data buoys have been around for many years, they continue to be developed and enhanced and therefore remain an integral part of our marine observing networks. With increased power generation and new battery technologies, a data buoy can take a significant instrument payload both in terms of power and size. For scientists to obtain a more complete understanding of the aquatic ecosystem, measurements throughout the water column are necessary. Historically, this has been achieved using an array of sensors suspended at different depths through the water column, although this method has many complications in terms of power, communications, sensor calibration and comparability. As sensors become less power hungry and battery technology advances, deployment times can be extended and, with the use of underwater modems, good communications can also be achieved. However, because these arrays comprise sensors that are suspended at fixed depths, intermediate water-column information is not obtained. To combat this problem a variety of profiling systems are becoming available for coastal and estuarine environments. These principally are in two forms: those deployed from a winch mounted on the surface buoy, or a winch deployed on the seabed and tethered to the surface buoy for power and communications. This added capability has the potential to further increase the contribution made by scientific data buoys.

Obviously, a moored scientific data buoy can only measure properties at a fixed geographic location, whereas to understand the inhomogeneous nature of estuarine and coastal systems scientists require data covering larger geographical areas. This has traditionally been covered using remote sensing solutions; however, these can only offer a limited amount of information from the surface waters and are also limited by weather conditions (e.g., optical and infrared sensors are unable to see through clouds and microwave instruments are affected by rain). Nevertheless, remote sensing continues to play a significant role in our understanding of the seas (see Chapter 11) and new in situ technologies help to build on this.

As the global population is expected to keep rising, with more and more people living in coastal regions, the reliance on our seas is ever increasing. With this comes the requirement to understand both the ecosystems that we depend on and the impacts that long-term anthropogenic changes will have on them. Scientists are now able to deploy several key technologies to aid this understanding, as described in this chapter, including data buoys, buoyancy gliders, autonomous underwater vehicles and autonomous surface vehicles. These technologies all offer different sampling strategies and together comprise a very powerful observing capability.

References

Burt, S. D., Mansfield, D. A., 1988. The great storm of 15–16 October 1988. *Weather* 43, 90–110. doi:10.1002/j.1477-8696.1988.tb03885.

Delauney, L., Compère, C., Lehaitre, M., 2010. Biofouling protection for marine environmental sensors. *Ocean Science* 6, 503–511.

EUMETNET, 2016. www.eumetnet.eu/e-surfmar [accessed April 2016].

Fishwick, J. R., Mason, P., Gallienne, C., 2011. Long-term autonomous data buoys for monitoring the English Channel. Recently overhauled buoy systems measure parameters missed by boat surveys, enhance collection of biogeochemical data. *Sea Technology* 52, 17–22.

Freeland, H. J., et al., 2010. Argo – a decade of progress. In: Hall, J., Harrison, D. E., Stammer, D. (eds.), *Proceedings of OceanObs'09: Sustained Ocean Observations and Information for Society*. Frascati, Italy: ESA Publications.

Ingleby, B., 2010. Factors affecting ship and buoy data quality: A data assimilation perspectiv. *Journal of Atmospheric and Oceanic Technology* 27, 1476–1489.

Manov, D. V., Chang, G. C., Dickey, T. D., 2003. Methods for reducing biofouling of moored optical sensors. *Journal of Atmospheric and Oceanic Technology* 21, 958–968.

Roemmich, D., and the Argo Steering Team, 2009. Argo: The challenge of continuing 10 years of progress. *Oceanography* 22, 46–55. http://dx.doi.org/10.5670/oceanog.2009.65.

Rusby, J. S. M., 1976. Background to the DB1 data buoy project. *Proceedings of the UK Data Buoy (DB1) Symposium*, Wormley, UK, 23 November 1976. IOS Report no. 44.

Smyth, T. J., Fishwick, J. R., Gallienne, C. P., Stephens, J. A., Bale, A. J., 2010. Technology, design, and operation of an autonomous buoy system in the Western English Channel. *Journal of Atmospheric and Oceanic Technology* 27, 2056–2064.

Smyth, T., Atkinson, A., Widdicombe, S., Frost, M., Allen, I., Fishwick, J., Queiros, A., Sims, D., Barange, M., 2015. The UK Western Channel Observatory: Integrating pelagic and benthic observations in a shelf sea ecosystem. *Progress in Oceanography* 137, Part B, 335–341.

Swail, V., Jensen, R. E., Lee, B., et al., 2010. Wave measurement needs and developments for the next decade. In: Hall, J., Harrison, D. E., Stammer, D. (eds.) *Proceedings of OceanObs'09: Sustained Ocean Observations and Information for Society*. 2, Venice, Italy, 21–25 September 2009. Frascati, Italy: ESA Publications. doi: http://dx.doi.org/10.5270/OceanObs09.cwp.87.

Turton, J., Pethica, C., 2010. Assessment of a new anemometry system for the Met Office's moored buoy network. *Journal of Atmospheric and Oceanic Technology* 27, 2031–2038.

Wilkerson, J. C., Earle, M. D., 1990. A study of differences between environmental reports by ships in the Voluntary Observing Program and measurements from NOAA buoys. *Journal of Geophysical Research* 95, C3, 3373–3385.

11 Satellite and Aircraft Remote Sensing

S. J. Lavender

1 Introduction

There are too many different definitions of remote sensing for the topic to be sufficiently covered within the confines of a single chapter; a search for 'remote sensing applications' within the internet, books and papers reveals that it can be applied to acoustic waves and satellite-based gravity missions, as well as the electromagnetic spectrum. This chapter reviews the data collected primarily from aircraft and satellites, with a focus on derived properties such as salinity, suspended particulate matter (SPM), temperature and surface water level. Therefore, what is being described here falls within the descriptive term 'Earth Observation' (EO); i.e., it applies to global or regional remote sensing that is used to study the Earth.

The derived remote sensing properties are our interpretation of the energy that is being transmitted and received within the electromagnetic spectrum, which includes passive and active optical, thermal, passive microwave and radar remote sensing. Even by narrowing the topic in this way it should still be recognised that this chapter can provide only an introduction, and so those wishing to gain a deeper knowledge should move on to more specific books, such as Njoku (2014), Robinson (2004) and Woodhouse (2006). Reviews of coastal remote sensing have also been written for aquatic environments in general (Miller et al., 2005), for the coastal management of tropical coastal resources (Green et al., 1996; Mumby et al., 1999) and for the European Seas (Barale and Gade, 2008).

2 Using the Electromagnetic Spectrum

We use only a small part of the electromagnetic spectrum for vision (illustrated in Figure 11-1); these visible wavelengths lie between 400 and 700 nm, where there is high solar irradiance (within the white area, Figure 11-1) and low atmospheric attenuation (dark grey shaded area under the curve, Figure 11-1). Going to longer wavelengths (lower frequencies) means that sensors use the infrared (IR), which includes thermal remote sensing, and then the microwave region, where measurements are in terms of frequencies with units defined by letters: e.g., a C-band instrument would operate between 4 and 8 GHz.

The shape of the dark grey shaded area (Figure 11-1) represents absorption by the Earth's atmosphere, with strong absorption by the permanent gases, including carbon

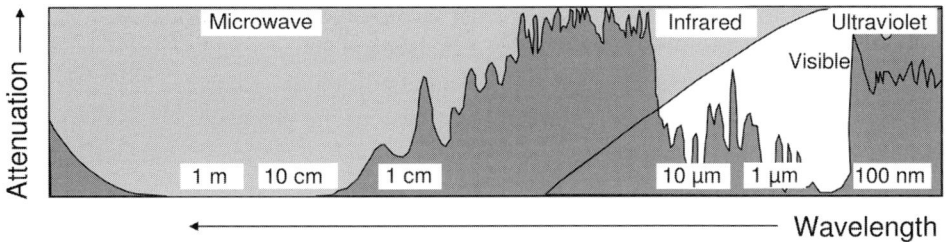

Figure 11-1 A representation of the electromagnetic spectrum (based on a figure originally produced by ESA). The wavelength nomenclature is shown and the wavelength dependence of atmospheric attenuation.

dioxide (CO_2) and oxygen (O_2), plus water vapour (H_2O) and ozone (O_3). Considering the shorter, ultraviolet (UV) wavelengths, then for these the atmospheric absorption increases rapidly due to ozone (shown in Figure 11-1), which is what protects life on Earth from the harmful UV rays emanating from the Sun.

The visible region sits in what is called an atmospheric window. As we move left (to longer wavelengths, Figure 11-1) into the IR, then there are strong absorption peaks caused by atmospheric gases, which need to be avoided if a sensor is designed to detect the properties of the Earth's surface rather than the atmosphere. The microwave region is also within an atmospheric window, but the signals are weak because there is low solar irradiance.

Sensors can be considered as either active or passive, with the difference being:

- Passive sensors detect the radiation naturally emitted or reflected by an object or area of interest – the land or water in the case of coastal remote sensing
- Active sensors emit energy towards the object or area of interest and then measure the return signal strength and its time delay

Examples of passive remote sensing include optical remote sensing where sunlight is reflected off the object, i.e., ocean colour, and thermal remote sensing where there is an emission from the earth or atmosphere rather than reflection. Passive microwave radiometers are used for the remote sensing of salinity and temperature. Active microwave examples include altimetry and Synthetic Aperture Radar (SAR), with Lidar being an active instrument within the optical region.

Because passive sensors are not emitting a signal, the platform power usage is reduced compared with active sensors and so, often, for satellites, passive imaging systems are global missions (i.e., they are always on and so collecting data across the whole globe), whereas active imaging systems may not be; e.g., SAR has historically collected data over predefined areas. The other limitation for the platform/sensor is data storage, and so there is also a trade-off between the amount of data collected for different sites, with commercial satellites often being targeted; e.g., high-resolution optical missions such as GeoEye only collect imagery where requested.

Aircraft mounted systems do not generally have power constraints, but aerial Unmanned Autonomous Vehicles (UAVs) do have power, weight and data storage restrictions. UAVs are becoming increasingly popular platforms for quickly and efficiently mapping areas such as estuaries and hence are considered as valuable scientific tools for environmental monitoring applications (Anderson and Gaston, 2013). This approach is developing rapidly because of the potential cost savings, but currently is still in its infancy with the development both of associated aviation regulations and the technology itself.

3   Optical Remote Sensing

Ocean colour satellites primarily operate on sun-synchronous platforms that provide global imagery plus an improved spatial resolution compared with geostationary satellites (related to altitude). Geostationary remote sensing platforms also exist, primarily for meteorological observations that can also be used for thermal remote sensing; their importance for ocean colour has been recognised, i.e., improved temporal coverage for the area being imaged. The first (and currently only, as of March 2016) geostationary ocean colour mission is the Geostationary Ocean Colour Imager (GOCI) launched by South Korea. It provides images with a spatial resolution of 500 m at hourly intervals of up to eight times a day, allowing observations of short-term changes in the northeast Asian region (Ryu et al., 2012).

Optical remote sensing utilises the electromagnetic radiation spectrum between 400 and 1000 nm, with visible radiation in the range 400 to 700 nm and the Near Infra-Red (NIR) occupying 700 to 1000 nm (Figure 11-1). Organic and inorganic material suspended in the water column can be quantified if it has a detectable effect on the water-leaving radiance, $L_w(\lambda)$; i.e., the signal coming from the sun that interacts with the water column and then travels to the sensor (Figure 11-2). An Atmospheric Correction (AC) is applied to remove the atmospheric signal (e.g., Gordon, 1978), which includes the path radiance, $L_p(\lambda)$ and transmittance, $T_d(\lambda)$, and allows the processing software to convert the imagery from a Top of Atmosphere (TOA) to a Bottom of Atmosphere (BOA) signal; i.e., converts from sensor radiance, $L_s(\lambda)$, to $L_w(\lambda)$, the water-leaving radiance (Figure 11-2).

Once the imagery has been atmospherically corrected it can then be used to derive quantitative variables; i.e., biogeochemical constituents (e.g., Gower and Borstadt, 1990). In estuarine and coastal waters the SPM or seston can be divided into two main groups of particles: inorganic and biological particles, with the latter (and sometimes the former; e.g., diatoms because they have a siliceous skeleton – frustule) dominated by phytoplankton in terms of their optical signature. In addition, absorption of electromagnetic radiation by coloured detrital matter and coloured dissolved organic matter (CDOM, also called gelbstoff or yellow substance) results from material derived from land runoff or biological degradation and collectively are referred to as CDM (i.e., absorption due to coloured dissolved and detrital organic matter; e.g., IOCCG, 2006; Shanmugam et al., 2016)

The effect of biogeochemical constituents on the water-leaving radiance depends on their absorption and scattering properties and varies in a nonlinear way in terms of both

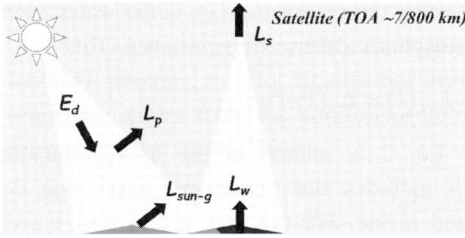

Figure 11-2 Optical signal paths through the atmosphere. The sky-glitter radiance (Eq. 1) is not drawn here; it would be a path of light, originating from the sun, which is first scattered within the atmosphere and then scattered by the sea surface.

wavelength and changes in concentration. Therefore, extracting accurate information from imagery requires the careful application of algorithms and an understanding of optical theory. In this chapter we will first review the different AC approaches and then consider how algorithms and optical modelling are applied to understand and quantify the content of estuarine and coastal waters.

3.1   Atmospheric Correction

Removing the atmospheric signal in ocean colour remote sensing is very important because only 5 to 10 per cent of the TOA signal originates from the water; over land the signal is much larger (10 to 50 per cent), but AC is increasingly commonly applied. Because it is not feasible to measure constantly all atmospheric variables, models must make rather broad assumptions, in particular about aerosols – the particles in the atmosphere.

In the oceanic situation, we have Case I waters (dominated by photosynthetic pigments), where the marine optical properties are gradually varying, and as the air mass is a long way from the land, the aerosol contribution can normally also be considered gradually varying and of marine origin. Case II waters (dominated by CDOM and/or SPM), which are present in estuarine and coastal environments, are more complex because both the water and aerosol properties exhibit significant variations over small spatial scales. Depending on the wind direction (and location) the aerosols will be a combination of anthropogenic, marine and terrestrial particles.

The TOA sensor spectral radiance (i.e., the signal received by the sensor, Figure 11-2) can be written as a linear sum of the various contributions, which all need to be estimated and taken into account:

$$L_s(\lambda) = L_p(\lambda) + L_{sky-g}(\lambda) + L_{sun-g}(\lambda) + L_{ws}(\lambda) \qquad (1)$$

Eq. 1 includes the path radiance and sun-glint terms, sky-glitter radiance (not drawn in Figure 11-2), $L_{sky-g}(\lambda)$, and sun-glitter radiance, $L_{sun-g}(\lambda)$, with the difference being whether the reflections from the ocean surface are from a direct or diffuse source; e.g., sun glitter can be seen at the sea surface on a sunny day when the water appears to be

is assumed that $\varepsilon(768, 865)$ will fall between the two models in the same manner. Then another LUT is used to derive the aerosol plus Rayleigh-aerosol reflectance from the Rayleigh-aerosol reflectance:

$$\varepsilon(765, 865) = \frac{1}{M} \sum_{k=1}^{M} \varepsilon_k(765, 865) \tag{14}$$

The AC algorithm for the Moderate Resolution Imaging Spectroradiometer (MODIS, Terra and Aqua missions, launched in 1999 and 2002) was developed from the SeaWiFS AC algorithm, using wavebands 15 (745–785 nm) and 16 (845–885 nm). The latter waveband was reduced in width (SeaWiFS was 857–872 nm) so that the MODIS AC would not have to account for the oxygen absorption that overlapped the SeaWiFS waveband (Wang, 1999).

Both MODIS and SeaWiFS employ a whitecap removal algorithm, whereby the normalized (N) whitecap reflectance $\lfloor R_{wc} \rfloor_N$ is assumed to be independent of wavelength (Eq. 15) and is calculated from the wind speed, W (Eq. 16; Gordon and Wang, 1994); the TOA whitecap reflectance $T \cdot R_{wc}(\lambda)$ is subtracted from the TOA reflectance before the AC is applied. Whitecaps (also called foam) occur in rough seas and will create an additional 'white' signal (i.e., making the water surface appear brighter):

$$R_{wc}(\lambda) = \lfloor R_{wc} \rfloor_N \cdot T(\theta, \lambda) \tag{15}$$

$$\lfloor R_{wc} \rfloor_N = 6.49 \times 10^{-7} W^{3.52} \tag{16}$$

The MEdium Resolution Imaging Spectrometer (MERIS, launched in 2002) Case I AC differs from SeaWiFS and MODIS by abandoning the assumption that the Rayleigh and aerosol radiances can be separated. Instead, the combined contributions to the multiple scattering are assessed using Eq. 17, where the atmospheric path reflectance, R_{path}, is derived from the reflectances due to single and multiple scattering by molecules only, R_r^* (but in the presence of aerosols), the equivalent term for aerosols, R_a^*, and that part of R_{path} that strictly results from successive aerosol and molecule scattering R_{ra}^* (Antoine and Morel, 1998, 1999):

$$R_{path} = R_r{}^* + R_a{}^* + R_{ra}{}^* \tag{17}$$

The MERIS AC also accounts for absorbing aerosols by assuming that the water-leaving radiance at 510 nm (the chlorophyll absorption hinge point) is insensitive to changes in chlorophyll and hence phytoplankton concentration. MERIS also utilised a 'Bright Pixel' (BP) component, run before the DP AC, where any significant NIR water-leaving radiance (due to the presence of high concentrations of SPM or coccolithophores, which have external calcium carbonate plates) can be estimated and removed (Moore et al., 1999).

Since the original MERIS implementation, a number of algorithms have been developed with varying hypotheses to account for the nonnegligible marine NIR signal (e.g., Ruddick et al., 2000; Stumpf et al., 2003; Lavender et al., 2005; Wang and Shi, 2007; Bailey et al., 2010). In addition, authors have moved the DP assumption into the shortwave Infra-Red (SWIR), where the assumption can remain valid for higher SPM concentrations; of the ocean colour sensors mentioned, only MODIS has suitable wavebands for this approach (Gao et al., 2007; Wang et al., 2009; Werdell et al., 2010).

MERIS has a second atmospheric correction approach that is applied to Case II waters; the Case 2 Neural Network, NNet (Schroeder et al., 2007). The underlying approach is to fit the atmospheric signal and a bio-optical reflectance model to the TOA reflectance throughout the visible spectrum, instead of, or in addition to, the NIR wavebands, and hence separate them. These approaches (including NNet and spectral optimization) have been employed by an increasing number of authors (e.g., Kuchinke et al., 2009; Shanmugam, 2012) because coastal waters are significantly more optically complicated than Case I waters; the aerosols are also more likely to be absorbing due to the proximity of the land and, additionally, anthropogenic sources of pollution.

More recent sensors to be launched include the Visible Infrared Imaging Radiometer Suite (VIIRS), launched on the Suomi National Polar-orbiting Partnership (NPP) platform in 2011, and the Hyperspectral Imager for the Coastal Ocean (HICO), which has been on the International Space Station since 2009. HICO has a full spectral coverage (88 channels between and covering 400 to 900 nm) with a high signal-to-noise ratio (SNR), and a spatial resolution of approx. 90 m (Corson et al. 2008). A complete list of missions (past, present and future) is available via the International Ocean Colour Coordinating Group website (IOCCG, 2016).

3.2   Application of Optical Theory to the Derivation of Water Constituents

We are interested in both the upwelling and downwelling light fields because changes in the solar irradiance will have a significant effect on the water-leaving radiance. In marine optics we use the ratio of the upwelling irradiance $E_u(\lambda, z)$ to downwelling irradiance $E_d(\lambda, z)$ to define reflectance:

$$R(\lambda, z) = \frac{E_u(\lambda, z)}{E_d(\lambda, z)} \tag{18}$$

In Eq. 18 the reflectance describes the flux of light across a boundary that is not a solid surface, but a certain point in space, e.g., a depth (z) of 5 m; the term *reflectivity* is used for light that is reflected off a solid surface. From Eq. 18 it is seen that reflectance is dimensionless, so we can therefore convert it to a percentage value. However, a remote sensor will often measure the upwelling radiance, $L_u(\lambda, z)$, rather than irradiance, because the signal returning is much weaker and more directionally dependent. Therefore, we can convert Eq. 18 to an equation for remote sensing reflectance (R_{rs}), which is the ratio of upwelling radiance to downwelling irradiance (see Eq. 19). Here the anisotropy factor (Q) converts the upwelling radiance to upwelling irradiance; it can be given a value of π steradians (sr) in an isotropic upwelling radiance distribution with approximately $3.5 sr$ for turbid waters (Bukata et al., 1988; Morel and Gentili, 1996). Sometimes authors neglect to include the Q factor, which means the units are then sr^{-1} rather than dimensionless.

$$R_{rs}(\lambda, z) = \frac{L_u(\lambda, z) \cdot Q}{E_d(\lambda, z)} \tag{19}$$

The most appropriate parameter to model or measure for the light field just above the water surface is the above-surface reflectance, $R_{+0}(\lambda)$. This can be determined from the ratio of upwelling to downwelling light (as seen in Eq. 19) or from the Inherent Optical Properties (IOPs) that are a function of the subsurface reflectance (Gordon et al., 1975), $R_{-0}(\lambda)$:

$$R_{+0}(\lambda) = 0.33 \cdot R_{-0}(\lambda) = 0.33 \cdot \frac{b_b(\lambda)}{b_b(\lambda) + a(\lambda)} \tag{20}$$

In this simplified version, the value of 0.33 (often given the symbol f) accounts for the loss of upwelling radiance as it passes through the sea-surface interface; therefore, the maximum possible above-surface reflectance is 33 per cent. In reality it is not a constant but will vary with the radiance distribution and volume scattering function; it has a value of 0.3244 when the sun is at the zenith (vertically in the sky with no clouds) and 0.3687 for a uniform (cloud covered) sky, but the variation from 0.33 never exceeds 5 per cent and only slightly varies throughout the commonly used part of the electromagnetic spectrum.

Historically, the angular dependence of R_{rs} has been ignored, with constant values used for f and Q when processing satellite or airborne ocean colour imagery, regardless of the positions of the sun and sensor (i.e., the solar zenith angle, satellite zenith angle and relative azimuth angle). For open ocean waters, Morel and Gentili (1996), updated by Morel et al. (2002), parameterized the f/Q factor as a function of Chlorophyll-a (*Chl*). Park and Ruddick (2005) showed that the particle scattering phase function is critical in determining the directional distribution. Since then, Gleason et al. (2012) have shown that a fixed phase function agreed well with in situ validation data (measurements made using a multispectral camera, Voss and Chapin, 2005) suggesting a single, average, particle scattering phase function could be used.

Lee et al. (1998) proposed an alternative form of Eq. 20 that is better adapted to turbid coastal waters:

$$R_{+0}(\lambda) = \pi \cdot \left(0.070 + 0.155 \cdot X^{0.752}\right) \cdot X \tag{21}$$

where:

$$X = \frac{b_b(\lambda)}{b_b(\lambda) + a(\lambda)}$$

The total absorption $a(\lambda)$ and backscattering $b_b(\lambda)$ coefficients are calculated from the absorption and scattering groups of biogeochemical constituents expressed as a function of their concentration and corresponding specific absorption and/or scattering (i.e., for one unit of concentration). Each group of biogeochemical constituents is made up of several substances (often present in differing proportions) with slightly different properties, represented by the routinely measured variable; e.g., the concentration of Chl is used to represent the phytoplankton, SPM is used to represent particles (inorganic and organic) and CDM represents the dissolved component and detrital absorption. Therefore, the specific absorption curves for each biogeochemical constituent group will be subject to variability (Prieur and Sathyendranath, 1981).

The total coefficients are calculated by summing the contributions from each of the specific in-water constituents (Eq. 22 and 23):

$$a(\lambda) = a_w(\lambda) + a_{CDM}(\lambda) \cdot CDM + a_{Chl}(\lambda) \tag{22}$$

$$b_b(\lambda) = 0.5 \cdot b_w(\lambda) + b_{b,Chl}(\lambda) + b_{b,SPM}(\lambda) \cdot SPM \tag{23}$$

The factor of 0.5 in the water (subscript 'w') backscattering term (Eq. 23) converts the water scattering ($b_w(\lambda)$) values to backscattering values and can be applied because of the symmetry of the volume scattering function (Morel, 1974). The Chl absorption and backscattering terms (Eq. 22 and 23) do not have multiplicative concentration values because they are nonlinear and such that concentration values are built into their full defining equations.

Chl and CDM fluorescence are not covered by this model because their effects on reflectance are of a much smaller magnitude than the suspended sediment backscattering in turbid estuarine environments. However, narrow wavebands (MERIS waveband 9 at 709 nm and MODIS waveband 14 at 676.7 nm) situated over the Chl fluorescence signal have been used to create products such as the Fluorescence Line Height (Gower et al., 1999) and Maximum Chlorophyll Index (Gower et al., 2005) in coastal waters with high Chl concentrations.

The number of constituent groups is sometimes increased with CDM split into the tripton (or Non-Algal Particles, NAP) and CDOM components; e.g., Belanger et al. (2008) developed an empirical algorithm to distinguish the contribution of CDOM to CDM, and Bukata et al. (1981) split the two components when modelling the optics of Lake Ontario. NAP and CDOM have similar exponentially decreasing (with increasing wavelength) absorption spectra and so can be difficult to distinguish, but NAP has a scattering contribution that CDOM does not have. In the open ocean, CDM is dominated by CDOM and has been found to have factors that do not covary (Siegel et al., 2005), whilst in estuarine waters and river plumes, CDOM can be inversely related to salinity when conservative mixing occurs (Chapter 4), allowing optical mapping of salinity (e.g., Palacios et al., 2009, for the Columbia River plume).

3.3   Application of Algorithms to Derive Quantitative Variables and Parameters

The most basic retrieval approach is to correlate a single spectral waveband with the biogeochemical constituent of interest; e.g., the natural logarithm of SPM and the reflectance in the red wavelength region. Retrieval variables can also include waveband ratios because they suppress solar angle and atmospheric effects and may also cancel out effects caused by the sensor tilt angle and Field-of-View (FOV).

Despite their usefulness, empirical statistical algorithms cannot provide detailed information on scattering and absorption and hence cannot discriminate between different absorbing agents, so that multispectral methods are needed if algorithms are to be extended spatially or temporally from the data set they were developed for. Often, semi-analytical algorithms (SAA) are used for detecting and separating different substances from multichannel measurements, using inverse modelling based on known absorption and scattering relationships (Eq. 20 or 21, 22 and 23). A further development of this approach is the 'optimisation' of inversion algorithms (e.g., Doerffer and Fischer, 1994; Lee et al., 1999). For MERIS, Schiller and Doerffer (1999) developed a NNet as a multiple nonlinear regression technique to parameterize the inverse relationship between concentration and reflectance within coastal waters. However, even these approaches perform better in some geographical regions and/or optical water types than others because of the difficulty of applying universally applicable algorithms for Case II waters, due to the variety of biogeochemical-specific IOPs (IOCCG, 2000, 2006 and 2010).

3.4   Penetration Depth and the Remote Sensing of Bathymetry

The subsurface downwelling irradiance diminishes in an approximately exponential manner with increasing depth, which is wavelength dependent because of the wavelength dependence of the absorption and scattering functions (see Eq. 22 and 23). The rate of decrease can be quantified by calculating the downwelling diffuse attenuation coefficient, $K_d(\lambda)$. In turn, the penetration depth, $z_{90}(\lambda)$, which is the depth above which 90 per cent of the diffusely reflected irradiance originates (Gordon and McCluney, 1975), is given by:

$$z_{90}(\lambda) = \frac{1}{K_d(\lambda)} \tag{24}$$

The penetration depth defined in Eq. 24 is considered to be the maximum depth from which an optical sensor receives a detectable signal, and so the signal originating from between this depth and the surface is integrated to create the water-leaving radiance. However, water that is closer to the surface contributes a greater percentage of the signal; e.g., if we have a subsurface *Chl* maximum, then the water below this depth will not be visible to an ocean colour sensor. Penetration depth is important because there will be no contamination of the signal by the bottom reflectance if the water depth is greater than this; conversely, in this case and in an approximate sense, a Lidar instrument is unlikely to be able to map bathymetry.

Lidar, the name of which can be interpreted as 'light detection and ranging' or, alternatively, 'light radar', is (like radar) an active remote sensing technique (Section 6 and Chapter 3). A laser pulse is transmitted towards the target with the time and strength of the return pulse measured (see Figure 11-3, where return signals are generated by both the sea surface and seabed). The use of Lidar for terrain elevation mapping began in the late 1970s, whereas modern systems map both terrestrial terrain and bathymetry. Traditional Lidars measure one or more discrete returns, whereas modern systems are full-waveform Lidars; i.e., there is a continuous return signal at a uniform sampling frequency. By analysing this signal it is possible to obtain a more detailed characterisation of the surface being remotely sensed, akin to water level detection from altimetry (Section 6.1 and Chapters 4 and 5).

The first free-flying satellite Lidar was launched by NASA in January 2003 (Winker et al., 2003); NASA's ICESat (Ice, Cloud and Land Elevation Satellite) carried the

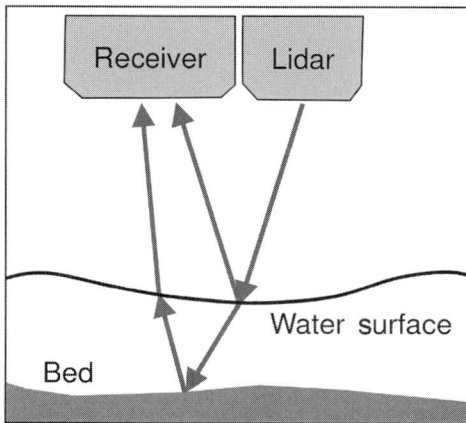

Figure 11-3 Lidar signals through the atmosphere and water column.

Geoscience Laser Altimeter System (GLAS), operating in the NIR and visible at 1.064 μm and 0.532 μm, respectively. Since then, atmospheric Lidars have been launched; e.g., CALIOP (Winker et al., 2003) in 2007, operating at wavelengths of 355 nm and 532 nm with dual polarisations designed for studying atmospheric aerosols and phytoplankton biomass. ICESat-2 is scheduled for launch in 2018; it will emit visible laser pulses at 532 nm with a pulse repetition rate of 10 kHz (current design estimate), which will generate dense, along-track sampling resolution of 0.70 m (Abdalati et al., 2010). The main focus of the data collection will be on polar ice coverage and vegetation canopy height; estuarine and catchment applications may also be possible.

4   Thermal Imagery

The temperature of estuarine and riverine environments is an important indicator of water quality, and this is influenced by groundwater and surface-water freshwater inputs, alongside tidally driven coastal saltwater inputs to estuaries. Although hydrologists and ecologists are often (ultimately) interested in the total water-column measurements, the radiant temperature emitted by the water surface and detected using thermal Infra-Red (TIR) remote sensing provides an attractive alternative to in situ measurements; this measurement is often called the 'skin' layer because it is a measurement that is only applicable to the top 100 μm for TIR. The NIR is often defined as the wavelengths between 760 nm and 1000 μm (1 mm), with TIR remote sensing typically occurring between 7 μm and 18 μm. Derived products are called Water Surface Temperature (WST) or, in marine environments, Sea Surface Temperature (SST).

All objects at temperatures above absolute zero emit thermal radiation, as specified by the radiation laws. A blackbody is a theoretical object that acts as a perfect absorber and emitter of radiation and is useful when describing and modelling the thermal behaviour of actual objects, such as the Earth. As the temperature of a blackbody increases, the wavelength of peak emission decreases, in accordance with Wien's displacement law:

$$\lambda_{max} = A/T \qquad (25)$$

In Eq. 25, A is 2898 μm $^\circ$K and T is the absolute temperature (in $^\circ$K) of the blackbody. The emissivity, $\varepsilon'(\lambda)$, is the ratio between the estimate of an object's emittance, e, and a blackbody at the same temperature, e_{bb}, which varies between 0 and 1:

$$\varepsilon'(\lambda) = e/e_{bb} \qquad (26)$$

where the Stefan-Boltzman law (in units of watt m^{-2}) is:

$$e_{bb} = 5.67 \times 10^{-8} \cdot T^4$$

As with all passive remote sensing approaches, TIR WST measurements are sensitive to the presence of clouds, scattering by aerosols, and atmospheric water vapour; i.e., meteorological conditions have a strong influence. For wavebands around 3.7 μm and/ or near 10 μm, fog and cloud are essentially opaque, and the signal-derived temperatures can be biased by aerosols by as much as 2°C or more. Therefore, as for ocean colour imagery, the data require a correction for the atmospheric influence that can only be derived from cloud-free pixels. Torgersen et al. (2001) utilised thermal imagery in the 8 to 12 μm waveband from the Thermovision-1000 forward-looking Infra-Red (FLIR) system, with a typical above-ground altitude of 300–700 m. The data were converted to temperatures using Planck's radiation law and sensor calibration curves (Atwell et al., 1971). The radiant water temperatures were then corrected for the emissivity of water (0.96) and atmospherically corrected using LOWTRAN-7 (Kneizys et al., 1988).

The Daedalus Airborne Thematic Mapper (ATM) has 11 wavebands in the visible, NIR, SWIR and thermal regions of the electromagnetic spectrum. Wavebands 2 (450–520 nm), 3 (520–600 nm), 5 (630–690 nm) and 7 (760–900 nm) are often used as simulations of Landsat-5 Thematic Mapper (TM) wavebands 1 to 4, and (as with Landsat) the ATM has been used extensively to quantify SPM within estuaries and in the near-shore coastal zone. Hedger et al. (2007) utilised a time series (approx. 10-min intervals) of ATM thermal images of Kirkcudbright Bay, on the northern side of the Solway Firth, southwest Scotland, to characterise and quantify the spatial dynamics of estuarine WST. The thermal imagery exhibited both banding and speckle features, attributed to sensor noise, which were removed by use of a mean-spatial filter. Marmorino and Smith (2008) extended the research to review the impact of bottom-generated vertical mixing on thermal features.

An important element of TIR remote sensing is the time of day because many surfaces have diurnal radiant temperature variations. During the daytime, direct sunlight will differentially heat an object according to its thermal characteristics and sunlight absorption, principally in the visible and NIR region of the electromagnetic spectrum. In the open ocean this heating would create a diurnal cycle (i.e., diurnal variability) of SST variations; near-surface, ocean diurnal warm-layer events have been observed to exceed 6°K (Merchant et al., 2008). In estuarine and coastal waters we would expect to see less (or no) diurnal impact in places where there is strong tidal or wind-driven mixing of the water column, but in inlets and estuaries with shallow waters and less vertical mixing it can become significant (e.g., the upper Plym Estuary, UK; Uncles et al., 2015).

A detailed review of thermal remote sensing for WST is available by Handcock et al. (2012) and by Robinson (2004) for SST.

5   Examples: Deriving Biogeochemical Constituents and Mapping WST

5.1   Airborne Remote Sensing

Airborne remote sensing (ARS) started in the early 1980s with the Daedalus ATM and then, in the late 1980s, began to include the Compact Airborne Spectrographic Imager (CASI), which was developed from the Fluorescence Line Imager (Borstadt et al., 1985). CASI initially operated over a 430- to 870-nm wavelength range with a resolution of 2.8 to 2.9 nm (termed hyperspectral), and the swathe width had 512 pixels spread over the FOV; CASI-1500 (Itres, 2014) now operates from 380 to 1050 nm and has 1550 pixels. The spatial resolution depends on the altitude, ground speed and integration time; it is typically operated in the 2–10 m range when flown between 2000 and 10000 feet (approx. 600 and 3000 m). Originally, CASI could operate in either a spatially or spectrally optimised mode. In the spectrally optimised mode the operator selects 39 separate pixels across the FOV or portion thereof (Wilson, 1995), and in the spatial mode the operator selects nonoverlapping spectral wavebands, specifying wavelength and bandwidth (Anger et al., 1994).

An example of CASI and Daedalus AADS-1268 ATM imagery, flown over the Tamar Estuary (southwest UK) on 13 June 2003, provides an illustration of estuarine remotely sensed data (Figure 11-4). The first two images (Figure 11-4a, left, and Figure 11-4b, middle) are pseudo-true colour composites from a CASI flightline (670 nm as red, 560 nm as green, and 490 nm as red blue) that show the before and after images following an AC that has been applied according to Lavender and Nagur (2002) and Lavender et al. (2005), to account for the signal in the NIR caused by optically turbid waters. The uncorrected CASI flightline shows an image without cloud contamination, but with significant variation in brightness across the image caused by sunglint (Figure 11-2). Once the image has been atmospherically corrected (middle, Figure 11-4b) the variations in water colour become more obvious, with a good example being the Devonport dockyard, where one of the dock areas has its dock gates open while the other two have theirs closed, resulting in a different water colour (area enclosed by the ellipse in the up-estuary part of Figure 11-4b) compared with the adjacent estuarine waters.

The right-most image (Figure 11-4c) shows ATM waveband 11 operating in the TIR region (8.4–11.5 μm). The in-water frontal features visible in the CASI imagery, which are due to variations in the biogeochemical constituents, are often visible in the thermal imagery, despite this being a raw brightness temperature image; it is not thermally

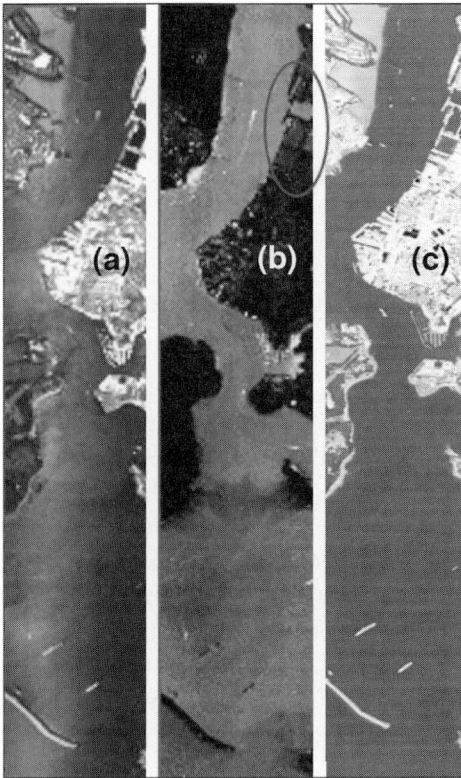

Figure 11-4 An illustration of CASI and ATM imagery from the Tamar Estuary, southwest UK, captured during 13 June 2003. The coloured images (a, b), rendered greyscale here, represent pseudo-true colour composites of wavebands 11, 6 and 3 as red, green and blue. The images are, from left to right: (a) uncorrected (sensor) CASI image; (b) CASI water-leaving reflectance; and (c) thermal band from ATM. The dark grey ellipse in the up-estuary part of (b) highlights the varying water colour (light and dark here) within the basins of Devonport dockyard. The data were supplied courtesy of the NERC ARSF.

calibrated or atmospherically corrected, but analytical and sea-truth approaches are still possible, as detailed by, e.g., Callison et al. (1987) and Uncles et al. (1999).

There are both advantages and disadvantages to using airborne as opposed to satellite remote sensing. The advantages of ARS include the ability to fly at user-specified times and with repeat cycles of minutes, rather than days, and the increased spatial resolutions when compared with satellite sensors (Collins and Pattiaratchi, 1984). Aircraft sensors can also be upgraded and regularly calibrated, whereas satellite sensors must rely on solar or lunar calibration measurements and 'vicarious adjustment' (i.e., on a complementary calibration using ground-truth measurements). The main disadvantage of ARS is that the sensors are on an unstable platform within the atmosphere, which causes additional complexities for both the atmospheric and geometric corrections. The Tamar imagery presented (Figure 11-4) has no geometric correction applied, and so there are distortions within the image caused by the yaw, pitch and roll of the aircraft.

5.2   Satellite Remote Sensing

The Landsat series of satellites, starting in 1972, have included the Multiple Spectral Scanner (MSS) that had two visible and an NIR waveband (500–600, 600–700 and 700–800 nm), and the TM and Enhanced Thematic Mapper plus (ETM+) that has three visible and an NIR waveband (450–520, 520–600, 630–690 and 760–900 nm). The TM and ETM sensors also have a thermal waveband (6), which operates in the 10.4–12.5 μm region. In 2013, Landsat 8 was launched with the Operational Land Imager (OLI) and Thermal Infrared Sensor (TIRS) aboard; OLI has nine wavebands from the visible to SWIR (430 nm–2.3 μm) with a spatial resolution of 15 m for the panchromatic waveband and 30 m for the multispectral wavebands, whilst TIRS has two wavebands in the TIR at 100-m spatial resolution.

The advantage of using Landsat for optically and thermally mapping estuaries and near-shore coastal waters, in comparison with the global ocean colour missions, is its spatial resolution, which is approximately 30 m for the optical wavebands and 120 m for the thermal waveband of the TM/ETM sensors, compared with several hundred metres to 1 km for typical ocean colour missions. However, the Landsat MSS/TM/ETM/OLI sensors are not optimised for ocean colour applications because they do not have narrow wavebands and have a poorer SNR compared with an ocean colour sensor; water targets have much lower reflectance. Landsat also has a large swathe (FOV in the across track direction), with no tilting mechanism to avoid sunglint and a poor repeat cycle of 14–17 days.

High-resolution satellite imagery (such as GeoEye, IKONOS, and Quickbird) can also be used for optical mapping, but suffers from the same types of limitations as Landsat, and most are currently only offered commercially; i.e., the data needs to be purchased and can become expensive for time-series analyses or large geographical regions. The WorldView-2 sensor has enhanced capabilities in that it has an approximately 0.5-m spatial resolution, a panchromatic waveband and eight multispectral wavebands at approximately 2-m resolution, four standard wavebands (blue, green, red and NIR1) plus four additional wavebands that are coastal (400–450 nm), yellow (585–625 nm), vegetation red edge (705–745 nm) and NIR2 (860–1040 nm, which overlaps with NIR1 but has a longer wavelength). This means that WorldView-2 has stimulated the generation of marine-orientated commercial products such as satellite bathymetry in optically shallow waters.

The SPOT (Satellite Pour l'Observation de la Terre) series of satellites were first launched in 1986 with SPOT 4 and 5 currently being active. SPOT 4 and 5 have identical specifications for the High Resolution Visible (HRV) sensor, including two visible and a NIR waveband (490–610, 610–680 and 780–890 nm) at 10-m spatial resolution with a 26-day repeat cycle. It has been followed by the French Pleiades-HR (High-Resolution Optical Imaging Constellation) that is formed from two satellites operating at a spatial resolution of 0.7 m for the panchromatic waveband and 2.8 m for the multispectral wavebands (blue, green, red and NIR). Other countries (such as India, Korea and Thailand) have also launched high-resolution satellite missions for their own purposes.

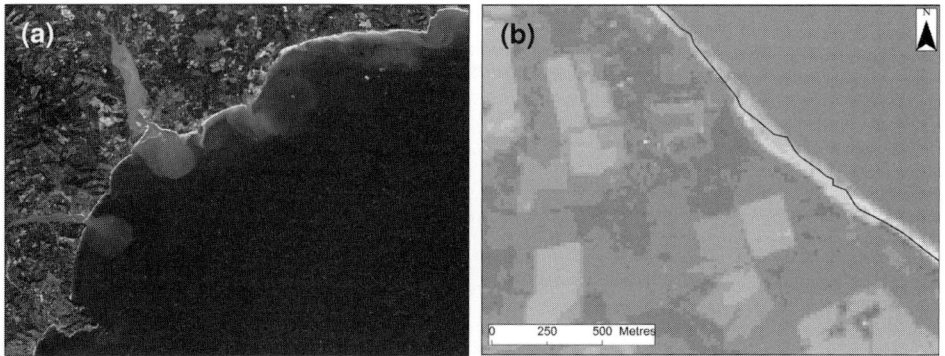

Figure 11-5 Landsat 8 imagery showing: left, (a) coastal plumes from estuaries along the south coast of Devon (UK), seen using the red waveband on 4 November 2013; and right, (b) panchromatic image for Happisburgh in Norfolk (UK) from 26 February 2014, overlaid with a black line indicating the cliff edge in 2003. The data were supplied by courtesy of USGS with processing undertaken by B. Hanlon (pers. comm.).

As previously mentioned, due to their complexity and seasonal variability, estuarine and coastal waters can be poor candidates for optical empirical algorithm approaches. Also, SAA performance is affected by the extreme signals found in these waters; e.g., very small water-leaving radiance values in the blue wavelength region due to high CDOM concentrations. Therefore, a large body of research has utilised the empirical statistical approach, especially for satellites with limited spectral capabilities, such as Landsat and SPOT. Empirical algorithms for estimating SPM concentration include the following approaches: the red waveband using an exponential function (Pavelsky and Smith, 2009); NIR reflectance (Sterckx et al., 2007); NIR/green ratio (Doxaran et al., 2002; Doxaran et al., 2006).

The optimum wavelength for SPM quantification (no matter the size or type) often depends on the concentration range; increasing SPM concentrations will create signal saturation at lower wavelengths, and so the algorithm/approach needs to move to higher wavelengths to obtain a detectable relationship between reflectance and concentration. Sources of uncertainty include variations in sediment colour, grain size, and mineralogy within the mapped study region (Novo et al., 1989; Bale et al., 1994), bottom reflectance in shallow areas (Tolk et al., 2000) and errors resulting from the AC (Stumpf and Pennock, 1989).

Two examples of Landsat 8 imagery are shown that illustrate its use in the coastal zone (Figure 11-5). The image on the left (Figure 11-5a) shows coastal plumes originating from estuaries (including the Exe and Teign) along the south coast of Devon, UK, on 4 November 2013. The image is the OLI red waveband 4 (655 nm) that correlates well with turbidity, as discussed previously. The image on the right (Figure 11-5b) is for Happisburgh in Norfolk, UK, from 26 February 2014, overlain with a black line indicating the cliff edge in 2003. On 11 December 2013, the BBC reported that, following the tidal surge along the east coast, 6 feet (1.8 m) of land in the area of Beach Road was lost to the sea (BBC, 2014). The panchromatic waveband 8 (500–680 nm) was used to investigate whether this could be seen using Landsat 8, given its higher spatial resolution than the multispectral wavebands (15 m as opposed to 30 m).

The launch of Landsat 8 in 2013 and the Copernicus Sentinel-2 Multi-Spectral Instrument (MSI) in 2017 are increasing the opportunities for improved processing approaches. MSI has 13 wavebands in the visible/NIR and SWIR and a spatial resolution that varies between 10, 20, and 60 m. Researchers have already been activity-processing Landsat with ocean colour missions so that the Landsat AC can use variables and parameters, especially related to the aerosols, which are derived using optimal wavebands (e.g., Hu et al., 2001).

6   Microwave Remote Sensing

Microwave satellite remote sensing has tended to be underutilised in estuarine environments because of the relatively large pixel sizes for satellite microwave image products and footprint for sounding products; also there has been the limited availability of airborne systems. However, reductions in satellite pixel size and the introduction of improved processing capabilities have started to create satellite products that could be of significant importance.

As with the optical and TIR regions of the electromagnetic spectrum, the microwave region has both atmospheric absorption bands and windows. The passive return signal is sensitive to temperature, roughness, salinity and moisture content, which can be both positive and negative; these are variables of interest that, depending on the application, can be corrected. The longer microwave wavelengths are largely unaffected by clouds and generally easier to correct for atmospheric effects, thus providing an improved frequency of temporal coverage, particularly in tropical environments.

Passive microwave SST instruments primarily operate near 7 GHz, with an additional waveband at 21 GHz for atmospheric water vapour correction. The ocean emissivity (Eq. 26) is fairly constant, and so variations in the brightness temperature (T_b) primarily relate to changes in the physical water temperature; salinity only influences emissivity at lower frequencies (< 6 GHz). Therefore, for an idealised ocean it would be possible to infer SST or salinity from T_b alone, but (as with optical remote sensing) there is a need to account for a nonideal surface; e.g., roughening effects from the wind. Also, the large footprint causes side lobes that can have a significant effect; e.g., for inland and coastal waters the footprint will often include 'bright' land areas in addition to 'darker' water areas. Plus, the lobes can pick up stray noise from other sources; e.g., mobile phones operate in the L-band (1.4 GHz), which is the same frequency as the Microwave Imaging Radiometer with Aperture Synthesis (MIRAS) on the Soil Moisture and Ocean Salinity satellite (SMOS), so there are radio frequency interference (RFI) sources (Hallikainen et al., 2010).

For an active sensor there is additional information in the timing of the return pulse, which can be used to measure distance. Therefore, active systems provide both the timing and properties of the echo, including strength and polarisation. Instruments making accurate measurements of distance are normally called altimeters, whereas instruments focusing on the echo properties are normally called scatterometers. A traditional radar altimeter is not an imaging device, but rather a nadir-pointing instrument that continuously records average surface 'spot' height directly below the satellite.

Radar is an acronym for 'radio detection and ranging' and was developed as a means of determining the positions of objects using radio waves. Short bursts, or pulses, of microwave energy are transmitted in the direction of interest, and the strength and origins of the echoes or reflections received are recorded. The value of a radar altimeter is the extremely high precision with which it can measure the average height of a surface; given the low frequencies, it is also unaffected by cloud cover. From an altitude of approximately 880 km an echo takes approximately 5 ms to travel to the Earth's surface and back. There are four major sources of error: quality of the instrument, accurate knowledge of the platform location, estimate of the speed of travel, and variations in surface topography. The speed of travel is influenced by the atmosphere, which causes a delay, and (as with optical remote sensing) AC is required.

Remote sensing techniques, such as SAR Interferometry (InSAR) from passive and active microwave observations, can offer important information; e.g., on the changing areal extent of large wetlands (e.g., Wdowinski et al., 2008). InSAR images are of the same geographical area, but acquired from two different positions; they are therefore akin to stereo photography. As SAR is an active system the transmitted signal is known, so the imaged area can be interpreted in terms of variations in both returned signal brightness and scattered signal phase. The interaction between the phases of the transmitted and scattered signals results in an interferogram, which can be translated into differences in terrain elevation via a process termed phase unwrapping.

Smooth water surfaces act as specular reflectors and so yield no radar returns, but a roughened surface will return a signal of varying strength that is dependent on the roughness. For interferometry, a flat water surface can be mapped where there is vegetation (e.g., reed beds) to cause a return signal. The return signal from a roughened water surface can be inverted to provide estimates of local wind speed. Variations in surface roughness can also be interpreted to map slicks created by biogenic films. Modern SAR instruments have a resolution of up 1 m; e.g., TerraSAR-X operates in the X-band (9.6 GHz) and has a choice of resolutions (1, 3, or 18.5 m). It has been in operational service since January 2008 and, with its twin satellite TanDEM-X launched in June 2010, acquires data for WorldDEM^{TM}, which is a worldwide and homogeneous Digital Elevation Model (DEM) that is available at 12-m spatial resolution. The COnstellation of small Satellites for the Mediterranean basin Observation (COSMO-SkyMed) is a constellation of four satellites providing X-band data for a variety of configurations, including 1-m spatial resolution data over a 10 km by 10 km area for HH or VV polarisation (see later), which is Spotlight-2 mode.

Polarisation denotes the orientation of the electric field vector component of the electromagnetic field that is emitted and received. Thus VV denotes Vertical emitted and Vertical received, and VH is Vertical emitted and Horizontal received. Different polarisations can be used to differentiate between surfaces of different roughness. As an example, alternating polarisation mode Envisat Advanced Synthetic Aperture Radar (ASAR) imagery is illustrated for Morecambe Bay (northwest England, Figure 11-6)

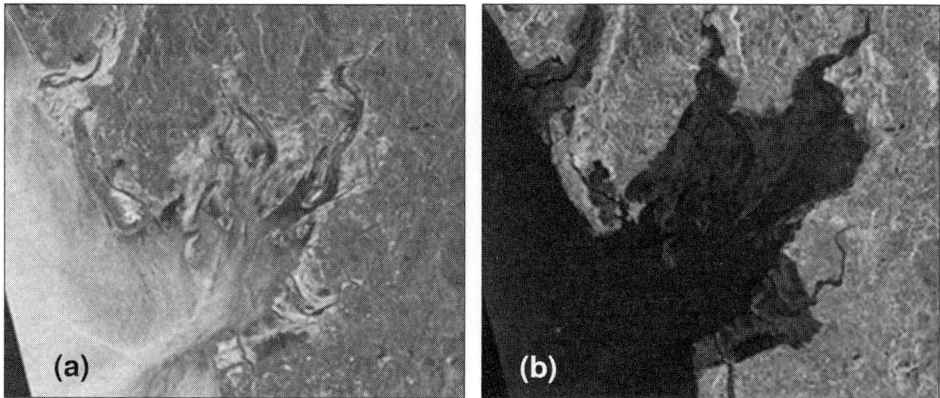

Figure 11-6 ASAR imagery of Morecambe Bay (northwest UK) from 27 December 2003 for Low Water + 0.8 hour, shown as: left, (a) V-V; and right, (b) V-H alternating polarisation modes (RSACUoP, 2005). Data provided by courtesy of ESA. This project was undertaken as part of the British National Space Centre (BNSC, now UK Space Agency) 'Government Information from the Space Sector' (GIFTSS) initiative, in collaboration with Lancaster City Council, UK (led by Remote Sensing Applications Consultants Ltd and Plymouth University, 2005).

during 27 December 2003, captured at Low Water + 0.8 hour. This imagery was taken as part of a project that investigated the applicability of EO and GIS methods for identifying positions and time-dynamics of low-water channels in Morecambe Bay, UK (RSACUoP, 2005).

6.1   Example: Water-Level Measurement

Satellite radar altimetry has been used to measure inland water-level variations over large river basins (e.g., Birkett, 1998). In these cases it has been argued that 'retracking' is not needed; retracking is the determination of the leading edge of a waveform when the waveform does not conform to the Brown (1977) 'standard' waveform model, which is applicable over the ocean and hence also over large freshwater bodies. However, retracking is needed for smaller water bodies, and there also needs to be careful data editing and/or filtering to ensure that only waveforms for which the leading edge can be determined accurately are selected (Crétaux et al., 2011). Therefore, a vital ingredient for smaller lakes and rivers is an accurate inland water locations mask, which allows short segments of altimetry data that contain the water surface to be selected from the overall data set that includes land. Once extracted, the altimetry data for a location are then often averaged to obtain a mean height and its associated uncertainty.

Validation is performed via comparison with gauge data; often the satellite-derived values (levels) are shifted in the y-axis by a constant offset to produce a best match with the gauge. Alternatively, an arbitrary reference datum (e.g., the mean over an observation period) is compared with 24-h-averaged gauge-stage data taken at a single latitude-longitude location and based on a local datum or mean sea level. However, it should be

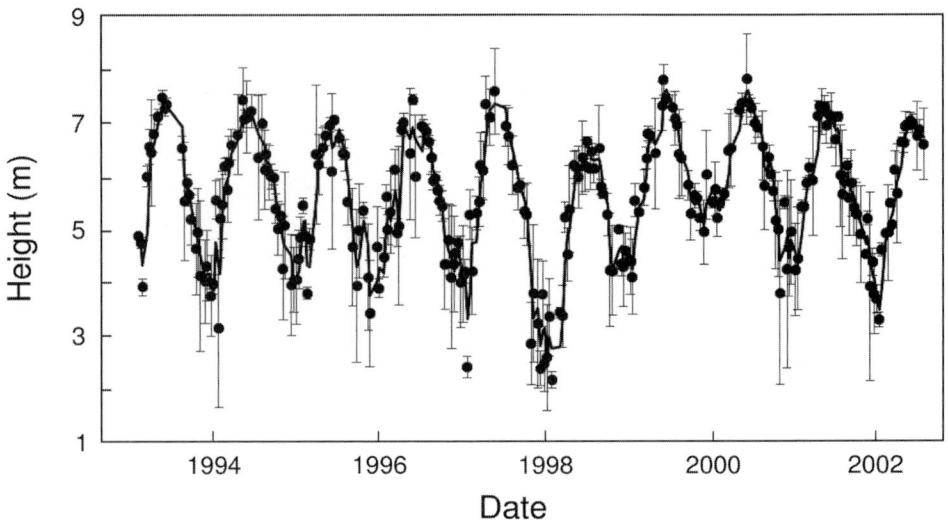

Figure 11-7 Amazon River water-level data from Topex/Poseidon, centred on 1.98°S, 53.85°W. Data provided by courtesy of the HYDROWEB site (HYDROWEB, 2016) with processing according to Crétaux et al. (2011).

remembered that the altimetry values differ from traditional gauge measurements in that they provide 'spot' heights, where spot refers to an area rather than a point.

Berry et al. (2005) used a rule-based expert system, which classifies each echo into one of eleven categories of waveform shape and picks an appropriate retracking algorithm. As the land is a relatively poor reflector of Ku-band energy (12–18-GHz portion of the electromagnetic spectrum) compared with inland water, the response from the 'bright' water target will often dominate the altimeter return (Berry et al., 2005); the outputs are included in the European Space Agency (ESA) River & Lake project website (ESA, 2016). In addition, experimental Coastal and Hydrology products are also available from the Pistach project via the AVISO+ website (AVISO, 2014), and the HYDROLARE project (Crétaux et al. 2011) provides water level variations for about 150 rivers, global lakes and reservoirs (HYDROWEB, 2016).

Water-level accuracy ranges from centimetres to decimetres, depending on the water-body size and also, to a lesser extent, the location (Crétaux et al., 2011); e.g., mountain lakes are more difficult to track. Data for water levels extracted from Topex/Poseidon, centred on 1.98°S 53.85°W, which corresponds to the Amazon River, show substantial temporal variability (plotted as height (m) and its standard deviation versus year in Figure 11-7); the data were downloaded from HYDROWEB (2016), with processing according to Crétaux et al. (2011). Water-level data are also available from sensors such as Europe's ENVISAT (2002–2012), Jason-1 (2001–2013), Jason-2 (2008-) and SARAL (2013-), which is onboard an Indian Space Research Organization (ISRO) satellite.

With the launch of the Cryosat-2 mission in 2010, an approach called SAR altimetry (Raney, 1998) was also developed. The instrument differs from a conventional radar altimeter in that it exploits coherent processing of groups of transmitted pulses. Its advantage is that it has an increased spatial resolution in the along-track dimension that allows for the accumulation of more data, leading to improved speckle reduction, and hence an improved precision for the altimetry measurements. The Copernicus Sentinel-3A Radar Altimeter, with Sentinel-3B scheduled for 2017 and Sentinel-3C before 2020 will further extend the availability of these data.

Further information on altimetry can be found within sources such as Chelton et al. (2001) and Rosmorduc et al. (2011).

7   New Developments

The most striking of the new developments in the two-year period up to March 2016 is the increase in availability of satellite remotely sensed data. Commercial satellite missions have been mentioned in this chapter, and with increased launches these will provide an increasing wealth of high spatial resolution data sets. Worldview-2 has now been joined by Worldview-3, with WorldView-4 due to launch in September 2016. Worldview-3 has the standard panchromatic and multispectral wavebands at spatial resolutions of 0.31 and 1.24 m, but also SWIR bands at 3.7 m resolution and CAVIS (Clouds, Aerosols, Vapors, Ice, and Snow) bands at 30-m resolution to increase the applications for which it can be used. In addition, companies like Planet Inc. operate these larger satellite missions in parallel with smaller cubesat-based instruments, e.g., the triple-cubes at Doves that aim to provide high temporal coverage. This increase in missions is also being matched by noncommercial providers. For example, there are plans to launch the European (ESA) Copernicus Sentinel-3D satellite unit after 2020 to provide continuous coverage up until at least 2030. NASA, and the USGS have also begun work on Landsat-9, due for launch in 2023, with missions such as the Hyperspectral Infrared Imager (HyspIRI) being studied for a future launch.

The increase in data availability is leading to a shift in how satellite data is handled, with a move towards using remote computing-based resources (often termed cloud computing), so that the large data sets do not need to be downloaded. It is also leading to an increased uptake of the data by nonspecialists, who are using the analytical power available from the interrogation of large data sets to extract valuable information.

8   Final Remarks

Estuarine and coastal waters and their adjacent environments are complex physical and biological environments with important ecological systems, natural habitats and vital assets that include centres of human population and their associated infrastructure. With the advent of remote sensing the scientific community saw the advantages it had to offer (i.e., snapshots of these complex environments), but they also encountered the difficulties of both high spatial variability within the land/water target and the need to remove the atmospheric signal. Therefore, although some applications have existed for many years, the science and technology continue to develop and mature, and as they do, new and improved products continue to emerge.

Remote sensing products are increasingly available to download from the internet, but without having some understanding of the physics behind the measurements, it is difficult to understand the assumptions and uncertainties. Eleveld et al. (2014) derived SPM concentrations for the Westerschelde Estuary and encountered and reported biases in the remote sensing observations that were due to tidal aliasing and the effects of variable weather conditions (e.g., cloudy skies); for this estuary, tidal aliasing resulted from the fact that during sun-synchronous satellite overpasses, high water occurred concurrently with spring tides, and low water occurred concurrently with neap tides.

Retrieving ocean colour products from raw data is complicated by several factors, including turbidity, which can make the 'Dark Pixel' assumptions invalid, as well as complex aerosol mixtures, that are not fully represented by global approaches and/or Look-Up Tables. Inherent Optical Properties' retrievals can be improved through improved sensor capabilities that allow for new and improved data approaches, including ultraviolet, hyperspectral and polarisation data. There is also an increasing focus on assimilating the data into hydrodynamic models. In situ data (e.g., mounted on fixed platforms, ships, buoys and floats, and unmanned autonomous vehicles, Chapter 10) remain key to validation, despite the cost of acquiring it. Finally, the improved spatial resolution and/or data coverage of microwave remote sensing data make it increasingly useful in estuarine and coastal environments.

References

Abdalati, W., Zwally, H. J., Bindschadler, R., et al., 2010. The ICESat-2 Laser Altimetry Mission, *Proceedings of the IEEE* 98, 735–751.

Anderson, K., Gaston, K. J., 2013. Lightweight unmanned aerial vehicles will revolutionize spatial ecology. *Frontiers in Ecology and the Environment* 11, 138–146.

Anger, C. D., Mah, S., Babey, S. K., 1994. Technological enhancements to the Compact Airborne Spectrographic Imager (CASI). *First International Airborne Remote Sensing Conference and Exhibition*. Strasbourg, France, 11–15 Sept. 1994. 205–213.

Antoine, D., Morel, A., 1998. Relative importance of multiple scattering by air molecules and aerosols in forming the atmospheric path radiance in the visible and near-infrared parts of the spectrum. *Applied Optics* 37, 2245–2259.

Antoine, D., Morel, A., 1999. A multiple scattering algorithm for atmospheric correction of remotely-sensed ocean colour (MERIS instrument): principle and implementation for atmospheres carrying various aerosols including absorbing ones. *International Journal of Remote Sensing* 20, 1875–1916.

Atwell, B. H., McDonald, R. B., Bartolucci, L. A., 1971. Thermal mapping of streams from airborne radiometric scanning. *Water Resources Bulletin* 7, 228–243.

AVISO. 2014. Experimental Coastal and Hydrology products. www.aviso.altimetry.fr/en/data/products/sea-surface-height-products/global/coastal-and-hydrological-products.html [accessed August 2016].

Bailey, S. W., Franz, B. A., Werdell, P. J., 2010. Estimations of near-infrared water-leaving reflectance for satellite ocean color data processing. *Optics Express* 18, 7521–7527.

Bale, A. J., Tocher, M. D., Weaver, R., Hudson, S. J., Aiken, J., 1994. Laboratory measurements of the spectral properties of estuarine suspended particles. *Netherlands Journal of Aquatic Ecology* 28, 237– 244.

Barale, V., Gade, M., 2008. *Remote Sensing of the European Seas.* New York: Springer.

BBC, 2014. British Broadcasting Corporation, UK. www.bbc.co.uk/news/uk-england-norfolk-25322214 [accessed August 2016].

Belanger, S., Babin, M., Larouche, P., 2008. An empirical ocean color algorithm for estimating the contribution of chromophoric dissolved organic matter to total light absorption in optically complex waters. *Journal of Geophysical Research* 113 (C04027). doi:10.1029/2007JC004436.

Berry, P. A. M., Garlick, J. D., Freeman, J. A., Mathers, E. L., 2005. Global inland water monitoring from multi-mission altimetry. *Geophysical Research Letters* 32 (L16401). doi:10.1029/2005GL022814.

Birkett, C. M., 1998. Contribution of the TOPEX NASA radar altimeter to the global monitoring of large rivers and wetlands. *Water Resources Research* 34, 1223–1239.

Borstadt, G. A., Edel, H. R., Gower, J. F. R., Hollinger, A. B., 1985. Analysis of test and flight data from the Fluorescence Line Imager. *Canadian Special Publication of Fisheries and Aquatic Sciences* 83, 38.

Brown, G. S., 1977. The average impulse response of a rough surface and its applications. *IEEE Transactions on Antennas and Propagation*, AP–25, 67–74.

Bukata, R. P., Bruton, J. E., Jerome, J. H., Jain, S., Zwick, H. H., 1981. Optical water quality model of Lake Ontario. 2: Determination of chlorophyll-*a* and suspended mineral concentrations of natural waters from submersible and low altitude optical sensors. *Applied Optics* 20, 1704–1714.

Bukata R. P., Jerome, J. H., Bruton, J. E., 1988. Particulate concentrations in Lake St. Clair as recorded by a shipborne multispectral optical monitoring system. *Remote Sensing of Environment* 25, 201–229.

Callison, R. D., Blake, P., Anderson, J. M., 1987. The quantitative use of Airborne Thematic Mapper thermal infrared data. *International Journal of Remote Sensing* 8, 113–126.

Chelton, D. B., Ries, J. C., Haines, B. J., Fu, L-L., Callahan, P. S., 2001. Satellite altimetry. In: Fu, L.-L., Cazenave, A. (eds.), *Satellite Altimetry and Earth Sciences: A Handbook of Techniques and Applications*, Vol. 69. San Diego, CA: Academic Press, 1–131.

Collins, M., Pattiaratchi, C., 1984. Identification of suspended sediment in coastal waters using airborne thematic mapper data. *International Journal of Remote Sensing* 5, 635–657.

Corson, M. R., Korwan, D. R., Lucke, R. L., Snyder, W. A., Davis, C. O., 2008. The Hyperspectral Imager for the Coastal Ocean (HICO) on the International Space Station. *IEEE Proceedings of the International Geoscience and Remote Sensing Symposium*, 978-1-4244-2808-3/08.

Crétaux, J-F., Jelinski, W., Calmant, S., et al., 2011. SOLS: A Lake database to monitor in near real time water level and storage variations from remote sensing data. *Journal of Advances in Space Research* 44 (9), 1497–1507. doi:10.1016/j.asr.2011.01.004.

Doerffer, R., Fischer, J., 1994. Concentrations of chlorophyll, suspended matter, and gelbstoff in Case II waters derived from satellite coastal zone color scanner data with inverse modeling methods. *Journal of Geophysical Research* 99, 7457–7466.

Doxaran, D., Castaing, P., Lavender, S. J., 2006. Monitoring the maximum turbidity zone and detecting fine-scale turbidity features in the Gironde estuary using high spatial resolution satellite sensor (SPOT HRV, Landsat ETM+) data. *International Journal of Remote Sensing* 27, 2303–2321.

Doxaran, D., Froidefond, J. M., Castaing, P., 2002. Remote-sensing reflectance of turbid sediment-dominated waters. Reduction of sediment type variations and changing illumination conditions effects by use of reflectance ratios. *Applied Optics* 42, 2623–2634.

Eleveld, M. A., van der Wal, D., van Kessel, T., 2014. Estuarine suspended particulate matter concentrations from sun-synchronous satellite remote sensing: Tidal and meteorological effects and biases. *Remote Sensing of Environment* 143, 204–215. doi:10.1016/j.rse.2013.12.019.

ESA, 2016. River & Lake. http://earth.esa.int/riverandlake/ [accessed August 2016].

Gao, B. C., Montes, M. K., Li, R. R., Dierssen, H. M., Davis, C. O., 2007. An atmospheric correction algorithm for remote sensing of bright coastal waters in MODIS land and ocean channels in the solar spectral region. *IEEE Transactions on Geoscience and Remote Sensing* 45, 1835–1843.

Gleason, A. C. R., Voss, K. J., Gordon, H. R., et al., 2012. Detailed validation of the bidirectional effect in various Case I and Case II waters. *Optics Express* 20, 7630–7645.

Gordon, H. R., 1978. Removal of atmospheric effects from satellite imagery of the oceans. *Applied Optics* 17, 1631–1636.

Gordon, H. R., 1994. Equivalence of the point and beam spread functions of scattering media: a formal demonstration. *Applied Optics* 33, 1120–1122.

Gordon, H. R., Brown, O. B., Jacobs, M. M., 1975. Computed relationships between the inherent and apparent optical properties of a flat homogeneous ocean. *Applied Optics* 14, 417–427.

Gordon, H. R., Castano, D. J., 1987. Coastal Zone Color Scanner atmospheric correction algorithm: multiple scattering effects. *Applied Optics* 26, 2111–2122.

Gordon, H. R., Clark, D. K., Brown, J. W., Brown, O. B., Evans, R. H., Broenkow, W. W., 1983. Phytoplankton pigment concentrations in the Middle Atlantic Bight: Comparison of ship determinations and CZCS estimates. *Applied Optics* 22, 20–36.

Gordon, H. R., McCluney, W. R., 1975. Estimation of the depth of sunlight penetration in the sea for remote sensing. *Applied Optics* 14, 413–416.

Gordon, H. R., Wang, M., 1994. Retrieval of water-leaving radiances and aerosol optical thickness over the oceans with SeaWiFS: A preliminary algorithm. *Applied Optics* 33, 443–452.

Gower, J. F. R., Borstadt, G. A., 1990. Mapping of phytoplankton by solar-stimulated fluorescence using an imaging spectrometer. *International Journal of Remote Sensing* 11, 313–320.

Gower, J., Doerffer, R., Borstad, G. A., 1999. Interpretation of the 685 nm peak in water-leaving radiance spectra in terms of fluorescence, absorption and scattering, and its observation by MERIS. *International Journal of Remote Sensing* 20, 1771–1786.

Gower, J., King, S., Borstad, G., Brown, L., 2005. Detection of intense plankton blooms using the 709 nm band of the MERIS imaging spectrometer. *International Journal of Remote Sensing* 26, 2005–2012.

Green, E. P., Mumby, P. J., Edwards, A. J., Clark, C. D., 1996. A review of remote sensing for the assessment and management of tropical coastal resources. *Coastal Management* 24, 1–40. doi:10.1080/08920759609362279.

Gregg, W. W., Carder, K. L., 1990. A simple spectral solar irradiance model for cloudless maritime atmospheres. *Limnology and Oceanography* 35, 1657–1675.

Hallikainen, M., Kainulainen, J., Seppanen, J., Hakkarainen, A., Rautiainen, K., 2010. Studies of radio frequency interference at L-band using an airborne 2-D interferometric radiometer. In: *Geoscience and Remote Sensing Symposium (IGARSS)*, 2490–2491. doi:10.1109/IGARSS.2010.5651866.

Handcock, R. N., Torgersen, C. E., Cherkauer, K. A., Gillespie, A. R., Tockner, K., Faux, R. N., Tan, J., 2012. Thermal infrared remote sensing of water temperature in riverine landscapes. In: Carbonneau, P. E., Piegay, H. (eds.), *Fluvial Remote Sensing for Science and Management.* Chichester, UK: John Wiley & Sons Ltd, 85–115.

Hanlon, B., Personal Communication. Pixalytics Ltd, 1 Davy Road, Plymouth Science Park, Plymouth, Devon PL6 8BX, UK.

Hedger, R. D., Malthus, T. J., Folkard, A. M., Atkinson, P. M., 2007. Spatial dynamics of estuarine water surface temperature from airborne remote sensing, *Estuarine, Coastal and Shelf Science* 71, 608–615.

Hu, C., Muller-Karger, F. E., Andrefouet, S., Carder, K. L., 2001. Atmospheric correction and cross-calibration of LANDSAT-7/ETM+ imagery over aquatic environments: A multiplatform approach using SeaWiFS/MODIS. *Remote Sensing of Environment* 78, 99–107.

HYDROWEB, 2016. Hydrology by altimetry. www.legos.obs-mip.fr/soa/hydrologie/hydroweb/Page_2.html [accessed August 2016].

IOCCG, 2000. Remote sensing of ocean colour in coastal, and other optically-complex, waters. In: Sathyendranath, S. (ed.), *Reports of the International Ocean-Colour Coordinating Group (IOCCG)*, No. 3, Dartmouth, Canada.

IOCCG, 2006. Remote sensing of inherent optical properties: Fundamentals, tests of algorithms, and applications. In: Lee, Z. (ed.), *Reports of the International Ocean-Colour Coordinating Group (IOCCG)*, No. 5, Dartmouth, Canada.

IOCCG, 2010. Atmospheric correction for remotely-sensed ocean-colour products. In: Wang, M. (ed.), *Reports of the International Ocean-Colour Coordinating Group (IOCCG)*, No. 10, Dartmouth, Canada.

IOCCG, 2016. Ocean-colour sensors. www.ioccg.org/sensors_ioccg.html/ [accessed August 2016].

Itres, 2014. Imagers. www.itres.com//imagers/ [accessed August 2016].

Kneizys, F. X., et al., 1988. *Atmospheric Transmittance/Radiance: Computer Code LOWTRAN-7.* AFGL-TR, 88–0177 (Air Force Geophysics Lab., Hanscom AFB, MA 01731, USA).

Kuchinke, C. P., Gordon, H. R., Franz, B. A., 2009. Spectral optimization for constituent retrieval in Case 2 waters I: Implementation and performance. *Remote Sensing of Environment* 113, 610–621.

Lavender, S. J., Nagur, C. R. C., 2002. Mapping coastal waters with high resolution imagery: Atmospheric correction of multi-height airborne imagery. *Journal of Optics A: Pure and Applied Optics* 4, S50–S55.

Lavender, S. L., Pinkerton, M. H., Moore, G. F., Aiken, J., Blondeau-Patissier, D., 2005. Modification to the atmospheric correction of SeaWiFS ocean colour images over turbid waters. *Continental Shelf Research* 25, 539–555.

Lee, Z., Carder, K. L., Mobley, C. D., Steward, R. G., Patch, J. S., 1998. Hyperspectral remote sensing for shallow waters 1: A semianalytical model. *Applied Optics* 37, 6329–6338.

Lee, Z., Carder, K. L., Mobley, C. D., Steward, R. G., Patch, J. S., 1999. Hyperspectral remote sensing for shallow waters 2: Deriving bottom depths and water properties by optimization. *Applied Optics* 38, 3831–3843.

Marmorino, G. O., Smith, G. B., 2008. Thermal remote sensing of estuarine spatial dynamics: Effects of bottom-generated vertical mixing. *Estuarine, Coastal and Shelf Science* 78, 587–591.

Merchant, C. J., Filipiak, M. J., Le Borgne, P., Roquet, H., Autret, E., Piollé, J.-F., Lavender, S., 2008. Diurnal warm-layer events in the western Mediterranean and European shelf seas. *Geophysical Research Letters* 35 (L04601). doi:10.1029/2007GL033071.

Miller, R. L., del Castillo, C. E., McKee, B. A., 2005. *Remote Sensing of Coastal Aquatic Environments*. Dordrecht, The Netherlands: Springer.

Moore, G. F., Aiken, J., Lavender, S. J., 1999. The atmospheric correction of water colour and the quantitative retrieval of suspended particulate matter in Case II waters: application to MERIS. *International Journal of Remote Sensing* 20, 1713–1733.

Morel, A., 1974. Optical properties of pure seawater. In: Jerlov, N. G., Nielsen, S. E. (eds.), *Optical aspects of Oceanography*. London: Academic Press, 1–24.

Morel, A., Antoine, D., Gentili, B., 2002. Bidirectional reflectance of oceanic waters: Accounting for Raman emission and varying particle scattering phase function. *Applied Optics* 41, 6289–6306.

Morel, A., Gentili, B., 1996. Diffuse reflectance of oceanic waters. III. Implication of bidirectionality for the remote-sensing problem. *Applied Optics* 35, 4850–4862.

Mumby, P. J., Green, E. P., Edwards, A. J., Clark, C. D., 1999. The cost-effectiveness of remote sensing for tropical coastal resources assessment and management. *Journal of Environmental Management* 55, 157–166.

Njoku, E. G., 2014. *Encyclopedia of Remote Sensing*. New York: Springer.

Novo, E. M. M., Hansom, J. D., Curran, P. J., 1989. The effect of sediment type on the relationship between reflectance and suspended sediment concentration. *International Journal of Remote Sensing* 10, 1283 – 1289. doi:10.1080/01431168908903967.

Palacios, S. L., Peterson, T. D., Kudela, R. M., 2009. Development of synthetic salinity from remote sensing for the Columbia River plume. *Journal of Geophysical Research* 114 (C00B05). doi:10.1029/2008JC004895.

Park, Y., Ruddick, K., 2005. Model of remote-sensing reflectance including bidirectional effects for Case 1 and Case 2 waters. *Applied Optics* 44, 1236–1249.

Pavelsky, T. M., Smith, L. C., 2009. Remote sensing of suspended sediment concentration, flow velocity, and lake recharge in the Peace-Athabasca Delta, Canada. *Water Resources Research* 45, W11417. doi:10.1029/2008WR007424.

Prieur, L., Sathyendranath, S., 1981. An optical classification of coastal and oceanic waters based on the specific spectral absorption curves of phytoplankton pigments dissolved organic matter and other particulate materials. *Limnology and Oceanography* 26, 671–689.

RSACUoP, 2005. *Implementation Test on the Application of Earth Observation and GIS Methods for Identifying Positions and Time-Dynamics of Low-Water Channels in Morecambe Bay*. Remote Sensing Applications Consultants Ltd and University of Plymouth Final Report, BNSC Project Ref CPBL/002/00136C.

Raney, R. K., 1998. The Delay / Doppler Radar Altimeter. *IEEE Transactions on Geoscience and Remote Sensing* 36, 1578–1588.

Robinson, I. S., 2004. *Measuring the Oceans from Space: The Principles and Methods of Satellite Oceanography*. Berlin: Springer/Praxis Publishing.

Rosmorduc, V., Benveniste, J., Bronner, E., Dinardo, S., Lauret, O., Maheu, C., Milagro, M., Picot, N., 2011. *Radar Altimetry Tutorial* (eds. J. Benveniste and N. Picot). www.altimetry.info [accessed August 2016].

Ruddick, K., Ovidio, F., Rijkeboer, M., 2000. Atmospheric correction of SeaWiFS imagery for turbid coastal and inland waters. *Applied Optics* 39, 897–912.

Ryu, J.-H., Han, H.-J., Cho, S., Park, Y.-J., Ahn, Y.-H., 2012. Overview of geostationary ocean color imager (GOCI) and GOCI data processing system (GDPS). *Ocean Science Journal* 47, 1738–5261.

Schiller, H., Doerffer, R., 1999. Neural network for emulation of an inverse model – operational derivation of Case II water properties from MERIS data. *International Journal of Remote Sensing* 20, 1735–1746.

Schroeder, T., Behnert, I., Schaale, M., Fischer, J., Doerffer, R., 2007. Atmospheric correction algorithm for MERIS above Case-2 waters. *International Journal of Remote Sensing* 28, 1469–1486.

Shanmugam, P., 2012. CAAS: An atmospheric correction algorithm for the remote sensing of complex waters. *Annales Geophysicae* 30, 203–220. doi:10.5194/angeo-30-203-2012.

Shanmugam, P., Varunan, T., Nagendra Jaiganesh, S. N., Sahay, A., Chauhan, P., 2016. Optical assessment of colored dissolved organic matter and its related parameters in dynamic coastal water systems. *Estuarine, Coastal and Shelf Science* 175, 126–145. http://dx.doi.org/10.1016/j.ecss.2016.03.020.

Siegel, D. A., Maritorena, S., Nelson, N. B., Behrenfeld, M. J., 2005. Independence and interdependencies among global ocean color properties: Reassessing the bio-optical assumption. *Journal of Geophysical Research* 110 (C7), C07011.

Sterckx, S., Knaeps, E., Bollen, M., Trouw, K., Houthuys, R., 2007. Retrieval of suspended sediment from advanced hyperspectral sensor data in the Scheldt Estuary at different stages in the tidal cycle. *Marine Geodesy* 30, 97–108. doi:10.1080/01490410701296341.

Stumpf, R. P., Arone, R. A., Gould, R. W., Ransibrahmanakul, V., 2003. A partially coupled ocean-atmosphere model for retrieval of water-leaving radiance from SeaWiFS in coastal waters. In: Hooker, S. B., Firestone, E. R. (eds.), *SeaWiFS Postlaunch Technical Report Series*, Vol. 22, chap. 9, NASA/TM-2003–206892. Greenbelt, MD: NASA Goddard Space Flight Center, 51–59.

Stumpf, R. P., Pennock, J. R., 1989. Calibration of a general optical equation for remote sensing of suspended sediments in a moderately turbid estuary. *Journal of Geophysical Research* 94, (C10), 14363–14371. doi:10.1029/JC094iC10p14363.

Tolk, B. L., Han, L., Rundquist, D. C., 2000. The impact of bottom brightness on spectral reflectance of suspended sediments. *International Journal of Remote Sensing* 21, 2259–2268. doi:10.1080/01431160050029558.

Torgersen, C. E., Faux, R. N., McIntosh, B. A., Poage, N. J., Norton, D. J., 2001. Airborne thermal remote sensing for water temperature assessment in rivers and streams. *Remote Sensing of Environment* 76, 386–396.

Uncles, R. J., Morris, K. P., Stephens, J. A., Robinson, M.-C., Murphy, R. J., 1999. Aircraft and sea-truth observations of salinity and temperature within the Tweed Estuary and coastal-zone frontal system. *International Journal of Remote Sensing* 20, 609–625.

Uncles, R. J., Stephens, J. A., Harris, C., 2015. Estuaries of southwest England: Salinity, suspended particulate matter, loss-on-ignition and morphology. *Progress in Oceanography* 137, Part B, 385–408. http://dx.doi.org/10.1016/j.pocean.2015.04.030.

Voss, K. J., Chapin, A. L., 2005. Upwelling radiance distribution camera system, NURADS. *Optics Express* 13, 4250–4262.

Wang, M., 1999. Validation study of the SeaWiFS oxygen A-band absorption correction: Comparing the retrieved cloud optical thicknesses from SeaWiFS measurements. *Applied Optics* 38, 937–944.

Wang, M., Shi, W., 2007. The NIR-SWIR combined atmospheric correction approach for MODIS ocean color data processing. *Optics Express* 15, 15722–15733.

Wang, M., Son, S., Shi, W., 2009. Evaluation of MODIS SWIR and NIR–SWIR atmospheric correction algorithms using SeaBASS data. *Remote Sensing of Environment* 113, 635–644.

Wdowinski, S., Kim, S. W., Amelung, F., Dixon, T. H., Miralles-Wilhelm, F., Sonenshein R., 2008. Space-based detection of wetlands' surface water level changes from L-band SAR interferometry. *Remote Sensing of Environment* 112, 681–696.